Applied Chemometrics
for Scientists

Applied Chemometrics
for Scientists

Richard G. Brereton
University of Bristol, UK

John Wiley & Sons, Ltd

Other Wiley Editorial Offices

John Wiley & Sons Inc., 111 River Street, Hoboken, NJ 07030, USA

Jossey-Bass, 989 Market Street, San Francisco, CA 94103-1741, USA

Wiley-VCH Verlag GmbH, Boschstr. 12, D-69469 Weinheim, Germany

John Wiley & Sons Australia Ltd, 42 McDougall Street, Milton, Queensland 4064, Australia

John Wiley & Sons (Asia) Pte Ltd, 2 Clementi Loop #02-01, Jin Xing Distripark, Singapore 129809

John Wiley & Sons Canada Ltd, 6045 Freemont Blvd, Mississauga, Ontario, L5R 4J3, Canada

Wiley also publishes its books in a variety of electronic formats. Some content that appears
in print may not be available in electronic books.

Library of Congress Cataloging-in-Publication Data:

Brereton, Richard G.
Applied chemometrics for scientists / Richard G. Brereton.
    p. cm.
Includes bibliographical references and index.
ISBN-13: 978-0-470-01686-2 (cloth : alk. paper)
ISBN-10: 0-470-01686-8 (cloth : alk. paper)
1.   Chemometrics.   I. Title.
QD75.4.S8B735 2007
543.01'5195 – dc22

                                 2006030737

British Library Cataloguing in Publication Data

A catalogue record for this book is available from the British Library

ISBN 978-0-470-01686-2

Typeset in 10/12pt Times by Laserwords Private Limited, Chennai, India.

Contents

Preface   xiii

1 Introduction   1
  1.1 Development of Chemometrics   1
    1.1.1 Early Developments   1
    1.1.2 1980s and the Borderlines between Other Disciplines   1
    1.1.3 1990s and Problems of Intermediate Complexity   2
    1.1.4 Current Developments in Complex Problem Solving   2
  1.2 Application Areas   3
  1.3 How to Use this Book   4
  1.4 Literature and Other Sources of Information   5
    References   7

2 Experimental Design   9
  2.1 Why Design Experiments in Chemistry?   9
  2.2 Degrees of Freedom and Sources of Error   12
  2.3 Analysis of Variance and Interpretation of Errors   16
  2.4 Matrices, Vectors and the Pseudoinverse   20
  2.5 Design Matrices   22
  2.6 Factorial Designs   25
    2.6.1 Extending the Number of Factors   28
    2.6.2 Extending the Number of Levels   28
  2.7 An Example of a Factorial Design   29
  2.8 Fractional Factorial Designs   32
  2.9 Plackett–Burman and Taguchi Designs   35
  2.10 The Application of a Plackett–Burman Design to the Screening of Factors Influencing a Chemical Reaction   37
  2.11 Central Composite Designs   39
  2.12 Mixture Designs   44
    2.12.1 Simplex Centroid Designs   45
    2.12.2 Simplex Lattice Designs   47
    2.12.3 Constrained Mixture Designs   47

2.13 A Four Component Mixture Design Used to Study Blending
      of Olive Oils   49
2.14 Simplex Optimization   51
2.15 Leverage and Confidence in Models   53
2.16 Designs for Multivariate Calibration   58
      References   62

3 Statistical Concepts   **63**
  3.1 Statistics for Chemists   63
  3.2 Errors   64
       3.2.1 Sampling Errors   65
       3.2.2 Sample Preparation Errors   66
       3.2.3 Instrumental Noise   67
       3.2.4 Sources of Error   67
  3.3 Describing Data   67
       3.3.1 Descriptive Statistics   68
       3.3.2 Graphical Presentation   69
       3.3.3 Covariance and Correlation Coefficient   72
  3.4 The Normal Distribution   73
       3.4.1 Error Distributions   73
       3.4.2 Normal Distribution Functions and Tables   74
       3.4.3 Applications   75
  3.5 Is a Distribution Normal?   76
       3.5.1 Cumulative Frequency   76
       3.5.2 Kolmogorov–Smirnov Test   78
       3.5.3 Consequences   79
  3.6 Hypothesis Tests   80
  3.7 Comparison of Means: the t-Test   81
  3.8 F-Test for Comparison of Variances   85
  3.9 Confidence in Linear Regression   89
       3.9.1 Linear Calibration   90
       3.9.2 Example   90
       3.9.3 Confidence of Prediction of Parameters   92
  3.10 More about Confidence   93
       3.10.1 Confidence in the Mean   93
       3.10.2 Confidence in the Standard Deviation   95
  3.11 Consequences of Outliers and How to Deal with Them   96
  3.12 Detection of Outliers   100
       3.12.1 Normal Distributions   100
       3.12.2 Linear Regression   101
       3.12.3 Multivariate Calibration   103
  3.13 Shewhart Charts   104
  3.14 More about Control Charts   106
       3.14.1 Cusum Chart   106
       3.14.2 Range Chart   108
       3.14.3 Multivariate Statistical Process Control   108
      References   109

4 Sequential Methods   **111**
  4.1 Sequential Data   111
  4.2 Correlograms   112
      4.2.1 Auto-correlograms   113
      4.2.2 Cross-correlograms   115
      4.2.3 Multivariate Correlograms 115
  4.3 Linear Smoothing Functions and Filters   116
  4.4 Fourier Transforms   120
  4.5 Maximum Entropy and Bayesian Methods   124
      4.5.1 Bayes' Theorem   124
      4.5.2 Maximum Entropy   125
      4.5.3 Maximum Entropy and Modelling 126
  4.6 Fourier Filters   128
  4.7 Peakshapes in Chromatography and Spectroscopy   134
      4.7.1 Principal Features   135
      4.7.2 Gaussians   136
      4.7.3 Lorentzians   136
      4.7.4 Asymmetric Peak Shapes   137
      4.7.5 Use of Peak Shape Information 138
  4.8 Derivatives in Spectroscopy and Chromatography   138
  4.9 Wavelets   142
      References   143

5 Pattern Recognition   **145**
  5.1 Introduction   145
      5.1.1 Exploratory Data Analysis 145
      5.1.2 Unsupervised Pattern Recognition 146
      5.1.3 Supervised Pattern Recognition 146
  5.2 Principal Components Analysis   147
      5.2.1 Basic Ideas   147
      5.2.2 Method   150
  5.3 Graphical Representation of Scores and Loadings   154
      5.3.1 Case Study 1   154
      5.3.2 Case Study 2   154
      5.3.3 Scores Plots   156
      5.3.4 Loadings Plots   157
      5.3.5 Extensions   159
  5.4 Comparing Multivariate Patterns   159
  5.5 Preprocessing   160
  5.6 Unsupervised Pattern Recognition: Cluster Analysis 167
  5.7 Supervised Pattern Recognition   171
      5.7.1 Modelling the Training Set 171
      5.7.2 Test Sets, Cross-validation and the Bootstrap 172
      5.7.3 Applying the Model   174
  5.8 Statistical Classification Techniques   174
      5.8.1 Univariate Classification 175
      5.8.2 Bivariate and Multivariate Discriminant Models 175

	5.8.3 SIMCA	178
	5.8.4 Statistical Output	182
5.9	K Nearest Neighbour Method	182
5.10	How Many Components Characterize a Dataset?	185
5.11	Multiway Pattern Recognition	187
	5.11.1 Tucker3 Models	188
	5.11.2 PARAFAC	189
	5.11.3 Unfolding	190
	References	190

6 Calibration — **193**

6.1	Introduction	193
6.2	Univariate Calibration	195
	6.2.1 Classical Calibration	195
	6.2.2 Inverse Calibration	196
	6.2.3 Calibration Equations	198
	6.2.4 Including Extra Terms	199
	6.2.5 Graphs	199
6.3	Multivariate Calibration and the Spectroscopy of Mixtures	202
6.4	Multiple Linear Regression	206
6.5	Principal Components Regression	208
6.6	Partial Least Squares	211
6.7	How Good is the Calibration and What is the Most Appropriate Model?	214
	6.7.1 Autoprediction	214
	6.7.2 Cross-validation	215
	6.7.3 Test Sets	215
	6.7.4 Bootstrap	217
6.8	Multiway Calibration	217
	6.8.1 Unfolding	217
	6.8.2 Trilinear PLS1	218
	6.8.3 N-PLSM	219
	References	220

7 Coupled Chromatography — **221**

7.1	Introduction	221
7.2	Preparing the Data	222
	7.2.1 Preprocessing	222
	7.2.2 Variable Selection	224
7.3	Chemical Composition of Sequential Data	228
7.4	Univariate Purity Curves	230
7.5	Similarity Based Methods	234
	7.5.1 Similarity	234
	7.5.2 Correlation Coefficients	234
	7.5.3 Distance Measures	235
	7.5.4 OPA and SIMPLISMA	236
7.6	Evolving and Window Factor Analysis	236

	7.6.1 Expanding Windows	237
	7.6.2 Fixed Sized Windows	238
	7.6.3 Variations	239
7.7	Derivative Based Methods	239
7.8	Deconvolution of Evolutionary Signals	241
7.9	Noniterative Methods for Resolution	242
	7.9.1 Selectivity: Finding Pure Variables	242
	7.9.2 Multiple Linear Regression	243
	7.9.3 Principal Components Regression	244
	7.9.4 Partial Selectivity	244
7.10	Iterative Methods for Resolution	246

8 Equilibria, Reactions and Process Analytics   **249**

8.1	The Study of Equilibria using Spectroscopy	249
8.2	Spectroscopic Monitoring of Reactions	252
	8.2.1 Mid Infrared Spectroscopy	253
	8.2.2 Near Infrared Spectroscopy	254
	8.2.3 UV/Visible Spectroscopy	255
	8.2.4 Raman Spectroscopy	256
	8.2.5 Summary of Main Data Analysis Techniques	256
8.3	Kinetics and Multivariate Models for the Quantitative Study of Reactions	257
8.4	Developments in the Analysis of Reactions using On-line Spectroscopy	261
	8.4.1 Constraints and Combining Information	261
	8.4.2 Data Merging	262
	8.4.3 Three-way Analysis	262
8.5	The Process Analytical Technology Initiative	263
	8.5.1 Multivariate Tools for Design, Data Acquisition and Analysis	263
	8.5.2 Process Analysers	264
	8.5.3 Process Control Tools	264
	8.5.4 Continuous Improvement and Knowledge Management Tools	264
	References	265

9 Improving Yields and Processes Using Experimental Designs   **267**

9.1	Introduction	267
9.2	Use of Statistical Designs for Improving the Performance of Synthetic Reactions	269
9.3	Screening for Factors that Influence the Performance of a Reaction	271
9.4	Optimizing the Process Variables	275
9.5	Handling Mixture Variables using Simplex Designs	278
	9.5.1 Simplex Centroid and Lattice Designs	278
	9.5.2 Constraints	280
9.6	More about Mixture Variables	283
	9.6.1 Ratios	283
	9.6.2 Minor Constituents	285
	9.6.3 Combining Mixture and Process Variables	285
	9.6.4 Models	285

10 Biological and Medical Applications of Chemometrics   **287**
    10.1 Introduction   287
            10.1.1 Genomics, Proteomics and Metabolomics   287
            10.1.2 Disease Diagnosis   288
            10.1.3 Chemical Taxonomy   288
    10.2 Taxonomy   289
    10.3 Discrimination   291
            10.3.1 Discriminant Function   291
            10.3.2 Combining Parameters   293
            10.3.3 Several Classes   293
            10.3.4 Limitations   296
    10.4 Mahalanobis Distance   297
    10.5 Bayesian Methods and Contingency Tables   300
    10.6 Support Vector Machines   303
    10.7 Discriminant Partial Least Squares   306
    10.8 Micro-organisms   308
            10.8.1 Mid Infrared Spectroscopy   309
            10.8.2 Growth Curves   311
            10.8.3 Further Measurements   311
            10.8.4 Pyrolysis Mass Spectrometry   312
    10.9 Medical Diagnosis using Spectroscopy   313
    10.10 Metabolomics using Coupled Chromatography and Nuclear Magnetic
            Resonance   314
            10.10.1 Coupled Chromatography   314
            10.10.2 Nuclear Magnetic Resonance   317
            References   317

11 Biological Macromolecules   **319**
    11.1 Introduction   319
    11.2 Sequence Alignment and Scoring Matches   320
    11.3 Sequence Similarity   322
    11.4 Tree Diagrams   324
            11.4.1 Diagrammatic Representations   324
            11.4.2 Dendrograms   325
            11.4.3 Evolutionary Theory and Cladistics   325
            11.4.4 Phylograms   326
    11.5 Phylogenetic Trees   327
            References   329

12 Multivariate Image Analysis   **331**
    12.1 Introduction   331
    12.2 Scaling Images   333
            12.2.1 Scaling Spectral Variables   334
            12.2.2 Scaling Spatial Variables   334
            12.2.3 Multiway Image Preprocessing   334
    12.3 Filtering and Smoothing the Image   335
    12.4 Principal Components for the Enhancement of Images   337

12.5 Regression of Images 340
12.6 Alternating Least Squares as Employed in Image Analysis 345
12.7 Multiway Methods In Image Analysis 347
 References 349

13 Food **351**
13.1 Introduction 351
 13.1.1 Adulteration 351
 13.1.2 Ingredients 352
 13.1.3 Sensory Studies 352
 13.1.4 Product Quality 352
 13.1.5 Image Analysis 352
13.2 How to Determine the Origin of a Food Product using Chromatography 353
13.3 Near Infrared Spectroscopy 354
 13.3.1 Calibration 354
 13.3.2 Classification 355
 13.3.3 Exploratory Methods 355
13.4 Other Information 356
 13.4.1 Spectroscopies 356
 13.4.2 Chemical Composition 356
 13.4.3 Mass Spectrometry and Pyrolysis 357
13.5 Sensory Analysis: Linking Composition to Properties 357
 13.5.1 Sensory Panels 357
 13.5.2 Principal Components Analysis 359
 13.5.3 Advantages 359
13.6 Varimax Rotation 359
13.7 Calibrating Sensory Descriptors to Composition 365
 References 368

Index **369**

Preface

The seed for this book was planted in 1999 when I undertook to write a series of fortnightly articles on chemometrics for ChemWeb, the aim being to summarize the main techniques and applications to a wide audience. In all, 114 articles were written over a period of nearly 5 years, until the end of 2003. These articles were a popular general reference. One important feature was that each article was around 1000 words in length plus figures and tables where appropriate, meaning that each topic was described in 'bite-sized' chunks: many people trying to keep up with chemometrics are busy people who have other duties and so having reasonably short and finite descriptions of a specific topic helps fit into a hectic schedule. Another feature, for the Web-based articles, was extensive cross-referencing. Finally presenting material in this way encourages readers to dip in and out rather than follow a track from beginning to end.

With completion of the ChemWeb series, it was proposed that the articles could be reorganized and edited into book format, and this text is the result. Many errors in the original Webpages have been corrected. Material has been reorganized and updated, but the main original themes are retained, and I have resisted the temptation for radical changes. In this text, there is new material covered in the first (introductory) chapter. Chapters 2–6 cover general topics in chemometrics (experimental design, statistical methods, signal analysis, pattern recognition and calibration). Chapters 7–13 cover many of the main applications of chemometrics (coupled chromatography, equilibria/reactions/Process Analytical Technology, optimizing yields, biological/medical aspects, biological macromolecules, multivariate image analysis and food). The later chapters build on the principles of the earlier ones. This particular book differs from existing general texts in the area in that it is very much focused on applications and on how chemometrics can be used in modern day scientific practice. Not all possible application areas are covered, in order to be faithful to the spirit of the original ChemWeb articles and in order to develop a text of acceptable length within a reasonable timescale. However, most of the important growth points over the past decade are covered, as chemometrics has metamorphosed from a subject that originated in many people's minds as an extension of statistical analytical chemistry into a very diverse area applicable to many forms of modern science.

Chapters 2, 4, 5, 6 and 7 cover some common ground with an existing text [1] but in a more condensed format, and with several additional topics but without worked examples or problems. The current text is aimed primarily at the scientific user of chemometrics methods, and as a first course for applied scientists. However, it should also serve as a very valuable aid to lectures in this area, the majority of people who are introduced to chemometrics are primarily interested in how the methods can be applied to real world problems. Almost all existing books on chemometrics are methodological, rather than application oriented, and where there are applications the major emphasis is in the context of analytical chemistry. There is, therefore, a need for more application based texts, as chemometrics is primarily an applied subject. An analogy might be with organic chemistry, the early part of an organic chemistry course may involve learning the basis of reactions, mechanisms, stereochemistry and functional groups in analogy to the building blocks of chemometrics, experimental design, signal analysis, pattern recognition, calibration and general statistics. However, later in the development of an organic chemist it is necessary to learn about the reactions and synthesis of specific compounds for example porphyrins, carbohydrates, terpenes and so on: equally, the second half of a balanced practical course in chemometrics should focus on applications, as in this book, so it is hoped that this text sets a new goal for books in this area for the future.

It was decided to keep reasonably faithful to the original articles, which were widespread and popular, and in consequence there is some overlap between chapters, and a small amount of repetition, whilst cross-referencing to other parts of the text, in order to keep each chapter coherent and not overly fragmented. This is inevitable as the basic techniques are applied in different contexts. Each later chapter then remains reasonably coherent but builds on some of the techniques in the earlier chapters. For example, Chapter 9 (Improving Yields and Processes using Experimental Designs), builds on methods of Chapter 2 (Experimental Design), Chapter 10 (Biological and Medical Applications of Chemometrics) builds on many of the classification methods of Chapter 5 (Pattern Recognition), Chapter 11 (Biological Macromolecules) builds on unsupervised methods of Chapter 5 (Pattern Recognition), and Chapter 8 (Equilibria, Reactions and Process Analytics) builds on some of the approaches of Chapter 7 (Coupled Chromatography). Some approaches such as Support Vector Machines and Discriminant Partial Least Squares are discussed in detail in only one chapter, in the context they probably are most commonly used, but there is no reason why the general reader should be restricted and a typical user of a book is likely to dip in and out of sections as the need arises. On the whole the philosophy is that to have a good basic grasp of chemometrics the contents of Chapters 2–6 provide essential broad information whereas the contents of the later chapters may be more of interest in the specific application areas.

In addition to the normal reorganization and corrections that have taken place to bring the ChemWeb articles to book format, including numerous corrections and extensive cross-referencing, several additional topics within the original themes have been included and some of the orientation of the original has been modified. The following lists the main changes:

- There is a specific section (2.4) on matrices, vectors and the pseudoinverse, early in the text.
- The covariance and correlation coefficient are now explicitly introduced, in Section 3.3.3.
- In Section 3.14 there is mention of Normal Operating Conditions, Q and D charts, in the context of multivariate statistical process control.
- Savitzky–Golay coefficients are tabulated in Chapter 4 (Tables 4.2 and 4.4).
- One of the examples in the introduction to Principal Components Analysis (Section 5.3) is non-chromatographic, previously both examples came from chromatography. The emphasis on chromatography is reduced in this chapter.

- There is discussion of the bootstrap (Sections 5.7.2 and 6.7.4) in the context of validating classification models and Partial Least Squares predictive ability.
- Chapter 7 is focused exclusively on chromatography as an application area.
- In Chapter 7, redundancy between sections has been reduced, involving substantial tightening, and most figures redrawn in Matlab.
- In Section 7.5.4 the principles of OPA and SIMPLISMA are introduced.
- In Chapter 7 there is mention of CODA and baseline correction. Overlap with Chapter 5 (data preprocessing) has been reduced.
- Alternating Least Squares is introduced in both Chapter 8 which involves the analysis of equilibria and reactions and Chapter 7, in addition to Chapter 12. Iterative Target Transform Factor Analysis is also described in Section 7.10.
- The sections on analysis of reactions in Chapter 8 have been restructured to reduce the emphasis on kinetics, whilst still introducing kinetic models.
- A new section (8.5) has been written on Process Analytical Technology.
- The section on Bayesian classification (10.5) also introduces concepts such as likelihood ratios, contingency tables and false positives/false negatives.
- The section on Support Vector Machines (10.6) has been completely rewritten.
- A new section (10.10) has been written on metabolomics including the use of coupled chromatography and nuclear magnetic resonance, and discusses problems of peak detection and alignment.
- New sections on the use of near infrared (13.3) and other analytical techniques (13.4) combined with chemometrics for the characterization of food have been written.

Many people are to be thanked that have made this book possible. First of all, I should mention the very wide range of external collaborators with my group in Bristol over the past 5 years, who have helped move work in Bristol from fairly theoretical and analytical chemistry based to very applied: these have helped focus on the wide and ever expanding range of applications of chemometric methods, and so develop a driving force for a text of this nature. Liz Rawling of ChemWeb is thanked as the first editor of the *Alchemist* to encourage my short articles, and Tina Walton for taking this over and developing a very professional series of material over several years. ChemWeb are thanked for permission to produce a publication based on these articles. Katya Vines of Wiley is thanked for her original advocacy of my existing text in 1998, as is Jenny Cossham for steering both the original and this new book to the market and Lynette James for the final editorial handling of this manuscript.

Richard G. Brereton
Bristol
July, 2006

REFERENCE

1. R.G. Brereton, *Chemometrics: Data Analysis for the Laboratory and Chemical Plant*, John Wiley & Sons, Ltd, Chichester, 2003

1
Introduction

1.1 DEVELOPMENT OF CHEMOMETRICS

1.1.1 Early Developments

Chemometrics has developed over the past decade from a fairly theoretical subject to one that is applied in a wide range of sciences. The word chemometrics was coined in the 1970s, and the early development went hand in hand with the development of scientific computing, and primarily involved using multivariate statistical methods for the analysis of analytical chemistry data. Early chemometricians were likely to be FORTRAN programmers using mainframes and statistical libraries of subroutines. The earliest applications were primarily to analytical chemical datasets, often fairly simple in nature, for example a high performance liquid chromatography (HPLC) cluster of two or three peaks or a set of ultraviolet (UV)/visible mixture spectra.

1.1.2 1980s and the Borderlines between Other Disciplines

Chemometrics as a discipline became organized in the 1980s, with the first journals, meetings, books, societies, packages and courses dedicated to the subject. Many of the papers in this era were to relatively simple analytical problems such as peak deconvolution, which nevertheless provided substantial challenges in the development of theoretical methods. Historically HPLC and near infrared (NIR) spectroscopy provided particularly important growth points in the 1980s, partly due to the economic driving forces at the time and partly due to the research interest and contacts of the pioneers. Industrial applications were particularly important in the developing phase.

The borderline between chemometrics and other disciplines gradually became established, and although of historic interest, helps define the scope of this, and related, texts. There is a generally recognized distinction between chemometrics, which is primarily involved with the analysis and interpretation of instrumental data, and methods applied to organic chemistry, such as quantitative structure–activity relationships (QSAR), which often use many of the same computational approaches, such as Partial Least Squares (PLS) or discriminant analysis, but in many cases poses different challenges: in analytical instrumentation there is often a critical phase of handling and preparing instrumental data prior to (or in some case

Applied Chemometrics for Scientists   R. G. Brereton
© 2007 John Wiley & Sons, Ltd

simultaneously with) pattern recognition algorithms, which is missing in many other areas of science. The borderline with computational chemistry is also fairly well established, for example, chemometrics does not involve structure representation or database algorithms, and is not part of quantum chemistry. Indeed there are interesting differences between chemometrics and traditional computational or theoretical chemistry, one example being that theoretical chemists rarely if ever come across matrices that are not square and need to be inverted (via the pseudoinverse). And chemometricians nowadays in most part develop methods using Matlab, reflecting the vintage of the subject which became established in the 1980s, rather than FORTRAN, preferred by quantum chemists, whose discipline was seeded in the 1960s. Finally there is a much more blurred distinction between chemometrics and the application of computational algorithms to handle analytical data, such as Neural Networks, Genetic Algorithms, Support Vector Machines and Machine Learning in general. This boundary has never really been resolved, and in this text we will be quite restrictive, although Support Vector Machines are popular in the statistically based community because they are reproducible for a given kernel and description of how the data have been processed, and so will be introduced.

1.1.3 1990s and Problems of Intermediate Complexity

In the 1990s the application of chemometrics started expanding, with special emphasis on certain industries especially the pharmaceutical industry, where large companies provided significant funding for developments. The sort of problems being tackled by the most active groups were of what might be called intermediate complexity. An example is chromatographic pattern recognition of pharmaceuticals, a typical series of perhaps 50 chromatograms may contain around 20 peaks in each chromatogram, most peaks being common to several of the chromatograms, there being quite well identified features that allow, for example, the ability to distinguish two groups according to their origins. Another typical example of this era relates to Process Analysis where there may be a reaction with four or five identifiable reactants or products which is to be monitored spectroscopically requiring the development of approaches for quantifying the presence of these reactants in the mixture. These sort of problems require the development of new approaches and have posed significant challenges particularly in self modelling curve resolution and pattern recognition. In contrast to much of the work in the 1980s and early 1990s, the size of problem is much larger.

1.1.4 Current Developments in Complex Problem Solving

A new and exciting phase has emerged as from the late 1990s, involving very complex datasets. Many are chromatographic, for example in biological applications such as metabolomics or proteomics. These chromatograms can consist of several hundred detectable peaks. A typical dataset, in turn, could consist of several hundred or even a thousand or more chromatograms. This could result in hundreds of thousands of detectable peaks throughout the dataset, and in many real life situations only a portion are common to each chromatogram. Instead of dealing with a dataset of perhaps 50 chromatograms of around 20 peaks each (1000 detectable peaks, many common to most or a significant proportion of the chromatograms), we may be dealing with 1000 chromatograms of around 500 peaks each (500 000 detectable peaks, many unique to a small number of chromatograms) and have a far more severe data analysis problem. This new and exciting phase is possible due to the capacity of analytical instruments to acquire large amounts of data rapidly, for example via autosamplers, and chemometrics becomes a type of data mining. All the building blocks are available, for example signal processing, chromatographic alignment, data scaling and pattern recognition but the problems are now much more complex. This third and very applied wave is likely to be a driving force at the forefront of chemometrics research for several years. In addition many of the newer applications are biologically driven and so there is emerging a new interface between chemometrics and bioinformatics, especially the relationship between analytical data as obtained by spectroscopy or chromatography and genetics, requiring some appreciation of gene or protein sequence and similarities. Hence in this book we introduce some basics of bioinformatics at the level that will be useful to chemometricians, but do not discuss aspects such as database searching and management.

1.2 APPLICATION AREAS

Traditionally chemometrics was regarded as very much based around laboratory and process instrumental analytical chemistry. Using partial least squares to calibrate the ultraviolet (UV)/Visible spectrum of a mixture of pharmaceuticals to determine the concentrations of each constituent is a typical application. There are numerous similar types of application in a wide variety of spectroscopies such as infrared (of all types), atomic spectroscopy, mass spectrometry, and in chromatography such as HPLC and gas chromatography-mass spectrometry (GC-MS). This area, whilst perhaps not very glamorous, is still very widespread and still the majority of reports on the use of chemometrics, in the academic literature, are to instrumental analytical chemistry where it is very successful and often regarded as essential especially in both experimental design and the interpretation of multivariate spectroscopic data.

However, the applications have diversified substantially over the last few years. There were certain economic driving forces that related partially to the organizations that employed the first pioneers of the subject. In the 1970s and 1980s there were not many dedicated academic chemometrics positions, so the 'parents' of the subject would have described themselves as organic chemists or analytical chemists or pharmaceutical chemists, etc., rather than chemometricians. Only in the mid 1980s and 1990s came a small trickle of dedicated chemometrics posts in academia, with industry perhaps moving faster [industry tends to wait until the academic seed has been sown but then can move much faster as there is no need to maintain departments for long periods so new staff can be rapidly hired (and fired of course) as the need arises]. In the 1970s and 1980s, food chemistry was one of the main early growth areas, probably because of the tradition of sensory statistics and a good interface between data analysis and analytical chemistry, hence a large development especially in the use of NIR and food analysis, and an important economic driving force especially in Scandinavia and Southern Europe. Process chemistry, also has a vintage in the 1980s with the establishment, over two decades, of several centres, pioneered first in the University of Washington (Seattle). Chemical engineers and analytical chemists work closely with manufacturing industry, spectroscopy including NIR (again) and the monitoring of processes has led eventually to developments, such as the Process Analytical Technology (PAT) initiative. A third and important influence was pharmaceutical chemistry, with a good link to hospitals, pharmacy and clinical applications, especially the use of HPLC: analytical chemists in Benelux who were trained as chromatographers first and foremost were responsible for the development of chemometrics within the area of pharmaceutical chromatography. These major growth points seeded major research groups, in Scandinavia, Benelux and the USA in the late 1970s and early 1980s allowing substantial growth in applications especially to HPLC and NIR as applied to pharmaceuticals, food and process analysis, and also the development of the intellectual driving force, first commercially marketed software and main principles.

During this period other people in industry quietly followed the academic lead: in a subject such as chemometrics, it is hard to disentangle the 'name' which is often trumpeted by academic groups who try to obtain students, grants, papers, theses and conference presentations, and people in industry that have less need to invent a name or write papers but get on with the job. Several industries notably the petrochemical industry also applied chemometric principles in these early days. There also was a separate development in environmental chemistry, some impinging on chemometrics, some being published as environmetrics, so diluting the focus in the analytical chemistry literature.

With the 1980s wrapping up and the early 1990s commencing, chemometrics became much more widespread. Individual researchers in academic positions, some with prior training in some of the more established groups, and others who read papers and books that had been written over this early period, dabbled with chemometrics, and so the applications spread not just to dedicated research groups but to investigators working in different environments. The most notable expansion even to this day is the biological and medical area, with the enormous increase in development of chemical analyses such as GC-MS, mass spectrometry (MS), nuclear magnetic resonance (NMR), optical spectroscopy and pyrolysis, and the tradition of computationally oriented biologists. The interface between chemometrics and bioinformatics is also a more modern growth point over the past 5 or 6 years, especially as biological data (e.g. gene and protein structure) is increasingly correlated with metabolomic data (metabolomics) obtained from chemical analyses. Other niche but important growth points include forensics (the use of chemical and spectroscopic information to determine origins of samples), multivariate image analysis (especially of pharmaceutical tablets), chemical engineering (far beyond the original process analytics), and a small but expanding interest in materials (including thermal analysis). All these applications have in common that potentially a large amount of multivariate data can be obtained, usually very rapidly, from instruments and there are major data analytical problems in their interpretation which is aided by chemometrics.

In this book we focus on several of the most widespread applications, although cannot cover all for reasons of length, ranging from the classical area of hyphenated chromatography to rapidly developing fields including the application to biology.

1.3 HOW TO USE THIS BOOK

It is anticipated that the readership of this book is very broad, coming from many different disciplines, with differing background knowledge. It is well know that scientists with different knowledge bases like to tackle the subject in separate ways. A synthetic organic chemist (who may be very interested in chemometrics for experimental design and reaction monitoring, for example) will not regard a text that builds on mathematical principles as a good one, they will expect a book that provides nonmathematical explanations, and so will not feel comfortable with a requirement to master the t-test or normal distribution or matrix algebra as a prerequisite to understanding, for example, experimental design. A statistician, in contrast, would be incredulous that anyone could possibly contemplate understanding how experiments are designed without this sort of knowledge, and would struggle to understand principles unless they are related back to statistical fundamentals. An analytical chemist will often appreciate the need to understand errors and hypothesis tests before diving into chemometrics, but may be less keen on algorithms or matrices or computational principles. A chemical engineer will want to build on principles of linear algebra such as matrices, but be less keen on understanding about chromatographic resolution and not think in terms of probability distributions.

Hence, developing a book designed for a variety of application based scientists is a difficult task, and the main principle is that the text is not designed to be read in a linear fashion. Different readers will dip in and out of different sections. This is in fact how most people absorb knowledge in these days of the Web. Certainly if I or one of my students comes across a new principle or method, often we use a search engine or take out some books from the library and dip into the most appropriate paragraphs and examples, often in contexts outside our own, and pick up the snippets of information necessary for our application. What is useful is extensive cross-referencing, this book containing around 400 cross-references between sections. This allows the use of the text in different ways according to need.

Chapters 2–6 contain many basic principles that are built on in later chapters, although some methods that are applied primarily in a specific application area are introduced in later chapters. A good chemometrician should have a thorough grounding in the material of Chapters 2–6, with specialist knowledge of some of the later chapters. Not all application areas have been covered in this text for reason of time and length, notable exceptions being environmental and forensic science, however most of the currently recognized historical and modern growth points are described. It is anticipated that this book will be very useful in applications based courses, as it gives a good overview of how chemometrics is now much more than just an adjunct to instrumental analytical chemistry, a secret that the practitioners have known for many years.

Most chapters contain a few references. These are of two types. The first are to a few primary articles or books or Web sites that provide overviews. A few are to specialist papers chosen as examples of the application of specific methods: the choice is essentially arbitrary, but a good lecture, for example, might involve illustrating an application by one or more well chosen examples. These do direct the reader to the primary literature, and to get a fuller view it is recommended to use a bibliographic citation database such as the Institute of Scientific Information's, to search either forward or backward in time to see who cites the specific paper and whom the paper cites, to obtain related papers. In certain areas, such as biomedical or food analysis, there are many hundreds of articles that use basic approaches such as calibration and discriminant analysis and the aim of this book is mainly to provide pointers to the tip of the iceberg rather than overwhelm the reader with detail.

1.4 LITERATURE AND OTHER SOURCES OF INFORMATION

There is a well developed chemometrics literature, and the aim of this section is to point readers in the direction of more detailed material.

For a more detailed description of most of the methods discussed in Chapters 2, 4, 5, 6 and 7, readers are recommended to the companion book [1], which contains extensive numerical examples and problems together with their solutions (on the chemometrics channel of the Web site www.spectroscopynow.com). The current book contains an abbreviated introduction suitable for reference and a first course for users of chemometrics.

The most comprehensive books on the subject remain those by Massart and coworkers [2,3] which cover a wide range of methods in substantial detail, primarily aimed at analytical chemists. There are several general books available. Otto's [4] is a good introduction primarily aimed (as most are) at analytical chemists but also tackling areas in general analytical chemistry data analysis. Beeb *et al.* have published a book aimed very much at the practising industrial NIR spectroscopist, but which is full of useful information [5]. Kramer has published a basic text that minimizes mathematics and covers areas such as calibration very well [6]. Gemperline has edited a text covering many modern topics in chemometrics [7].

Historically there are several important texts which are also still quite relevant, perhaps not very applied but useful nevertheless. The first edition of Massart's text is much cited in the context of analytical chemical applications [8]. Sharaf *et al.* produced what was probably historically the first text dedicated to chemometrics [9]. Brereton produced a small introductory book in the area [10]. The proceedings of a key meeting held in Cosenza, Italy, in 1983 were edited into an early landmark volume [11] capturing many of the very earliest contributors to the field. Tutorial articles from the first few years of the journal *Chemometrics and Intelligent Laboratory Systems* have been edited into two volumes, covering a wide range of topics [12,13].

Some books that may appear from their scope to be slightly more specialist are nevertheless of general interest. There are several books based primarily on the use of statistical methods in analytical chemistry [14–16]. Miller and Miller's book contains an introduction to some topics in chemometrics in the final chapters [14]. Martens and Martens have written a very applied book on the multivariate analysis of quality [17], which contains good practical advice especially for those people using calibration in the food industry. Malinowski's book on factor analysis was one of the first chemometrics texts and has evolved into a third edition which is still very popular especially with physical chemists and spectroscopists [18]. In the area of multivariate calibration, Martens and Naes' book is a classic [19]. A recent text focused very much on NIR spectroscopy is quite popular for users of these techniques especially in the food industry [20]. Meloun and coworkers have published two volumes aimed at statistical methods as applied to analytical chemistry [21,22]. There are several books on experimental design that are useful to chemists [23–25], Deming and Morgan's being a well regarded first text in the field trying to outline the ideas to chemists based on many years' successful experience. In pattern recognition, Massart and Kaufmann wrote an early text which is very lucid [26] and Brereton edited a text [27] that covers several of the basic methods authored by a number of leading experts in the early 1990s and illustrated by several examples. For spectroscopists often interested in specific statistical methods, there are several excellent books. Mark and Workman [28] have produced a text based on many years' experience, that is well recommended. Gans' book is well regarded [29] containing introductions to several important methods. Adams has produced a tutorial text on chemometrics in spectroscopy [30]. Several companies have published books that are good introductions to chemometrics but are based very much around their software and commercial courses, but are useful for users of specific packages. Notable are CAMO [31] and Umetrics [32], whose books are often revised with new editions and authors, examples being cited here, and are valuable aids for users of their software.

In the area of chemometrics most of the highest cited work tends to be published in book form rather than as papers, because texts are a good way of summarizing methods, which have broad appeal. The majority of papers in chemometrics involve very specific applications or description of methods and most tend not to be cited as widely as the main books or a few very focused review type articles.

The number of journals that publish chemometrics papers is very large, papers being widespread throughout science. There are two dedicated journals, *Journal of Chemometrics* (John Wiley & Sons, Ltd), and *Chemometrics Intelligent Laboratory Systems* (Elsevier) that publish mainly articles about theoretical advances in methodology. The American Chemical Society's *Journal of Chemical Information and Modelling* also contains a number of relevant methodological articles. However, perhaps only 5–10 % of all chemometrics based papers are published in these core journals, as the application is very widespread. All the main analytical chemistry journals, for example, regularly publish chemometrics papers. In addition, journals in areas such as chromatography, food chemistry, spectroscopy, and some areas of biology regularly publish good application articles. Because of the enormous diversity of applications, chemometrics is not a unitary subject such as, for example, synthetic chemistry, where the vast majority of papers are published in a small number of dedicated core journals.

There are numerous sources of information on the Web also, but since Web sites change rapidly only a very small selection will be listed. The NIST handbook of statistical methods, although aimed primarily at engineering statistics, is a superb repository of information [33] developed over many years, and very relevant to chemometrics. The Spectroscopynow Web site is a well maintained general repository of information in chemometrics [34].

REFERENCES

1. R.G. Brereton, *Chemometrics: Data Analysis for the Laboratory and Chemical Plant,* John Wiley & Sons, Ltd, Chichester, 2003
2. D.L. Massart, B.G.M. Vandeginste, L.M.C. Buydens, S. De Jong, P.J. Lewi and J. Smeyers-Verbeke, *Handbook of Chemometrics and Qualimetrics Part A*, Elsevier, Amsterdam, 1997
3. B.G.M. Vandeginste, D.L. Massart. L.M.C. Buydens, S. De Jong, P.J. Lewi and J. Smeyers-Verbeke, *Handbook of Chemometrics and Qualimetrics Part B*, Elsevier, Amsterdam, 1998
4. M. Otto, *Chemometrics: Statistics and Computer Applications in Analytical Chemistry*, Wiley-VCH, Weinheim, 1998
5. K.R. Beebe, R.J. Pell and M.B. Seasholtz, *Chemometrics: a Practical Guide*, John Wiley & Sons, Inc., New York, 1998
6. R. Kramer, *Chemometrics Techniques for Quantitative Analysis*, Marcel Dekker, New York, 1998
7. P.J. Gemperline (editor), *Practical Guide to Chemometrics*, 2nd Edn, CRC, Boca Raton, FL, 2006
8. D.L. Massart, B.G.M. Vandeginste, S.N. Deming, Y. Michotte, and L. Kaufman, *Chemometrics: a Textbook*. Elsevier, Amsterdam, 1988
9. M.A. Sharaf, D.L. Illman and B.R. Kowalski, *Chemometrics*, John Wiley & Sons, Inc., New York, 1986
10. R.G. Brereton, *Chemometrics: Applications of Mathematics and Statistics to Laboratory Systems*, Ellis Horwood, Chichester, 1990
11. B.R. Kowalski (editor), *Chemometrics: Mathematics and Statistics in Chemistry*, Reidel, Dordrecht, 1984
12. D.L. Massart, R.G. Brereton, R.E. Dessy, P.K. Hopke, C.H. Spiegelman and W. Wegscheider (editors), *Chemometrics Tutorials*, Elsevier, Amsterdam, 1990
13. R.G. Brereton, D.R. Scott, D.L. Massart, R.E. Dessy, P.K. Hopke, C.H. Spiegelman and W. Wegscheider (editors), *Chemometrics Tutorials II*, Elsevier, Amsterdam, 1992
14. J.N. Miller. and J.C. Miller, *Statistics and Chemometrics for Analytical Chemistry*, 5th Edn, Pearson, Harlow, 2005

15. W.P. Gardiner, *Statistical Analysis Methods for Chemists: a Software-based Approach*, Royal Society of Chemistry, Cambridge, 1997

16. R. Caulcutt and R. Boddy, *Statistics for Analytical Chemists*, Chapman and Hall, London, 1983

17. H. Martens and M. Martens, *Multivariate Analysis of Quality*, John Wiley & Sons, Ltd, Chichester, 2000

18. E.R. Malinowski, *Factor Analysis in Chemistry*, 3rd Edn, John Wiley & Sons, Inc., New York, 2002

19. H. Martens and T. Næs, *Multivariate Calibration*, John Wiley & Sons, Ltd, Chichester, 1989

20. T. Naes, T. Isaksson, T. Fearn and T. Davies, *A User Friendly Guide to Multivariate Calibration and Classification*, NIR Publications, Chichester, 2002

21. M. Meloun, J. Militky and M. Forina, *Chemometrics for Analytical Chemistry Vol. 1*, Ellis Horwood, Chichester, 1992

22. M. Meloun, J. Militky and M. Forina, *Chemometrics for Analytical Chemistry Vol. 2*, Ellis Horwood, Chichester, 1994

23. S.N. Deming and S.L. Morgan, *Experimental Design: a Chemometric Approach*, Elsevier, Amsterdam, 1994

24. C.K. Bayne and I.B. Rubin, *Practical Experimental Designs and Optimisation Methods for Chemists*, VCH, Deerfield Beach, 1986

25. E. Morgan, *Chemometrics: Experimental Design*, John Wiley & Sons, Ltd, Chichester, 1995

26. D.L. Massart and L. Kaufmann, *The Interpretation of Analytical Chemical Data by the Use of Cluster Analysis*, John Wiley & Sons, Ltd, New York, 1983

27. R.G. Brereton (editor), *Multivariate Pattern Recognition in Chemometrics, Illustrated by Case Studies*, Elsevier, Amsterdam, 1992

28. H. Mark and J. Workman, *Statistics in Spectroscopy*, 2nd Edn, Academic Press, New York, 2003

29. P. Gans, *Data Fitting in the Chemical Sciences: by the Method of Least Squares*, John Wiley & Sons, Ltd, Chichester, 1992

30. M.J. Adams, *Chemometrics in Analytical Spectroscopy*, 2nd Edn, Royal Society of Chemistry, Cambridge, 2004

31. K.H. Esbensen, *Multivariate Data Analysis in Practice*, CAMO, Oslo, 2002

32. L. Eriksson, E. Johansson, N. Kettaneh-Wold and S. Wold, *Multi- and Megavariate Data Analysis: Principles and Applications*, Umetrics, Umeå , 2001

33. http://www.itl.nist.gov/div898/handbook/

34. http://www.spectroscopynow.com

2

Experimental Design

2.1 WHY DESIGN EXPERIMENTS IN CHEMISTRY?

Many chemists believe they are good at designing experiments. Yet few have a knowledge of formal statistical experimental design.

There are many reasons why the chemist can be more productive if he or she understands the basis of design. There are four main areas:

- *Screening.* Such experiments involve seeing which factors are important for the success of a process. An example may be the study of a chemical reaction, dependent on proportion of solvent, catalyst concentration, temperature, pH, stirring rate, etc. Typically 10 or more factors might be relevant. Which can be eliminated, and which studied in detail?
- *Optimization.* This is one of the commonest applications in chemistry. How to improve a synthetic yield or a chromatographic separation? Systematic methods can result in a better optimum, found more rapidly.
- *Saving time.* In industry, this is possibly the major motivation for experimental design. There are obvious examples in optimization and screening, but an even broader range of cases arise from areas such as quantitative structure–property relationships. From structural data, of existing molecules, it is possible to predict a small number of compounds for further testing, representative of a larger set of molecules. This allows enormous savings in time.
- *Quantitative modelling.* Almost all experiments, ranging from simple linear calibration in analytical chemistry to complex physical processes, where a series of observations are required to obtain a mathematical model of the system, benefit from good experimental design.

A simple example is that of the optimization of the yield of a reaction as a function of reagent concentration. A true representation is given in Figure 2.1. In reality this contour plot is unknown in advance, and the experimenter wishes to determine the pH and concentration (in mM) that provides the best reaction conditions. To within 0.2 of a pH and concentration unit, this optimum happens to be pH 4.4 and 1.0 mM. Many experimentalists will start by guessing one of the factors, say concentration, and then finding the best pH at that concentration. Consider an experimenter who chooses to start the experiment at 2 mM and wants to find the best pH. Figure 2.2 shows the yield at 2 mM. The best pH is undoubtedly

Applied Chemometrics for Scientists   R. G. Brereton
© 2007 John Wiley & Sons, Ltd

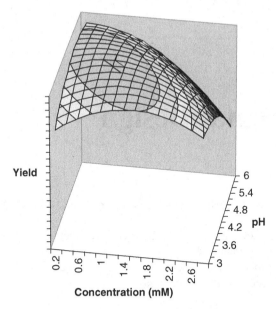

Figure 2.1 Graph of yield versus concentration and pH of an imaginary reaction

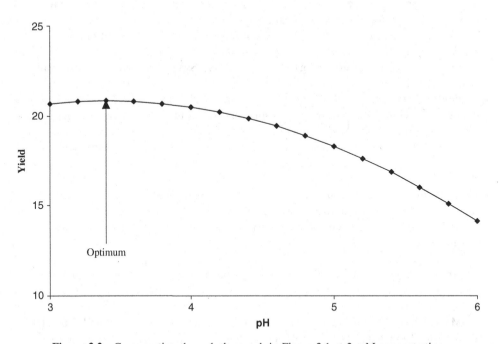

Figure 2.2 Cross-section through the graph in Figure 2.1 at 2 mM concentration

a low one, in fact pH 3.4. So the next stage is to perform the experiments at pH 3.4 and improve on the concentration, as shown in Figure 2.3. The best concentration is 1.4 mM. These answers, 1.4 mM and pH 3.4, are quite far from the true ones. The reason for this problem is that the influences of pH and temperature are not independent. In chemometric

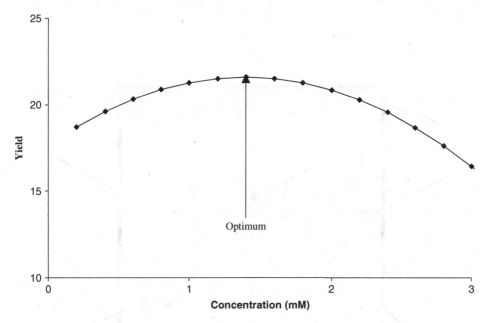

Figure 2.3 Graph of yield versus concentration at the optimum pH (3.4) obtained in Figure 2.2

terms, they 'interact'. In many cases, interactions are common sense. The best pH found in one solvent, may be different to that found in another solvent. Chemistry is complex, but how to find the true optimum, by a quick and efficient manner, and to be confident in the result? Experimental design provides the chemist with a series of rules to guide the optimization process which will be explored later.

Another example relates to choosing compounds for biological tests. Consider the case where it is important to determine whether a group of compounds are toxic, often involving biological experiments. Say there are 50 potential compounds in the group. Running comprehensive and expensive tests on each compound is prohibitive. However, it is likely that certain structural features will relate to toxicity. The trick of experimental design is to choose a selection of the compounds and then perform tests only on this subset. Chemometrics can be employed to develop a mathematical relationship between chemical property descriptors (e.g. bond lengths, polarity, steric properties, reactivities, functionalities) and biological functions, via a computational model. The question asked is whether it is really necessary to test all 50 compounds for this model? The answer is no. Choosing a set of 8 or 16 compounds may provide adequate information to predict not only the influence of the remaining compounds (and this can be tested), but any unknown in the group.

Figure 2.4 illustrates a simple example. Consider an experimenter interested in studying the influence of hydrophobicity and dipoles on a set of candidate compounds, for example, in chromatography. He or she finds out these values simply by reading the literature and plots them in a simple graph. Each circle in the figure represents a compound. How to narrow down the test compounds? One simple design involves selecting nine candidates, those at the edges, corners and centre of the square, indicated by arrows in the diagram. These candidates are then tested experimentally, and span the typical range of compounds. In reality there are vastly more chemical descriptors, but similar approaches can be employed,

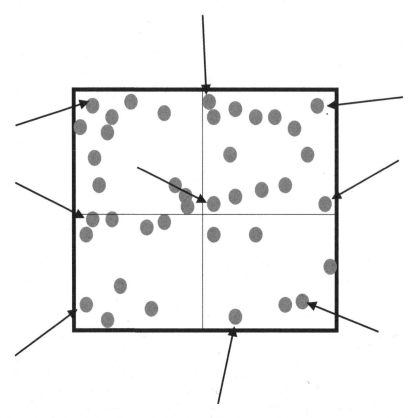

Figure 2.4 Example of choosing molecules from descriptors. The axes may refer to physical properties, e.g. dipole and hydrophobicity, and the nine compounds chosen are said to best span the space to be used for experiments to predict the behaviour of other compounds

using, instead of straight properties, statistical functions of these (for example Principal Components as discussed in Sections 5.2 and 5.3) to reduce the number of axes, typically to about three, and then choose a good and manageable selection of compounds.

The potential uses of rational experimental design throughout chemistry are large, and the basic principles of experimental design together with some of the most popular designs are described below.

2.2 DEGREES OF FREEDOM AND SOURCES OF ERROR

Every experimental procedure is subject to errors, that is an experiment is never perfectly repeatable. This is important to understand, especially how to study and take this problem into account. One of the most important ideas to understand is the concept of degrees of freedom and how they relate to sources of error.

Most experiments result in some sort of *model* which is a mathematical way of relating the experimental *response* to the *factors*. An example of a response is the yield of a synthetic reaction; the factors may be the pH, temperature and catalyst concentration. An experimenter wishes to run a reaction at a given set of conditions and predict what the yield will be.

How many experiments should be performed in order to provide confident predictions of the yield? Five, ten, or twenty? Obviously, the more the experiments, the better the predictions, but the greater the time, effort and expense. So there is a balance, and statistical experimental design helps to guide the chemist as to how many and what type of experiments should be performed.

The idea of *degrees of freedom* is a key concept. Consider a linear calibration experiment, for example, measuring the peak height in electronic absorption spectroscopy as a function of concentration, at five different concentrations, illustrated in Figure 2.5. This process can be called calibration whereby the peak height is calibrated to concentration, with the eventual aim of doing away with measuring the concentration directly after reference standards have been produced, and to use instead the spectroscopic intensity for estimation (see Chapter 6).

Often the chemist fits a straight line *model* (or mathematical relationship) to the response of the form $y = b_0 + b_1 x$, where y is the response (in this case the peak height), x is the factor (in this case concentration) and b_0 and b_1 are said to be the *coefficients* of the model. There are two coefficients in this equation, but five experiments have been performed. The degrees of freedom available for determining how well the data fits a straight line (often called the lack-of-fit: seeing how well or badly the theoretical model is fitted) are given by:

$$D = N - P$$

where N is the number of experiments (5 in this case) and P the number of coefficients (2 in this case), and equals 3 in this case. The degrees of freedom tell us how much information is available to check that the response is truly linear. The more degrees of freedom, the more information we have to be able to determine whether the data are consistent with an underlying linear process. Is it justifiable to fit a straight line to the data? If there are baseline problems or if the solution is too concentrated, a linear model may not be appropriate, at high

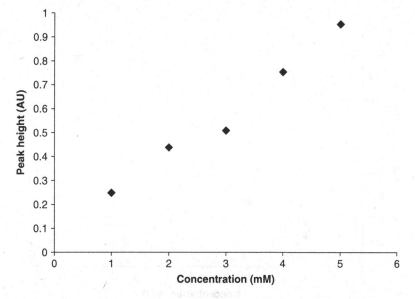

Figure 2.5 Graph of peak height (e.g. in absorbance units at a specific wavelength) versus concentration

concentrations the Beer–Lambert law is no longer obeyed, whereas at low concentrations the signal is dominated by noise.

From these data we can obtain an error which relates to how well the experiment obeys a linear model. This error, however, is in the form of a simple number in our example expressed in absorbance units (AUs). Physical interpretation is not so easy. Consider an error that is reported as 1000 mg/l: this looks large, but then express it as g/cm^3 and it becomes 0.001. Is it now a large error? The absolute value of the error must be compared with something, and hence the importance of repeating experiments under as close as possible to identical conditions, which is usually called *replication*, comes into play. It is useful in any design to repeat the experiment a few times under identical conditions: this gives an idea of the experimental error. The larger the error the harder it is to make good predictions. Figure 2.6 is of a linear calibration experiment with large errors: these may be due to many reasons, for example, instrumental performance, accuracy of volumetric flasks, accuracy of repeat weighings and so on. It is hard to see whether the results are linear or not. The experiment at the top right hand corner might be a 'rogue' experiment. Consider a similar experiment, but with lower experimental error (Figure 2.7). Now it is clear that a linear model is unlikely to be suitable, but this is only detectable because the experimental error is small compared with the deviation from linearity. In Figures 2.6 and 2.7, an extra five degrees of freedom (the five replicates) have been added to provide information on replication errors. The degrees of freedom available to test for the lack of fit to a linear model now become:

$$D = N - P - R$$

where R equals the number of replicates (5), and $N = 10$.

In many experiments it is important to balance the number of unique experimental conditions against the number of replicates. Each replicate provides a degree of freedom towards measuring experimental error. A degree of freedom tree is a method often used to represent

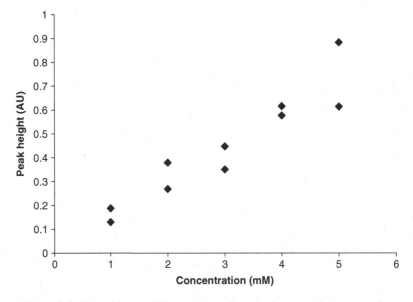

Figure 2.6   Experiments of Figure 2.5 replicated twice at each concentration

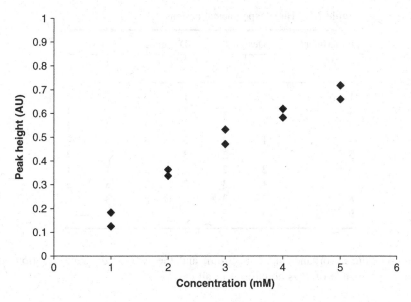

Figure 2.7 A similar experiment to that in Figure 2.6 except that the experimental errors are smaller

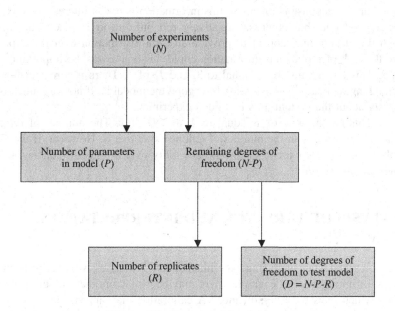

Figure 2.8 Degrees of freedom tree

this information; a simplified version is illustrated in Figure 2.8. A good rule of thumb is that the number of replicates (R) should equal approximately the number of degrees of freedom for testing for the lack-of-fit to the model (D) unless there is an overriding reason for studying one aspect of the system in preference to another. Consider three experimental designs in Table 2.1. The aim is to produce a linear model of the form $y = b_0 + b_1x_1 + b_2x_2$: y

Table 2.1 Three experimental designs

Experiment number	Design 1		Design 2		Design 3	
	A	B	A	B	A	B
1	1	1	1	2	1	3
2	2	1	2	1	1	1
3	3	1	2	2	3	3
4	1	2	2	3	3	1
5	2	2	3	2	1	3
6	3	2	2	2	1	1
7	1	3	2	2	3	3
8	2	3	2	2	3	1
9	3	3				

may represent the absorbance in a spectrum, and the two xs the concentrations of two compounds. The value of P is equal to 3 in all cases.

- *Design 1.* This has a value of R equal to 0, and D of 6. There is no information about experimental error and all effort has gone into determining the model. If it is known with certainty that the response is linear (or this information is not of interest) this experiment may be a good one, but otherwise relatively too little effort is placed in replication. Although this design may appear to provide an even distribution over the experimental domain, the lack of replication information could, in some cases, be important.
- *Design 2.* This has a value of R equal to 3, and D of 2. There is a reasonable balance between taking replicates and assessing how good the model is. If nothing much is known of certainty about the system, this is a good experiment.
- *Design 3.* This has a value of R equal to 4, and D of 1. The number of replicates is rather large compared with the number of unique experiments. However, if the main aim is simply to investigate experimental reproducibility over a range of concentrations, this approach might be useful.

2.3 ANALYSIS OF VARIANCE AND INTERPRETATION OF ERRORS

A key question asked when analysing experimental data is to assess whether a specific factor is significant or important. For example, there may be 10 factors that influence the yield of a reaction – which ones have significance? A calibration may appear curved – is it really curved, or can it be approximated by a straight line, is the squared term in the equation significant? It might appear that two factors such as pH and temperature interact – are these interactions important (see Section 2.1 for an introduction to interactions)?

A conventional approach is to set up a mathematical model linking the response to coefficients of the various factors. Consider our simple linear calibration experiment, where the height of a peak (y) in a spectrum is related to concentration (x) by the equation:

$$y = b_0 + b_1 x$$

The term b_0 represents an intercept term, which might be influenced by the baseline of the spectrometer, the nature of a reference sample (for a double beam instrument) or the solvent absorption. Is this second, intercept, term significant? Adding extra terms will *always* improve the fit to the straight line, so simply saying that a better straight line is obtained by adding the intercept term does not always help: it would be possible to include all sorts of terms such as logarithms and sinewaves and cubes and each time a term is added the fit gets better: this does not imply that they have any relevance to the underlying physical reality. One approach to answering this question is to see how important the intercept term is relative to the experimental (replicate) error. In Section 2.2 we discussed how experimental design errors be divided up into different types of degrees of freedom due to replication and lack of fit to the model, and we will use this information in order to obtain a measure of significance of the intercept term.

Results from two experiments are illustrated in Figure 2.9, the question being asked is whether there is a significant intercept term. This may provide an indication as to how serious a baseline error is in a series of instrumental measurements, for example, or whether there is a background interferent. The first step is to determine the number of degrees of freedom for determining the fit to the model (D). For each experiment:

- N (the total number of experiments) equals 10;
- R (the number of replicates) equals 4, measured at concentrations 1, 3, 4 and 6 mM.

Two models can be examined, the first without an intercept of the form $y = b_1 x$ and the second with an intercept of the form $y = b_0 + b_1 x$. In the former case:

$$D = N - R - 1 = 5$$

and in the latter case:

$$D = N - R - 2 = 4$$

because there is only one parameter in the first model, but two in the second.

The tricky part comes in determining the size of the errors. An intercept could be important if it contributes significantly more to improving the quality of the model than the replicate error. In other words, this would imply that there is good evidence that there is a baseline problem that is not just due to problems with reproducibility of the experiments. How can these errors be determined?

- The replicate error can be obtained from the difference between the experimental response when the measurement is repeated at identical concentrations. For example, if an experiment is performed at 1 mM twice, and the first time the response is 20 units, and the second time it is 24 units, then the average response is 22 units. The replicate error is 2 units for the 1 mM experiments. The replicate errors at concentrations 1, 3, 4 and 6 mM (in the case described here) are then squared and summed. The larger this sum, the bigger the error.
- The total error sum of squares is simply the sum of square difference between the observed readings and those predicted using a best fit model (for example obtained using standard regression procedures in Excel).

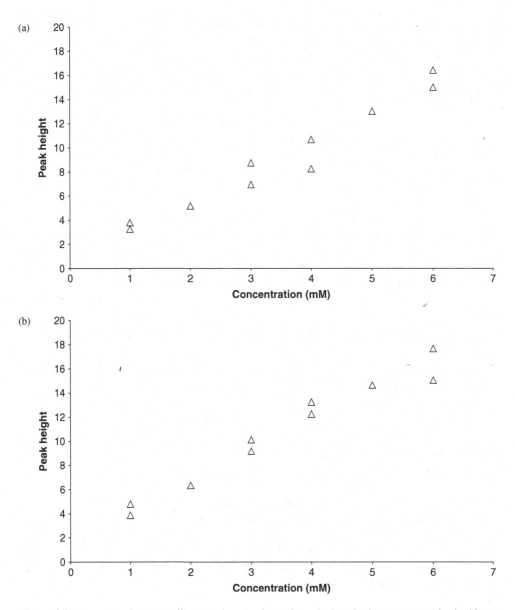

Figure 2.9   Two experiments to illustrate how to determine whether the intercept term is significant

There are, of course, two ways in which a straight line can be fitted, one with and one without the intercept. Each generates a different sum of squares according to the model. The values are given in Table 2.2.

Several conclusions can be drawn.

- The replicate sum of squares is obviously the same no matter which model is employed for a given experiment, but differs for each of the two experiments. Each experiment results in roughly similar replicate errors, suggesting that the experimental procedure (e.g.

Table 2.2 Sum of square replicate errors for the two datasets in Figure 2.9, using models with and without an intercept

	Figure 2.9(a)	Figure 2.9(b)
Model without intercept		
Total error sum of squares	9.11512	15.4694
Replicate error sum of squares ($d.f. = 4$)	5.66499	4.77567
Difference between sum of squares ($d.f. = 5$)	3.45013	**10.6937**[a]
Model with intercept		
Total error sum of squares	8.37016	7.24014
Replicate error sum of squares ($d.f. = 4$)	5.66499	4.77567
Difference between sum of squares ($d.f. = 4$)	2.70517	2.46447

d.f., degrees of freedom.
[a] Note the high error (indicated in bold) for the model without the intercept for the data in Figure 2.9(b).

dilutions, instrumental method) is similar in both cases. Only four degrees of freedom are used to measure this error so exact correspondence is unlikely.
- The total error reduces when an intercept term is added in both cases, as will always happen.
- The difference between the total error and the replicate error gives a term often called the 'lack-of-fit' error. The bigger this is, the worse the model. The lack-of-fit error is slightly smaller than the replicate error, in all cases, except for the model without the intercept in data of Figure 2.9(b). This suggests that adding the intercept term to the data of Figure 2.9(b) makes a big difference to the quality of the model and is significant.

The numbers in Table 2.2 can be statistically compared by dividing each error by their respective number of degrees of freedom, to give an 'average' error for each type of variability (replicate error with R degrees of freedom and lack-of-fit error with D degrees of freedom), often called a variance (see Section 3.3.1). The ratio between the average lack-of-fit and replicate error provides a numerical value of the significance of the error. If excluding a term, such as the intercept, greatly increases this ratio, then there is evidence that such a term is important. Often the value of this ratio is assessed using a statistic called the F-test (Section 3.8) which provides a probability that the term is significant. Most common statistical software performs these calculations automatically. This method is often called Analysis of Variance (ANOVA) and is described in much greater detail in many texts. A well known book is that by Deming and Morgan, pioneers of experimental design in chemistry [1].

ANOVA can be used in many situations to determine the significance of factors or terms in models. For example, if there are 10 possible factors in determining the success of a process, which ones are significant? Are the interactions between two factors significant? Are the squared terms important? Many designs used by chemists provide sufficient information for determination of significance of various different factors, allowing the answer to numerous questions. Most common statistical packages such as Minitab and SAS, make extensive use of ANOVA for determining significance of factors, the principles of which are quite straightforward, and involve comparing the size of errors, usually the correlation of a factor or term in an equation against a replicate or experimental error.

2.4 MATRICES, VECTORS AND THE PSEUDOINVERSE

At this point we will digress to give a short introduction to matrices. Matrix notation is generally useful throughout chemometrics and this small section has relevance to several parts of this book.

In this book we employ IUPAC recommended notation as follows.

- A scalar (or a single number for example a single concentration), is denoted in italics, e.g. x.
- A vector (or row/column of numbers for example a single spectrum) is denoted by lower case bold italics, e.g. $\boldsymbol{x}$.
- A matrix (or two-dimensional array of numbers for example a set of spectra) is denoted by upper case bold italics, e.g. $\boldsymbol{X}$. Scalars and vectors can be regarded as matrices, one or both of whose dimensions are equal to 1.

A matrix can be represented by a region of a spreadsheet and its dimensions are characterized by the number of rows (for example the number of samples or spectra) and the number of columns (for example the number of chromatographic peaks or biological measurements). Figure 2.10 is a 3×6 matrix.

The inverse of a matrix is denoted by $^{-1}$ so that $\boldsymbol{X}^{-1}$ is the inverse of $\boldsymbol{X}$. Only square matrices (where the number of rows and columns are equal) can have inverses. A few square matrices do not have inverses, this happens when the columns are correlated. The product of a matrix and its inverse is the identity matrix.

The transpose of a matrix is often denoted by $'$ and involves swapping the rows and columns: so, for example, the transpose of a 5×8 matrix is an 8×5 matrix.

Most people in chemometrics use the dot ('.') product of matrices (and vectors). For the dot product $\boldsymbol{X}.\boldsymbol{Y}$ to be viable, the number of columns in $\boldsymbol{X}$ must equal the number of rows in $\boldsymbol{Y}$ and the dimensions of the product equal the number of rows in the first matrix and number of columns in the second one. Note that in general $\boldsymbol{Y}.\boldsymbol{X}$ (if allowed) is usually different to $\boldsymbol{X}.\boldsymbol{Y}$. For simplicity, sometimes the dot symbol is omitted although it is always the dot product that is computed.

To illustrate the use of matrices, we will take a set of measurements y that are related to a variable x as follows:

$$y = bx$$

If there are several such measurements, there will be several y values and their corresponding x values, so y and x are no longer single numbers but vectors of numbers (the length corresponding to the number of measurements, normally expressed as row vectors), and the equation can be generalized to:

$$\boldsymbol{y} = \boldsymbol{x}\, b$$

24	23	20	19	17	19
56	51	52	51	45	50
31	31	30	27	25	25

Figure 2.10 A 3×6 matrix represented by a region of a spreadsheet

where b is a constant or scalar. Knowing x and y can we deduce b? The equation above is not normally exactly obeyed and includes some errors in it, and may alternatively be best expressed by:

$$y \approx x \, b$$

Note a slight quirk in that the scalar b should be placed after and not before the x. This can be understood by reference to the dimensions of each vector in the equation above. If 10 measurements are performed, then both y and x have dimensions 10×1, where b has dimensions 1×1, and the rules of matrix multiplication (a scalar can be considered as a 1×1 matrix) dictate that the number of columns in the first matrix to be multiplied (x) must equal the number of rows of the second one (b) ($=1$).

All that is required is to introduce the concept of an 'inverse' of a vector or matrix, which we denote x^+, then:

$$b \approx x^+ . y$$

The tricky bit is that only square matrices (where the number of columns and rows are equal) have mathematical inverses, but chemometricians use the idea of a *pseudoinverse*. There are two types, the right pseudoinverse:

$$x' . (x \, x')^{-1}$$

and the left pseudoinverse:

$$(x' x)^{-1} . x'$$

Some packages such as Matlab, calculate the pseudoinverse in one step, but it is quite easy to obtain these by manually multiplying matrices, e.g. in Excel. When using vectors, the pseudoinverse above simplifies still further but we will not discuss this here. In the case above we use the left pseudoinverse, so:

$$b \approx (x' x)^{-1} . x' . y$$

Note that $(x' x)^{-1}$ is actually a scalar or a single number in this case (when extending the calculation to a situation where there is more than one column in 'x' this is not so), equal to the inverse of the sum of squares of the elements of x. An easy way to understand this, is to put some simple numbers into Excel or some equivalent package and play with them to check that all the summations do indeed add up correctly.

Often there is more than one term in x for example:

$$y \approx b_1 x_1 + b_2 x_2 + b_3 x_3$$

Now the power of matrices and vectors becomes evident. Instead of using a single column vector for X we use a matrix with three columns of experimentally known x values, and b becomes a column vector of three elements, each element corresponding to one coefficient in the equation. The equation can now be expressed as:

$$y \approx X . b$$

so that:

$$b \approx X^+ . y$$

which is an easy way of calculating the coefficients computationally, and saves using complex equations.

These ideas can easily be extended so that Y and/or B become matrices rather than vectors.

2.5 DESIGN MATRICES

A key plank for the understanding of experimental designs is the concept of a design matrix. A design may consist of a series of experiments performed under different conditions, e.g. a reaction at differing pHs, temperatures, and concentrations. Table 2.3 illustrates a typical experimental set-up. Note the replicates in the final five experiments: in Section 2.11 we will discuss this type of design in greater detail. Further discussion about matrices is provided in Section 2.4.

It is normal to model experimental data by forming a relationship between the factors or independent variables such as temperature or pH and a response or dependent variable such as a synthetic yield, a reaction time or a percentage impurity. A typical model for three factors might be expressed as:

$$y =$$ (response)

b_0+ (an intercept or average)

$b_1x_1 + b_2x_2 + b_3x_3+$ (linear terms depending on each of the three factors)

$b_{11}x_1^2 + b_{22}x_2^2 + b_{32}x_3^2+$ (quadratic terms depending on each of the three factors)

$b_{12}x_1x_2 + b_{13}x_1x_3 + b_{23}x_2x_3$ (interaction terms between the factors)

An explanation for these terms is as follows:

Table 2.3 A typical experimental set-up, where a reaction is run varying three conditions

pH	Temperature (°C)	Concentration (mM)
6	60	4
6	60	2
6	20	4
6	20	2
4	60	4
4	60	2
4	20	4
4	20	2
6	40	3
4	40	3
5	60	3
5	20	3
5	40	4
5	40	2
5	40	3
5	40	3
5	40	3
5	40	3
5	40	3

- The intercept relates to the average experimental value. Most experiments study variation about an average. For example, if the yield of a reaction is 80% at pH 5 and 60% at pH 7, the yield at the lower pH is 10% greater than the average.
- The linear terms allow for a direct relationship between the response and a given factor. For some experimental data, there are only linear terms. If pH increases, does the yield increase or decrease, and if so by how much?
- In many situations, quadratic terms are also common. This allows curvature, and is one way of obtaining a maximum or minimum. Most chemical reactions have an optimum performance at a particular pH, for example. Almost all enzymic reactions work in this way. Quadratic terms balance out the linear terms.
- In Section 2.1 we discussed interaction terms. These arise because the influence of two factors on the response is rarely independent. For example, the optimum pH at one temperature may differ from that at a different temperature.

Some of these terms may not be very significant or relevant, but it is up to the experimenter to check which terms are significant using approaches such as ANOVA (see Section 2.3). In advance it is often hard to predict which factors are important.

There are 10 terms or parameters in the equation on page 22, and a design matrix involves setting up a matrix for which each column represents one of the parameters and each row an experiment. For readers of this book that are not familiar with matrix notation, there is further discussion in Section 2.4. A matrix is analogous to a portion of a spreadsheet. There are simple rules for multiplication of matrices, which can be performed easily using basic spreadsheet functions that are readily available in packages such as Excel.

Continuing with the example in Table 2.3, there are 10 parameters in the model implying 10 columns, and 20 experiments implying 20 rows, as illustrated in Figure 2.11. The design matrix can then be calculated (Table 2.4). The reader can check the numbers numerically: for example, the interaction term between pH and temperature for the first experiment is 360 which equals 6 (pH) × 60 (temperature), and appears in the eighth column of the first row, corresponding to the term x_1x_2 [term 8 in the equation on page 22] for experiment 1.

The design matrix depends on two features, namely:

- the number and arrangement of the experiments; and
- the mathematical model to be assessed.

It can be shown that:

- the 20 responses form a vector with 20 rows and 1 column, called y;
- the design matrix is a matrix with 20 rows and 10 columns as illustrated in Table 2.4, which we will call D;
- the 10 coefficients form a vector with 10 rows and 1 column, called b.

The ideal relationship between the responses, the coefficients and the experimental conditions can be expressed by:

$$y = D.b$$

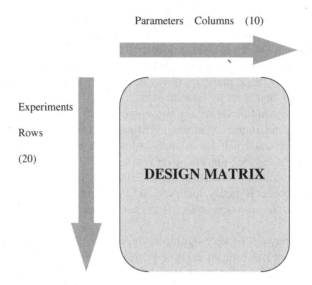

Parameters   Columns   (10)

Experiments

Rows

(20)

DESIGN MATRIX

Figure 2.11   Design matrix for experiments in Table 2.3 and model in Section 2.5

Table 2.4   Numerical calculation of design matrix illustrated in Figure 2.11

Intercept	Linear terms			Quadratic terms			Interaction terms		
1	6	60	4	36	3600	16	360	24	240
1	6	60	2	36	3600	4	360	12	120
1	6	20	4	36	400	16	120	24	80
1	6	20	2	36	400	4	120	12	40
1	4	60	4	16	3600	16	240	16	240
1	4	60	2	16	3600	4	240	8	120
1	4	20	4	16	400	16	80	16	80
1	4	20	2	16	400	4	80	8	40
1	6	40	3	36	1600	9	240	18	120
1	4	40	3	16	1600	9	160	12	120
1	5	60	3	25	3600	9	300	15	180
1	5	20	3	25	400	9	100	15	60
1	5	40	4	25	1600	16	200	20	160
1	5	40	2	25	1600	4	200	10	80
1	5	40	3	25	1600	9	200	15	120
1	5	40	3	25	1600	9	200	15	120
1	5	40	3	25	1600	9	200	15	120
1	5	40	3	25	1600	9	200	15	120
1	5	40	3	25	1600	9	200	15	120
1	5	40	3	25	1600	9	200	15	120

as is illustrated in Figure 2.12. Of course this relationship is not normally exactly obeyed because of experimental error but is the best approximation.

An aim of analysis of designed experiments is to find out b (the best fit coefficients) from y (the experimental responses) and D (the design matrix).

Response = **Design matrix** × **Coefficients**

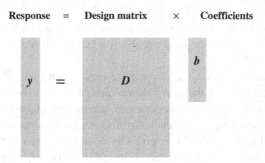

Figure 2.12 Illustration of how a response is the product of a design matrix and the coefficients in the equation

Three types of information are useful:

- The size of the coefficients can inform the experimenter how significant the coefficient is. For example, does pH significantly improve the yield of a reaction? Or is the interaction between pH and temperature significant? In other words does the temperature at which the reaction has a maximum yield differ at pH 5 and at pH 7? ANOVA (Section 2.3) can be employed for assessing the significance of these coefficients and many statistical packages can be used to automate these calculations.
- The coefficients can be used to construct a model of the response, for example, the yield of a reaction as a function of pH and temperature and so establish the optimum conditions for obtaining the best yield. The experimenter is often not so interested in the nature of the equation for the yield as an end in itself, but is very interested in the best pH and temperature.
- Finally, a quantitative model may be interesting. Predicting the concentration of a compound from the absorption in a spectrum requires an accurate knowledge of the relationship between the variables. Under such circumstances the precise value of the coefficients is important.

There are numerous ways of computing **b** from **y** and **D**. It is important to realize that most traditional approaches were developed before modern computers were available. Some early workers even calculated coefficients using pencil and paper, and some designs were even proposed to make these calculations easy. It might have taken a couple of days to determine the coefficients from the responses for complex designs before the years of digital computers – remember principles of experimental design are more than 100 years old. When reading the literature it is important to recognize this, and realize that the majority of modern approaches for calculation are matrix based which can easily be done using common packages on microcomputers. The simplest computational method is via the pseudoinverse (Section 2.4), but if **D** is a square matrix (the number of rows equalling the number of columns), the straight inverse can be employed instead.

2.6 FACTORIAL DESIGNS

Now we have discussed the basic building blocks of experimental design we will outline several of the most fundamental designs used by chemists.

One of the simplest is a factorial design.

The most common are two level factorial designs. These are mainly used for screening, that is to determine the influence of a number of factors on a response, and which are important.

Consider a chemical reaction, whose outcome may depend on pH or temperature, and possibly the interaction between these. A factorial design will consist of running the reaction when each of two factors (pH and temperature), are at one of two levels. This series of experiments can be studied by a two level, two factor experimental design. The number of experiments for a full factorial design is given by $N = l^f$ where l is the number of levels, and f the number of factors, so a four factor design will consist of 16 experiments.

The first step in setting up such series of experiments is to choose a high and low level for each factor, for example, $30°$ and $60°$, and pH 4 and 6. As in all chemometrics, some advance knowledge of the system is required, and these ranges have to be sensible ones, which can only realistically be determined from chemical (or domain) knowledge.

The next step is to use a standard design. In this case it will consist of 2^2 or four experiments. Each factor is 'coded' as '$-$' (low) or '$+$' (high). Note that some authors use -1 and $+1$ or even 0 and 1 for low and high. When reading different texts check the notation carefully. These four sets of experimental conditions can be represented as a table equivalent to four binary numbers 00 ($--$), 01 ($-+$), 10 ($+-$) and 11 ($++$) which represent the true pHs and temperatures under which each reaction is performed.

Third, perform the experiments and obtain the response. Table 2.5 illustrates the coded and true set of experiments plus the response, for example the percentage of a by-product, the lower the better. Something immediately appears strange from these results. Although it is obvious that the higher the temperature, the higher the percentage by-product, there does not, at first, seem to be any consistent trend as far as pH is concerned. Providing the experiments were performed correctly, this suggests that there must be an interaction between temperature and pH. At a lower temperature, the percentage decreases with increase in pH, but vice versa at a higher temperature. How can we interpret this?

The fourth step, of course, is to analyse the data, by setting up a design matrix (see Section 2.5). We know that interactions must be taken into account, and so set up a design matrix as given in Table 2.6. This can be expressed either in terms of the true or coded concentrations. Note that four coefficients are obtained from the four experiments. We will return to this issue later. Note also that each of the columns are different. This is an important property and allows each of the four possible terms to be distinguished, and relates to orthogonality: we will discuss this property further in Sections 2.8 and 2.9.

Table 2.5 Coded and true experimental conditions for the first design in Section 2.6 together with a response

Experiment number	Factor 1	Factor 2		Temperature (°C)	pH		Response
1	$-$	$-$		30	4		12
2	$-$	$+$	$\longrightarrow$	30	6	$\longrightarrow$	10
3	$+$	$-$		60	4		24
4	$+$	$+$		60	6		25

Table 2.6 Design matrix for the experiment in Table 2.5

Intercept	Temperature (°C)	pH	Temperature pH		1	x_1	x_2	x_1x_2
1	30	4	120		+	−	−	−
1	30	6	180	⟶	+	−	+	−
1	60	4	240		+	+	−	+
1	60	6	360		+	+	+	+

Fifth, calculate the coefficients. It is not necessary to employ specialist statistical software for this. There are simple manual approaches, or it is possible to use common spreadsheets such as Excel. The response is given by $y = D.b$ where b consists of four coefficients. In the case of this design, simply use the matrix inverse function, so that $b = D^{-1}.y$. The coefficients are listed below.

For raw values:

- intercept = 10;
- temperature coefficient = 0.2;
- pH coefficient = −2.5;
- interaction term = 0.05.

For coded values:

- intercept = 17.5;
- temperature coefficient = 6.75;
- pH coefficient = −0.25;
- interaction term = 0.75.

Sixth, interpret the coefficients. Note that for the raw values, it appears that pH is much more influential than temperature, also that the interaction is very small. In addition, the intercept term is not the average of the four readings. The reason why this happens is that the temperature range is numerically larger than the pH range. This can be understood by considering whether a length is recorded in m or cm. A length of 0.1 m is the same as a length of 10 cm. Which of these is large? Temperature has a significant influence over the reaction no matter whether we use degrees Celsius, degrees Fahrenheit, or kelvin, but the value of the calculated coefficient depends on the units of measurement, and the product of the coefficients (b) and the measurements (x) must be the same so a big value of x will result in a small value of b and vice versa. It is the physical significance of the factor that matters. A better measure of significance comes from using the coefficients of the coded factors, in which temperature appears to have a much larger influence, pH is seen to be quite small, and the interaction is close to 1. Using coded coefficients means that the range of each of the factors is the same, so we can compare significance on the same scale.

Such two level factorial designs can be used very effectively for screening to see which factors are most important but also have limitations:

- They only provide an approximation within the experimental region. Note that for the model above it is possible to obtain nonsensical predictions of negative percentages outside the experimental region.

- They cannot take quadratic terms into account in the model, as the experiments are performed only at two levels.
- There is no replicate information.

There are two common extensions of the simple design above.

2.6.1 Extending the Number of Factors

Most reactions depend on even more factors. Typical factors influencing the yield of a reaction may be: temperature; pH; whether there is stirring or not; volume of solvent; reaction time; two types of catalyst; reagent to substrate ratio, and so on. We have listed seven factors, not atypical of a real experiment. At two levels, this would lead to 128 possible experiments. A four factor, two level design (16 experiments) is illustrated in Table 2.7.

2.6.2 Extending the Number of Levels

It may be inadequate to simply study two levels. For example, is it really sufficient to select only two temperatures? A more detailed model can be obtained using three temperatures. These designs are called multilevel designs. A two factor, three level design is illustrated in Table 2.8, consisting of nine experiments ($= 2^3$). Note that in the text below we will use $+1$ rather than $+$ to denote a high level and similarly for other levels in a design. These experiments can be visualized pictorially as in Figure 2.13, where each axis represents the levels of one factor, and each experiment is represented by a circle. It is obviously possible to visualize two and three factor experiments this way, but not when there are more than three factors. Note that since there are three levels, quadratic terms can be used in the model.

It is possible to conceive of any number of factors and experiments. A four level, five factor design involves 1024 ($= 4^5$) experiments, for example. However, as the size of

Table 2.7 Four factor, two level design

Factor 1	Factor 2	Factor 3	Factor 4
−	−	−	−
−	−	−	+
−	−	+	−
−	−	+	+
−	+	−	−
−	+	−	+
−	+	+	−
−	+	+	+
+	−	−	−
+	−	−	+
+	−	+	−
+	−	+	+
+	+	−	−
+	+	−	+
+	+	+	−
+	+	+	+

Table 2.8 Two factor three level, design

Factor 1	Factor 2
−1	−1
−1	0
−1	+1
0	−1
0	0
0	+1
+1	−1
+1	0
+1	+1

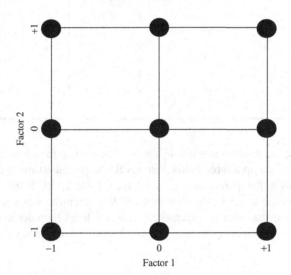

Figure 2.13 Illustration of the coded levels of a three level, two factor design; each axis represents one factor

the experiments increases it becomes impracticable to perform the design, and there are a number of tricks, outlined below, to reduce the volume of experimentation.

2.7 AN EXAMPLE OF A FACTORIAL DESIGN

In order to illustrate the uses and analysis of factorial designs we will employ a case study based on the paper recommended by Drava and colleagues of the University of Genova [2].

The example involves measuring the stability of a drug, diethylpropione, as measured by high performance liquid chromatography (HPLC) after 24 h. Various factors influence the stability, and the aim is to study three possible factors:

- moisture;
- dosage form;
- an additive, clorazepate.

Table 2.9   Levels of the three factors in the example in Section 2.7

Factor	Level (−)	Level (+)
Moisture (%)	57	75
Dosage form	Powder	Capsule
Clorazepate (%)	0	0.7

Table 2.10   Three factor, two level design and response (stability measured by percentage retained of the drug) for the example in Section 2.7

Experiment	Factor 1	Factor 2	Factor 3	Response
1	−1	−1	−1	90.8
2	1	−1	−1	88.9
3	−1	1	−1	87.5
4	1	1	−1	83.5
5	−1	−1	1	91.0
6	1	−1	1	74.5
7	−1	1	1	91.4
8	1	1	1	67.9

The first step is to determine the sensible levels of these three factors as given in Table 2.9.

The next step is to set up a three factor, two level design consisting of eight experiments, and then to determine the percentage retained (see Table 2.10). Note that the design is coded. For example, the second column implies 88.9 % retention when the moisture is at a high level (75 %), and the other two factors are at a low level (powder and no clorazepate). The design can be visualized as a box (Figure 2.14). The corners of the box represent

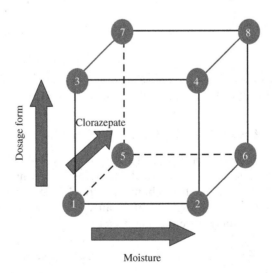

Figure 2.14   Graphical representation of design in Table 2.10

Table 2.11 Design matrix and coefficients for the experiment in Table 2.10

Intercept	Moisture	Dosage	Clorazepate	Interaction moisture/ dosage	Interaction moisture/ clorazepate	Interaction dosage/ clorazepate	Three way interaction
1	−1	−1	−1	1	1	1	−1
1	1	−1	−1	−1	−1	1	1
1	−1	1	−1	−1	1	−1	1
1	1	1	−1	1	−1	−1	−1
1	−1	−1	1	1	−1	−1	1
1	1	−1	1	−1	1	−1	−1
1	−1	1	1	−1	−1	1	−1
1	1	1	1	1	1	1	1
b_0	b_1	b_2	b_3	b_{12}	b_{13}	b_{23}	b_{123}
84.44	−5.74	−1.86	−3.24	−1.14	−4.26	0.31	−0.61

experimental conditions. For example, experiment 2 on the bottom, front and right-hand corner corresponds to a high level of moisture and a low level of the other factors.

There are numerous ways of looking at the data at the next stage. The simplest is to produce a design matrix (Table 2.11), and calculate the effect of the eight possible terms. Note that it is only sensible to use coded variables in this case as the dosage form has only two categories. Sometimes a mixture of easily measured numbers such as concentrations and temperatures, and categorical variables (was the reaction stirred – yes or no? was the colour blue – yes or no?), is employed in a typical experiment. Note that it is possible to conceive of a three parameter interaction term between all three factors, even though this may seem unlikely, but also that exactly eight terms can be used to model the eight experiments which implies that there are no degrees of freedom left to determine how well the model is fit to the data (so $D = 0$ in this case). However, in some cases this may not necessarily be a problem: in the case in question it is not really necessary to obtain an exact model for the behaviour of one factor, mainly it is important to determine whether one or more specific factors have significant influence over the response. Providing experiments are not too irreproducible, this design gives good clues.

The final stage is to interpret the results:

- All three main factors appear to have an influence, although dosage does not seem too important. As a rule of thumb, look first at the range of responses, in this experiment, the response y ranges over slightly more than 20 %. To interpret the size of the coefficients, a coefficient of +5, for example, implies that changing the level of the coded factor by 1 unit results in a 5 % increase in response: since the range of each coded factor is 2 units (from −1 to +1), this would imply that the response changes by 10 % (= 2 × 5 %) or around half the overall range (20%) between high and low levels, which is very significant. A coefficient of +1, implies that the factor exerts approximately a tenth of the influence over the response between high and low values, still quite important, since there are several factors involved.
- All the main (single factor) effects are negative, implying a lower value of stability at higher values of the factors. Moisture is the most significant.

- Interestingly the interaction between moisture and clorazepate is highly significant and negative. This 'hidden' factor reveals some very important results which may not be obvious to the experimenter at first. Without considering the interactions it might be suggested that the best conditions are at low values for all three factors, because of the negative coefficients for the single factor terms, but, in fact, the best results are for experiments 5 and 7, where moisture and clorazepate are at opposite levels. This factor is larger than the single factor terms, so suggests that the most important considerations are (a) a low moisture content and (b) a high clorazepate content in the presence of a low moisture content. These single out experiments 5 and 7 as the best.
- Note that a high clorazepate content in the presence of a high moisture content does not provide good results. Because of a significant interaction term, it is not possible to provide a good prediction of the effect of clorazepate independently of moisture.
- Finally some factors, such as the three factor interaction terms, are not very relevant. If further studies were to be performed, the next step would be to eliminate some of the less relevant factors.

In this section we have seen how to design and interpret the data from a real experiment. Without a systematic approach much experimentation could be wasted, and subtle but important trends (such as interaction effects) may be missed altogether.

2.8 FRACTIONAL FACTORIAL DESIGNS

In Sections 2.6 and 2.7 we discussed the use of factorial designs. A weakness of factorial designs is the large number of experiments that must be performed when the number of factors is large. For example, for a 10 factor design at two levels, 1024 experiments are required. This can be impracticable. Especially in the case of screening, where a large number of factors may be of potential interest, it is inefficient to run so many experiments in the first instance. There are, fortunately, numerous tricks to reduce the number of experiments.

Consider the case of a three factor, two level design. The factors, may, for example, be pH, temperature, and concentration, and the response the yield of a reaction. Eight experiments are listed in Table 2.12, the conditions being coded as usual. The design matrix (Section 2.5) can be set up as is also illustrated in the table, consisting of eight possible columns, equalling the number of experiments. Some columns represent interactions, such as a three factor interaction, that are not very likely. At first screening we primarily wish to say whether the three main factors have any real influence on the response, not to study the model in detail. In other typical cases, for example, when there may be 10 possible factors, reducing the number of factors to be studied further to three or four makes the next stage of experimentation easier.

How can we reduce the number of experiments? Two level fractional factorials reduce the number of experiments by 1/2, 1/4, 1/8 of the total and so on. Can we halve the number of experiments? A simple, but misguided, approach might be to perform the first four experiments of Table 2.12. These, however, leave the level of the first factor at $+1$ throughout. A problem is that the variation of factor 1 is now no longer studied, so we do not obtain any information on how factor 1 influences the response.

Can a subset of four experiments be selected that allows us to study the variation of all three factors? Rules have been developed, to produce fractional factorial designs. These are

Table 2.12 Experimental set-up and design matrix for a full factorial three factor design

Experiments			Design matrix							
Factor 1	Factor 2	Factor 3	1	x_1	x_2	x_3	$x_1 x_2$	$x_1 x_3$	$x_2 x_3$	$x_1 x_2 x_3$
1	1	1	1	1	1	1	1	1	1	1
1	1	−1	1	1	1	−1	1	−1	−1	−1
1	−1	1	1	1	−1	1	−1	1	−1	−1
1	−1	−1	1	1	−1	−1	−1	−1	1	1
−1	1	1	1	−1	1	1	−1	−1	1	−1
−1	1	−1	1	−1	1	−1	−1	1	−1	1
−1	−1	1	1	−1	−1	1	1	−1	−1	1
−1	−1	−1	1	−1	−1	−1	1	1	1	−1

Table 2.13 Fractional factorial design

Experiments			Matrix of effects							
Factor 1	Factor 2	Factor 3	x_0	x_1	x_2	x_3	$x_1 x_2$	$x_1 x_3$	$x_2 x_3$	$x_1 x_2 x_3$
1	1	1	1	1	1	1	1	1	1	1
1	−1	−1	1	1	−1	−1	−1	−1	1	1
−1	−1	1	1	−1	−1	1	1	−1	−1	1
−1	1	−1	1	−1	1	−1	−1	1	−1	1

obtained by taking a subset of the original experiments, but it must be the correct subset. Table 2.13 illustrates a possible fractional factorial design that enables all factors to be studied. There are a number of important features:

- Each of the four columns in the experimental matrix is different.
- In each column, there are an equal number of '−1' and '+1' levels.
- For each experiment at level '+1' for factor 1, there are an equal number of experiments for factors 2 and 3 which are at levels '+1' and '−1', and so on for every combination of factors.

This latter property is sometimes called *orthogonality*. It means that each factor is independent of each other. This is important, otherwise it is not always easy to distinguish the effect of two factors varying separately.

The properties of this design can be understood better by visualization (Figure 2.15): half the experiments have been removed. For the remainder, each face of the cube now represents two rather than four experiments, and every alternate corner corresponds to an experiment.

The design matrix of Table 2.13 is also interesting. Whereas the first four columns are all different, the last four each correspond to one of the first four columns. For example the $x_1 x_2$ column is exactly the same as the x_3 column. What does this imply? As the number of experiments is reduced, the amount of information is correspondingly reduced. Since only four experiments are now performed, it is only possible to measure four unique effects, and

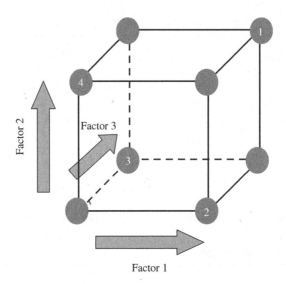

Figure 2.15   Visualization of fractional factorial design

so four out of the original eight effects can no longer be studied (it is not possible to study more than four effects if only four experiments are performed), and these effects are now confused with the effects we are most interested in. The interaction between factors 1 and 2 is said to be *confounded* with factor 3. This might mean, for example, that the interaction between temperature and pH is indistinguishable from the influence of concentration alone if these variables correspond to the three factors in the design. However, not all interactions are likely to be significant, and the purpose of preliminary screening is often simply to sort out which main factors should be studied in detail later, and by reducing the number of experiments we are banking on some of the interactions not being of primary significance or interest at this point in time.

In more complex situations, such as 10 factor experiments, it is almost certain that higher order interactions will not be significant, or if they are, these interactions are not measurable or do not have a physical meaning (we can often understand the meaning of a two or three factor interaction but not an eight factor interaction). Therefore, it is possible to select specific interactions that are unlikely to be of interest, and consciously reduce the experiments in a systematic manner by confounding these with simpler parameters. Table 2.14 is of a six factor, two level, eight experiment design. The full design would consist of 64 experiments, so only an eighth of the original possibilities have been included. Nearly all possible interactions are confounded: for example, the interaction between factors 1 and 2 is the same as the influence of factor 4 (this can be checked by multiplying the first two columns together).

There are obvious advantages in two level fractional factorial designs, but these do have some drawbacks:

- There are no quadratic terms, as the experiments are performed only at two levels, this also is a weakness of full factorial designs, and we will look at ways of overcoming this in Section 2.11.

Table 2.14 Six factor fractional factorial design

Factor 1	Factor 2	Factor 3	Factor 4	Factor 5	Factor 6
1	1	1	1	1	1
1	1	−1	1	−1	−1
1	−1	1	−1	1	−1
1	−1	−1	−1	−1	1
−1	1	1	−1	−1	1
−1	1	−1	−1	1	−1
−1	−1	1	1	−1	−1
−1	−1	−1	1	1	1

- There are no replicates.
- The number of experiments must be a power of two.

Nevertheless, this approach is very popular in many exploratory situations and has the additional advantage that the data are easy to analyse.

2.9 PLACKETT–BURMAN AND TAGUCHI DESIGNS

Fractional factorial designs have a limitation, in that the number of experiments must equal a power of two. Is it possible to set up a design to perform one experiment more than the number of factors being studied (this allows for an intercept or average), so that the minimum number of experiments required to study 19 factors should be 20? A fractional factorial design will require 32 experiments. Can this number be reduced further?

Plackett and Burman's classic 1946 paper [3] originated from the need for war-time testing of components in equipment manufacture. A large number of factors influenced the quality of these components and efficient procedures were required for screening. They propose a number of two level factorial designs, whose length (the number of experiments) is a multiple of four and whose width (the number of factors) is one less than the number of experiments. These designs reduce the number of experiments required.

One such design is given in Table 2.15 for 11 factors and 12 experiments. There are several features of the design. First, note its construction. The first row consists of an experiment at a single level. In the remaining 11 rows, the levels are related diagonally, and it is only necessary to know one row or column to obtain the entire design. The second row is a 'generator', and Plackett and Burman publish a number of such generators for various different numbers of factors.

There are as many high as low levels of each factor over the 12 experiments, as would be expected. The most important property of the design, however, is *orthogonality*, as discussed in Section 2.8. Consider factors 1 and 2:

- There are six instances in which factor 1 is at a high level, and six at a low level.
- For each of the six instances at which factor 1 is at a high level, in three cases factor 2 is at a high level, and in the other three cases it is at a low level. A similar relationship exists where factor 1 is at a low level.
- Any combination of two factors is related in a similar way.

Table 2.15 Plackett–Burman design for 11 factors

Experiments	Factors										
	1	2	3	4	5	6	7	8	9	10	11
1	−	−	−	−	−	−	−	−	−	−	−
2	+	−	+	−	−	−	+	+	+	−	+
3	+	+	−	+	−	−	−	+	+	+	−
4	−	+	+	−	+	−	−	−	+	+	+
5	+	−	+	+	−	+	−	−	−	+	+
6	+	+	−	+	+	−	+	−	−	−	+
7	+	+	+	−	+	+	−	+	−	−	−
8	−	+	+	+	−	+	+	−	+	−	−
9	−	−	+	+	+	−	+	+	−	+	−
10	−	−	−	+	+	+	−	+	+	−	+
11	+	−	−	−	+	+	+	−	+	+	−
12	−	+	−	−	−	+	+	+	−	+	+

Now we notice two important consequences of the design:

- The number of experiments must be a multiple of four.
- There are only certain very specific generators that have these properties: there are published generators for different numbers of experiments (see Table 9.3 for more details).

The importance of orthogonality can be illustrated by a simple example, in which factor 1 is always at a high level when factor 2 is low, and vice versa. Under such circumstances, it is impossible to separate the influence of these two factors. When the first factor increases, the second decreases. For example, if a reaction is always observed either at pH 5 and 30 °C or at pH 3 and 50 °C it is not possible to state whether a change in rate is a result of changing pH or temperature. The only way to be sure is to ensure that each factor is completely independent, or orthogonal, as above. Even small deviations from orthogonality can lead to the influence of different factors getting muddled up.

Standard Plackett–Burman designs exist for 3, 7, 11, 15, 19, etc., factors. If the number of experimental factors is less than the number in a standard design, the final factors can be set to *dummy* factors. Hence if we want to study just 10 real factors, use an 11 factor design, the final factor being a dummy one: this may be a variable that has no influence on the experiment, such as the technician that handed out the glassware, or the colour of laboratory furniture. In fact it is often good practice always to have one or more dummy factors when screening, because we can then feel satisfied that a real factor is not significant if its influence on the response is less than that of a dummy factor. More discussion of these designs is given in Section 9.3 in the context of improving synthetic yields.

An alternative approach comes from the work of Taguchi. His method of quality control was much used by Japanese industry, and only over the last few years was it recognized that certain aspects of the theory are very similar to Western practices. His philosophy was that consumers desire products that have constant properties within narrow limits. For example, a consumer panel may taste the sweetness of a product, rating it from 1 to 10. A good marketable product may result in a taste panel score of 8: above this the product is too sickly, and below, the consumer expects the product to be sweeter. There will be a huge number of factors in the manufacturing process that might cause deviation from the norm, including suppliers of raw materials, storage and preservation of the food and so on. Which factors are significant? Taguchi developed designs for screening a large number of potential factors.

His designs are presented in the form of table similar to that of Plackett and Burman, but using a '1' for a low and '2' for a high level. Superficially Taguchi designs might appear different to Plackett–Burman designs, but by changing the notation, and swapping rows and columns around, it is possible to show that both types of design are identical. There is a great deal of controversy surrounding Taguchi's work; while many statisticians feel that he has reinvented the wheel, he was an engineer, and his way of thinking had a major and positive effect on Japanese industrial productivity. Before globalization and the Internet, there was less exchange of ideas between different cultures. His designs are part of a more comprehensive approach to quality control in industry.

Taguchi designs can be extended to three or more levels, but construction becomes fairly complicated. Some texts do provide tables of multilevel screening designs, and it is also possible to mix the number of levels, for example having one factor at two, and another at three, levels. In many areas of chemistry, though, for screening, two levels are sufficient. Where multilevel designs have an important role is in calibration, in which each compound in a mixture is measured at several concentrations: this will be discussed in Section 2.16.

2.10 THE APPLICATION OF A PLACKETT–BURMAN DESIGN TO THE SCREENING OF FACTORS INFLUENCING A CHEMICAL REACTION

Although Plackett–Burman designs were first proposed for engineers, in fact, they have a large and significant impact in chemistry, and a quick review of the literature suggests that their classic paper is most cited within the chemical literature.

In order to illustrate the use of such designs, the yield of a reaction of the form:

$$A + B \longrightarrow C$$

is to be studied as influenced by 10 possible experimental conditions, listed in Table 2.16. The aim is to see which of the conditions have a significant influence over the reaction.

Table 2.16 Ten possible factors that influence the reaction in Section 2.10

Factor		Low	High
x_1	NaOH (%)	40	50
x_2	Temperature (°C)	80	110
x_3	Nature of catalyst	A	B
x_4	Stirring	Without	With
x_5	Reaction Time (min)	90	210
x_6	Volume of Solvent (ml)	100	200
x_7	Volume of NaOH (ml)	30	60
x_8	Substrate/NaOH ratio (mol/ml)	0.5×10^{-3}	1×10^{-3}
x_9	Catalyst/substrate ratio (mol/ml)	4×10^{-3}	6×10^{-3}
x_{10}	Reagent/substrate ratio (mol/mol)	1	1.25

Table 2.17 Results of performing a Plackett–Burman design on the reaction in Section 2.10

	x_1	x_2	x_3	x_4	x_5	x_6	x_7	x_8	x_9	x_{10}	x_{11}	Yield (%)
	−	−	−	−	−	−	−	−	−	−	−	15
	+	+	−	+	+	+	−	−	−	+	−	42
	−	+	+	−	+	+	+	−	−	−	+	3
	+	−	+	+	−	+	+	+	−	−	−	57
	−	+	−	+	+	−	+	+	+	−	−	38
	−	−	+	−	+	+	−	+	+	+	−	37
	−	−	−	+	−	+	+	−	+	+	+	74
	+	−	−	−	+	−	+	+	−	+	+	54
	+	+	−	−	−	+	−	+	+	−	+	56
	+	+	+	−	−	−	+	−	+	+	−	64
	−	+	+	+	−	−	−	+	−	+	+	65
	+	−	+	+	+	−	−	−	+	−	+	59
Coefficients	8.3	−2.3	0.5	8.8	−8.2	−2.2	1.3	4.2	7.7	9.0	4.8	

Table 2.18 Design matrix for Table 2.17

$$
\begin{pmatrix}
1 & -1 & -1 & -1 & -1 & -1 & -1 & -1 & -1 & -1 & -1 & -1 \\
1 & 1 & 1 & -1 & 1 & 1 & 1 & -1 & -1 & -1 & 1 & -1 \\
1 & -1 & 1 & 1 & -1 & 1 & 1 & 1 & -1 & -1 & -1 & 1 \\
1 & 1 & -1 & 1 & 1 & -1 & 1 & 1 & 1 & -1 & -1 & -1 \\
1 & -1 & 1 & -1 & 1 & 1 & -1 & 1 & 1 & 1 & -1 & -1 \\
1 & -1 & -1 & 1 & -1 & 1 & 1 & -1 & 1 & 1 & 1 & -1 \\
1 & -1 & -1 & -1 & 1 & -1 & 1 & 1 & -1 & 1 & 1 & 1 \\
1 & 1 & -1 & -1 & -1 & 1 & -1 & 1 & 1 & -1 & 1 & 1 \\
1 & 1 & 1 & -1 & -1 & -1 & 1 & -1 & 1 & 1 & -1 & 1 \\
1 & 1 & 1 & 1 & -1 & -1 & -1 & 1 & -1 & 1 & 1 & -1 \\
1 & -1 & 1 & 1 & 1 & -1 & -1 & -1 & 1 & -1 & 1 & 1 \\
1 & 1 & -1 & 1 & 1 & 1 & -1 & -1 & -1 & 1 & -1 & 1
\end{pmatrix}
$$

The first step is to set up a design, as given in Table 2.17. Note that an eleventh (dummy) factor (x_{11}) must be added as these designs (see Section 2.9) do not exist for 10 factors. This factor could, for example, be the colour of the shoes worn by the experimenter. A high level is represented by '+' and a low level by '−'. The yields or the reaction for the 12 experiments are also tabulated.

Table 2.18 is of the design matrix, often denoted as **D**. Columns 2–12 correspond to the 11 columns of Table 2.17. An extra, first, column is added to represent the intercept or average of the data, making a square matrix, with as many rows as columns. The coefficients $b_1 - b_{11}$ for the factors can be calculated by a variety of ways, two of which are as follows:

- The classical approach is simply to multiply the yields by the values of each factor for each experiment and divide by twelve. The result is the value of the coefficient. So, for factor 1, the coefficient becomes:

$$b_1 = (-15 + 42 - 3 + 57 - 38 - 37 - 74 + 54 + 56 + 64 - 65 + 59)/12 = 8.3$$

This approach involves simple summations and was often introduced in textbooks prior to the modern computer age.

- A more modern approach, involves calculating:

$$b = D^{-1}.y$$

where y is a vector of responses (synthetic yields) and D^{-1} is the inverse of the design matrix. This is feasible because the matrix D is a square matrix with as many rows as columns. Most packages such as Excel can readily determine inverse matrices.

These coefficients are placed in Table 2.17. In addition a coefficient for the intercept of 47 is calculated, which is the average of the entire dataset.

The next step is to interpret the coefficients. For example the value of b_1 ($= 8.3$) implies that on average over all 12 experiments, the yield is 8.3 % higher when the percentage of NaOH is increased from 40 to 50. Hence increasing NaOH in the reaction mixture improves the yield. The design can be relied on to give meaningful results because all the other factors are equally assigned to low and high levels corresponding to low and high levels of NaOH. If the design had been unbalanced it would not be possible to completely disentangle the effects of each factor, and so distinguish uniquely the different influences on the yield. Note that some coefficients are negative. This implies that increasing the value of the factor decreases the yield. An example is reaction time, implying that the highest yield is obtained earlier in the reaction, possibly due to degradation later.

The size of the coefficients relate to the significance of each factor. There are sophisticated statistical tests that can be employed to give an indication of the significance of any factor, often represented as a probability that the factor has influence. However, since all the factors are coded on the same scale, namely between $+1$ and -1, it is relatively straightforward to simply look at the size of the coefficients. The eleventh (dummy) factor has a value of $+4.8$, so, on the whole, factors whose magnitude is less than that are unlikely to have much influence on the reaction. Note that had the experiments been poorly designed it would not be possible to make these statements with much confidence.

From this preliminary study, the original 10 factors can be reduced to five that appear to have a significant influence on the yield of the reaction, namely, %NaOH, stirring rate, time of reaction, catalyst/substrate ratio and reagent/substrate ratio.

The next step would be to perform more detailed experiments on these remaining factors. Remember that the exploratory design reported in this section does not provide information about interactions or squared terms. What the design does do is give an indication of what factors are likely to be important when optimizing the experimental yield. Further work would be required to pinpoint those conditions that provide the best yield. Below we discuss some of these designs.

2.11 CENTRAL COMPOSITE DESIGNS

Most of the designs described above are used for exploratory purposes. After finding the main factors, it is often useful to obtain a more detailed model of a system. There are two prime reasons. The first is for optimization: to find the conditions that result in a maximum or minimum as appropriate. An example is when improving the yield of synthetic reaction. The second is to produce a detailed quantitative model: to predict mathematically how a response relates to the values of various factors. An example may be how the near infrared (NIR) spectrum of a manufactured product relates to the nature of the material and processing employed in manufacturing.

Most exploratory designs do not provide replicate information, nor any information on squared or interaction terms (Section 2.5). The degrees of freedom for the lack-of-fit for the model (D) are often zero, also. More informative models are best estimated using a modest number of factors, typically from two to five, to reduce the volume of experimentation.

We will illustrate such designs via a three factor experiment represented in Figure 2.16 and Table 2.19:

- A minimal three factor factorial design consists of four experiments, used to estimate the three linear terms and the intercept. Such design will not provide estimates of the interactions, replicates or squared terms.
- Extending this to eight experiments provides estimates of all interaction terms. When represented by a cube, these experiments are placed on the eight corners of the cube.
- Another type of design, often denoted as a star design, can be employed to estimate the squared terms. In order to do this, at least three levels are required for each factor, often denoted by $+1$, 0, and -1, with level '0' being in the centre. The reason for this is that there must be at least three points to fit a quadratic. For three factors, a star design consists of the centre point, and a point in the middle of each of the six faces of the cube.
- Finally it is often important to estimate the error, and this is typically performed by repeating the experiment in the centre of the design five times. Obviously other arrangements of replicates are possible.
- Performing a full factorial design, a star design and five replicates, results in twenty experiments. This design is often called a *central composite design*. The exact position of the star points relates to certain other properties of the design. Where the star points are at ± 1, the specific name is sometimes denoted as a 'face centred cube'.

The central features for three factor designs were briefly introduced previously (Section 2.5):

- A full model including all two factor interactions consists of 10 terms.
- A degree of freedom tree can be drawn up (see Section 2.2 for a discussion of the principles) as illustrated in Figure 2.17. We can see that:
  - there are 20 experiments overall;
  - 10 parameters in the model (Section 2.2);
  - 5 degrees of freedom to determine replication error;
  - 5 degrees of freedom for the lack-of-fit (see Section 2.3 for a reminder).

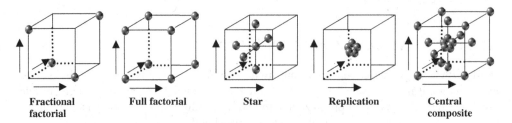

| Fractional factorial | Full factorial | Star | Replication | Central composite |

Figure 2.16 Construction of a three factor central composite design

Note that the number of degrees of freedom for the lack-of-fit equals that for replication, suggesting quite a good design.

In the example above, the position of the 'star' points is at ± 1. For statistical reasons it is more usual to place these at $\pm^4\sqrt{(2^f)}$ where f is the number of factors, equalling 1.41 for two factors, 1.68 for three factors and 2 for four factors. These designs are often referred to as *rotatable central composite designs* as all the points except the central points lie approximately on a circle or sphere or equivalent multidimensional surface, and are at equal distance from the origin. Table 2.20 is of a rotatable two factor central composite design with the factorial, star and replicate points shaded as in Table 2.19. It can be shown that the total number of terms including linear, quadratic and interactions is six, so there

Table 2.19 A simple three factor central composite design, consisting of three elements

Fractional factorial

1	1	1
1	−1	−1
−1	−1	1
−1	1	−1

Full factorial

1	1	1
1	1	−1
1	−1	1
1	−1	−1
−1	1	1
−1	1	−1
−1	−1	1
−1	−1	−1

Star

0	0	−1
0	0	1
0	1	0
0	−1	0
1	0	0
−1	0	0
0	0	0

Replication

0	0	0
0	0	0
0	0	0
0	0	0
0	0	0

Table 2.19 (*continued*)

Central composite

1	1	1
1	1	−1
1	−1	1
1	−1	−1
−1	1	1
−1	1	−1
−1	−1	1
−1	−1	−1
0	0	−1
0	0	1
0	1	0
0	−1	0
1	0	0
−1	0	0
0	0	0
0	0	0
0	0	0
0	0	0
0	0	0
0	0	0

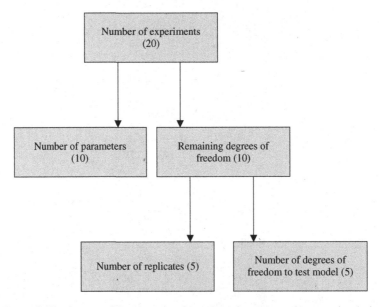

Figure 2.17 Degree of freedom tree for a three factor central composite design

Table 2.20 Rotatable two factor central composite design

Experiment	Factor 1	Factor 2
1	+1	+1
2	+1	−1
3	−1	+1
4	−1	−1
5	+1.41	0
6	−1.41	0
7	0	+1.41
8	0	−1.41
9	0	0
10	0	0
11	0	0
12	0	0
13	0	0
14	0	0

are three degrees of freedom remaining for determination of the lack-of-fit (D), compared with five replicate points (R).

Some authors use slightly different terminology and call only the rotatable designs true central composite designs. It is very important to recognize that the literature of statistics is very widespread throughout science, especially in experimental areas such as biology, medicine and chemistry and to check carefully an author's precise usage of terminology.

The true experimental conditions can be easily calculated from the design. For example, if coded levels +1, 0 and −1 correspond to temperatures of 30, 40 and 50 °C for the two factor design of Table 2.20, the star points correspond to temperatures of 25.9 and 54.1 °C (at coded positions ±1.41 from the centre), whereas for a rotatable four factor design these points correspond to temperatures of 20 and 60 °C (at coded positions ±2 from the centre).

A recommended four factor design consists of 30 experiments, namely:

- 16 factorial points at all possible combinations of ±1;
- nine star points, including a central point of (0,0,0,0) and eight points of the form (±2,0,0,0), etc.;
- typically five further replicates in the centre.

Sometimes, the number of replicates is increased somewhat as the number of factors increases, so that the number of degrees of freedom for lack-of-fit and replicate error are equal, but this is not essential.

Once the design is performed it is then possible to calculate the values of the terms using regression and design matrices (Section 2.5) or almost any standard statistical software including Excel and assess the significance of each term using ANOVA (Section 2.3). It is important to recognize that these designs are mainly employed for detailed modelling, and also to look at interactions and higher order (quadratic) terms. The number of experiments becomes excessive if the number of factors is large, and if more than about five significant factors are to be studied, it is usual to narrow down the problem first using exploratory designs as discussed in Sections 2.6–2.10.

2.12 MIXTURE DESIGNS

Chemists and statisticians often use the term 'mixture' to mean different things. To a chemist, any combination of several substances is a mixture. In more formal statistical terms, however, a mixture involves a series of factors whose total is a constant sum. Hence in statistics (and chemometrics) a solvent system in high performance liquid chromatography (HPLC) or a blend of components in a product such as paints, drugs or food, is a mixture, as each component can be expressed as a percentage and the total adds up to 1 or 100 %. The response could be a chromatographic separation, or the taste of a food, or physical properties of a manufactured material. Often the aim of experimentation is to find an optimum blend of components that taste best, or provide the best chromatographic separation, or are most durable.

The design of compositional mixture experiments involves different considerations to those employed in normal unconstrained experiments, since the value of each factor is connected or dependent on other factors. Take, for example, a three component mixture of acetone, methanol and water, e.g. as solvents used in the mobile phase for a chromatographic separation. If we know that there is 80 % of water in the mixture, there can be no more than 20 % of acetone or methanol. If there is 15 % acetone, the amount of methanol is fixed at 5 %. In fact, although there are three components in the mixtures, these translate to two independent factors.

A three component mixture can often be represented by a mixture triangle (Figure 2.18), which is a cross-section of the full three-dimensional mixture space, represented by a cube:

- the three corners correspond to single components;
- points along the edges correspond to binary mixtures;
- points inside the triangle correspond to ternary mixtures;
- the centre of the triangle corresponds to an equal mixture of all three components;
- all points within the triangle are physically allowable blends.

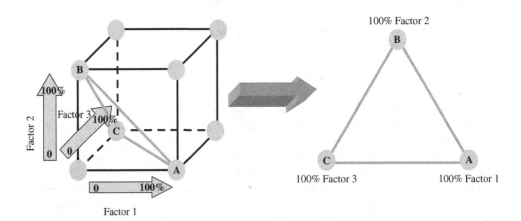

Figure 2.18   Mixture triangle

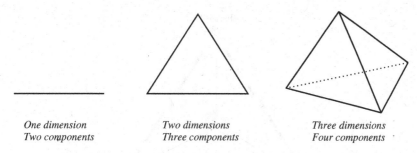

One dimension
Two components

Two dimensions
Three components

Three dimensions
Four components

Figure 2.19 Illustration of simplex

As the number of components increases so does the dimensionality of the mixture space:

- for two components the mixture space is simply a straight line;
- for three components a triangle;
- for four components a tetrahedron.

Each object (Figure 2.19) is called a *simplex* – the simplest possible object in space of a given dimensionality, where the dimensionality is one less than the number of factors or components in a mixture.

Experiments are represented as points in this mixture space. There are a number of common designs.

2.12.1 Simplex Centroid Designs

This is probably the most widespread design. For f factors this involves performing $2^f - 1$ experiments. It involves all possible combinations of the form $1, 1/2$ to $1/f$ and is best illustrated by an example. A three factor design consists of:

- three single factor combinations;
- three binary combinations;
- one ternary combination.

Graphically these experiments are represented in mixture space by Figure 2.20, and in Table 2.21.

Just as previously a model and design matrix can be set up (Section 2.5). However, the most common type of model contains all interaction terms but not quadratic or intercept terms. Since seven experiments have been performed, seven terms can be calculated, namely:

- three one factor terms;
- three two factor interactions;
- one three factor interaction.

The equation for the model is usually expressed by:

$$y = b_1x_1 + b_2x_2 + b_3x_3 + b_{12}x_1x_2 + b_{13}x_1x_3 + b_{23}x_2x_3 + b_{123}x_1x_2x_3$$

with the design matrix given in Table 2.22.

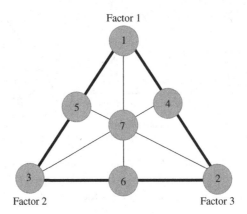

Figure 2.20   Simplex centroid design

Table 2.21   Three factor simplex centroid design

Experiment	Factor 1	Factor 2	Factor 3	
1	1	0	0	⎫
2	0	1	0	⎬ Single factor
3	0	0	1	⎭
4	1/2	1/2	0	⎫
5	1/2	0	1/2	⎬ Binary
6	0	1/2	1/2	⎭
7	1/3	1/3	1/3	⎫ Ternary

Table 2.22   Design matrix for three factor simplex centroid mixture design

x_1	x_2	x_3	$x_1 x_2$	$x_1 x_3$	$x_2 x_3$	$x_1 x_2 x_3$
1.000	0.000	0.000	0.000	0.000	0.000	0.000
0.000	1.000	0.000	0.000	0.000	0.000	0.000
0.000	0.000	1.000	0.000	0.000	0.000	0.000
0.500	0.500	0.000	0.250	0.000	0.000	0.000
0.500	0.000	0.500	0.000	0.250	0.000	0.000
0.000	0.500	0.500	0.000	0.000	0.250	0.000
0.333	0.333	0.333	0.111	0.111	0.111	0.037

Modelling of mixture experiments can be quite complex, largely because the value of each factor is related to other factors, so, for example, in the equation on page 45 $x_3 = 1 - x_1 - x_2$, so an alternative expression that includes intercept and quadratic terms but not x_3 could be conceived mathematically. However, it is simplest to stick to models of the type above, that include just single factor and interaction terms, which are called Sheffé models. Mixture models are discussed in more detail in Section 9.6.4.

2.12.2 Simplex Lattice Designs

A weakness of simplex centroid designs is the very large number of experiments required when the number of factors becomes large. For five factors, 31 experiments are necessary. However, just as for factorial designs (Section 2.6), it is possible to reduce the number of experiments. A full five factor simplex centroid design can be used to estimate all 10 possible three factor interactions, many of which are unlikely. An alternative class of designs called *simplex lattice* designs have been developed. These designs can be referenced by the notation $\{f, m\}$ where there are f factors and m fold interactions. The smaller m, the less experiments.

2.12.3 Constrained Mixture Designs

In chemistry, it is often important to put constraints on the value of each factor. By analogy, we might be interested in studying the effect of changing the proportion of ingredients in a

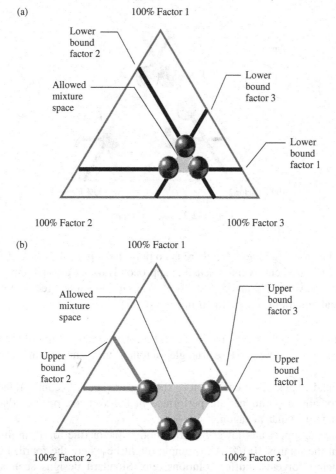

Figure 2.21 Constrained mixture designs. (a) Lower bounds defined; (b) upper bounds defined; (c) upper and lower bounds defined: fourth factor as a filler; (d) upper and lower bounds defined

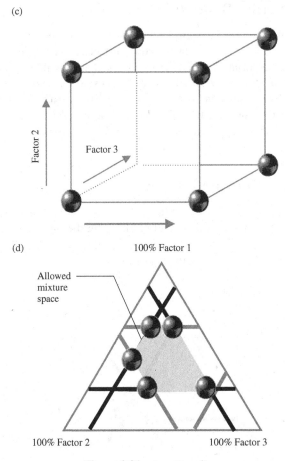

Figure 2.21   (*continued*)

cake. Sugar will be one ingredient, but there is no point baking a cake using 100 % sugar and 0 % of each other ingredient. A more sensible approach is to put a constraint on the amount of sugar, perhaps between 2 % and 5 % and look for solutions in this reduced mixture space. There are several situations, exemplified in Figure 2.21:

- A lower bound is placed on each factor. Providing the combined sum of the lower bounds is less than 100 %, a new mixture triangle is found, and standard mixture designs can then be employed.
- An upper bound is placed on each factor. The mixture space becomes rather more complex, and typically experiments are performed at the vertices, on the edges and in the centre of a new irregular hexagon.
- Each factor has an upper and lower bound. Another factor (the fourth in this example) is added so that the total is 100 %. An example might be where the fourth factor is water, the others being solvents, buffer solutions, etc. Standard designs such as the central composite design (Section 2.11) can be employed for the three factors in Figure 2.21, with the percentage of the final factor computed from the remainder.

Finally, it is possible to produce a mixture space where each factor has an upper and lower bound, although only certain combinations are possible. Often quite irregular polygons are formed by this mechanism. Note, interestingly, that in the case illustrated in Figure 2.21, the upper bound of factor 3 is never reached.

There is a great deal of literature on mixture designs, much of which is unfamiliar to chemists. A well known text is by Cornell [4]. Mixture designs are described further in Section 9.4.

2.13 A FOUR COMPONENT MIXTURE DESIGN USED TO STUDY BLENDING OF OLIVE OILS

In order to illustrate the use of mixture designs we will consider an example. Food chemistry is an especially important field for the study of mixtures, in which each factor is an ingredient. The result of the blending may be the quality of a finished product, for example as assessed by a taste panel.

Vojnovic *et al.* [5] report a four compound mixture design consisting of 14 experiments, studying up to three component interactions. The aim is to look at blends of olive oils. The response relates to the score of a taste panel when assessing a particular blend. The higher the score, the better the taste. Four fundamental tastes were scored (sweet, salt, sour and bitter) and four defects (bitter, marc, wined and rancid). The maximum possible in each category was 9. Extra virgin olive oil is defined as one with an average score of at least 6.5.

The design can be illustrated by points in a tetrahedron, each corner representing one of three pure components (see Figure 2.22). There are:

- four corners corresponding to single blends;
- six sides corresponding to binary blends;
- four faces corresponding to ternary blends.

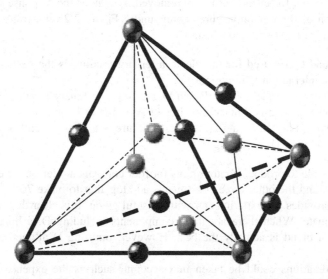

Figure 2.22 Experimental design for example in Section 2.13

Table 2.23 Design plus response for the olive oil taste panels discussed in Section 2.13

A	B	C	D	Response
1	0	0	0	6.86
0	1	0	0	6.5
0	0	1	0	7.29
0	0	0	1	5.88
0.5	0.5	0	0	7.31
0.5	0	0.5	0	6.94
0.5	0	0	0.5	7.38
0	0.5	0.5	0	7
0	0.5	0	0.5	7.13
0	0	0.5	0.5	7.31
0.33	0.33	0.33	0	7.56
0.33	0.33	0	0.33	7.25
0.33	0	0.33	0.33	7.31
0	0.33	0.33	0.33	7.38

The experimental matrix together with the responses is given in Table 2.23. A best fit equation can be calculated, involving:

- four single component terms;
- six two component interaction terms;
- four three component interaction terms.

Once the number of terms becomes large a good method for determining the optimum is to visualize the change in response by contouring.

One way of looking at these data is to see what happens when only three components (A–C) are employed, for example, if D is removed, how does the response vary according to the proportion of the remaining three compounds? Figure 2.23 illustrates the behaviour and we can make a number of conclusions:

- When A, B and C are used for the blend, a ternary mixture is the best, suggesting that the three way interaction is important.
- If the constituents include only A, B and D, the best is a binary mixture of D and B.
- In the case of A, C and D, binary mixtures of A and D or C and D are the best.
- When employing only B, C and D, the best mixture is a binary blend of D and C, with slightly more C than D.

It also is possible to produce contour plots for four component blends, for example, when there is 30 % D, and the other three constituents add together to make 70 % of the mixture.

The model provides the experimenter with a useful guide. It is clear that one component alone is inadequate. When there are three components including D, a binary rather than three component blend is actually the best. However, omitting D altogether provides the best samples.

Other considerations could be taken into account, such as the expense or availability of each ingredient. The overall optimum might, in practice, not be the most suitable if it

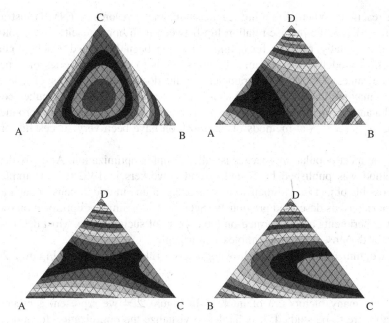

Figure 2.23 Contour response surface plot of the olive oils

contains a high proportion of an expensive ingredient. Another factor relates to the time of year: the blend may have to change according to seasonal considerations. One olive grower may have a limited supply and a competing manufacturer might have purchased the majority of the crop one year. A different blend might need to be used in January of one year as opposed to July.

Sometimes several different responses can be modelled, such as the expense, taste and appearance of a product, and contour plots compared for each response, to aid decision making. Potential applications of mixture designs are large, and may include, for example, fuels, paints, drugs and so on. Some chemists have to tackle quite difficult mixture problems even with 10 or more components, and the considerations of design, modelling and graphical visualization are quite complex. In practice, most properties of substances such as colour, hardness and taste result from complex interactions between the various ingredients which are most efficiently studied using rational designs, especially if each mixture is expensive to produce and test.

2.14 SIMPLEX OPTIMIZATION

Experimental designs can be employed for a large variety of purposes, one of the most widespread being optimization. Most statistical approaches normally involve forming a mathematical model of a process, and then, either computationally or algebraically, optimizing this model to determine the best conditions.

There are many applications, however, in which a mathematical relationship between the response and the factors that influence it is not of primary interest, and it is here that often methods developed first by engineers rather than statisticians can equally well aid the chemist. Is it necessary to model precisely how pH and temperature influence the yield of a reaction? When shimming a nuclear magnetic resonance (NMR) instrument, is it necessary to know the precise relationship between field homogeneity and resolution? In engineering, especially, methods for optimization have been developed which do not require a mathematical model of the system. The approach is to perform a series of experiments, changing the values of the control parameters, until a desired quality of response is obtained. Statisticians may not like this approach as it is not normally possible to calculate confidence in the model and the methods may fall down when experiments are highly irreproducible, but in practice these sequential methods of optimization have been very successful throughout chemistry.

One of the most popular approaches is called simplex optimization. An early description of the method was published by Spendley and coworkers in 1962 [6]. A simplex is the simplest possible object in f-dimensional space, e.g. a line in one dimension, and a triangle in two dimensions, as described previously (Section 2.12). Simplex optimization means that a series of experiments are performed on the corners of such a figure. Most descriptions are of two factor designs, where the simplex is a triangle.

Simplex optimization involves following a set of rules, illustrated in Figure 2.24.

The main steps are as follows:
- Define how many factors are of interest. In Figure 2.24 we represent a process where two factors are to be studied. It is harder to visualize the optimization for more factors.
- Code these factors. Clearly if we are going to study a reaction between pH 5 and 8 and between 50 °C and 80 °C one pH unit has roughly the same significance as 10 temperature units.
- Choose $f + 1$ starting conditions, where f is the number of factors (two in this example). These correspond to the conditions at which the initial experiments are performed. These are indicated in the top left triangle in Figure 2.24. The size of each side of the triangle in Figure 2.24 is often called the 'step size'. This triangle is a simplex. For a three factor design the simplex will be a tetrahedron and involves four rather than three initial experiments.

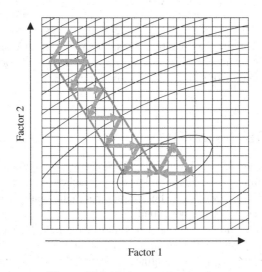

Figure 2.24 Progress of a simplex

- Next comes the tricky part. Look at the response (e.g. a reaction yield or chromatographic separation) in each of these three experiments, and order them (for three factors) as 'Best', 'Next' and 'Worst'. Rules then determine the conditions for the next experiment as indicated by the arrows in Figure 2.24.
- Continue performing experiments, which, ideally result in a better response, until an optimum is detected. There are a number of ways of determining an optimum, often called 'stopping rules'. Oscillation is common under such circumstances, but when detected this is a clue that we are near the optimum.

In many well behaved cases, simplex performs fine. It is quite an efficient approach for optimization. There are, however, a number of limitations:

- If there is a large amount of experimental error, the response is not very reproducible. This can cause problems, for example, when searching a fairly flat response surface.
- Sometimes there can be serious boundary conditions that have to be taken into account, i.e. impossible combinations of parameters even ones that create explosions. Rules can be modified, of course, but it is important not to allow the method to stray into these regions. Occasionally an optimum might be very close to an unacceptable boundary (analogous to catastrophes), so caution must be built in.
- Sensible initial conditions and scaling (coding) of the factors is essential. This can only come from empirical chemical knowledge.
- A major problem relates to the choice of original conditions and step size. The smaller the simplex, the higher the chance of reaching a true optimum, but the slower the optimization. A large step size may overlook the optimum altogether. To overcome this, certain modifications of the original method have been developed, the most well known being the 'modified simplex' method. In this the step size varies, being largest farthest away from and smallest when close to the optimum. This is analogous to the size of triangles changing in Figure 2.24, being quite small when close to the optimum. Even more elaborate is the 'super-modified simplex' method requiring mathematical modelling of the change in response at each step.

There is some controversy as to whether simplex methods are genuinely considered as experimental designs. Most statisticians will almost totally ignore these approaches, and, indeed, many books and courses on experimental design in chemistry will omit simplex methods, concentrating exclusively on approaches for mathematical modelling of the response surfaces. However, engineers and programmers have employed simplex and related approaches for optimization for many years, and these methods have been much used, for example, in spectroscopy and chromatography, so should be part of the battery of techniques available to the chemometrician. As a practical tool where the detailed mathematical relationship between response and underlying variables is not of primary concern, the methods described above are very valuable. They are also easy to implement computationally and to automate.

2.15 LEVERAGE AND CONFIDENCE IN MODELS

An important experimental problem relates to how well quantitative information can be predicted. For example, if an experiment is performed between 40 °C and 80 °C, can we make predictions about the outcome of an experiment at $90\,^\circ\text{C}$? It is traditional to sharply cut-off the model outside the experimental region, so that predictions are made only within the experimental limits. However, this misses much information. In most cases, the ability to make a prediction reduces smoothly from the centre of a set of experiments, being best at $60\,^\circ\text{C}$ and worse the farther away from the centre in our example. This does not imply that it is impossible to make any statement as to what happens at $90\,^\circ\text{C}$, simply that there is less confidence in the prediction than at $80\,^\circ\text{C}$, which, in turn is predicted less well than at $60\,^\circ\text{C}$. It is important to be able to visualize how the ability to predict a response (e.g. a synthetic yield or a concentration) varies as the independent factors (e.g. pH, temperature) are changed.

In univariate (single variable) calibration (Section 6.2), the predictive ability is often visualized by confidence bands. The 'size' of these confidence bands depends on the magnitude of the experimental error. The 'shape' however, depends on the experimental design, and can be obtained from the design matrix (Section 2.5) and is influenced by the arrangement of experiments, replication procedure and mathematical model. Leverage can be used as a measure of such confidence and has has certain properties:

- The value is always greater than 0.
- The lower the value the higher the confidence in the prediction. A value of 1 indicates a very poor prediction.
- The sum of the values for leverage at each experimental point adds up to the number of coefficients in the model.

To show how leverage can help, three designs (Table 2.24) will be illustrated. Each involves performing 11 experiments at five different concentration levels, the only difference being the arrangement of the replicates. The aim is simply to perform linear calibration to produce a model of the form $y = b_0 + b_1 x$, where x is an independent factor, such as concentration. The leverage can be calculated from the design matrix D (see Section 2.5 for more discussion about design matrices) which consists of 11 rows (corresponding to each experiment) and two columns (corresponding to each coefficient) in our experiments. In matrix notation a 'hat matrix' can be calculated, given by $H = D.(D'.D)^{-1}.D'$, where

Table 2.24 Three designs for illustration of leverage calculations

Concentrations			Leverage		
Design A	Design B	Design C	Design A	Design B	Design C
1	1	1	0.23377	0.29091	0.18
1	1	1	0.23377	0.29091	0.18
1	2	1	0.23377	0.14091	0.18
2	2	1	0.12662	0.14091	0.18
2	3	2	0.12662	0.09091	0.095
3	3	2	0.09091	0.09091	0.095
4	3	2	0.12662	0.09091	0.095
4	4	3	0.12662	0.14091	0.12
5	4	3	0.23377	0.14091	0.12
5	5	4	0.23377	0.29091	0.255
5	5	5	0.23377	0.29091	0.5

the $'$ stands for a transpose and the $^{-1}$ for an inverse (Section 2.4). The hat matrix consists of 11 rows and 11 columns, the numbers on the diagonal being the values of leverage for each experimental point. Although these calculations may appear complicated, in fact, using modern spreadsheets it is easy to perform simple matrix operations, this being a good way to learn about matrix algebra if you are unfamiliar. The leverage for each experimental point is given in Table 2.24. It is also possible to obtain a graphical representation of the equation as shown in Figure 2.25 for the three designs.

What does this tell us?

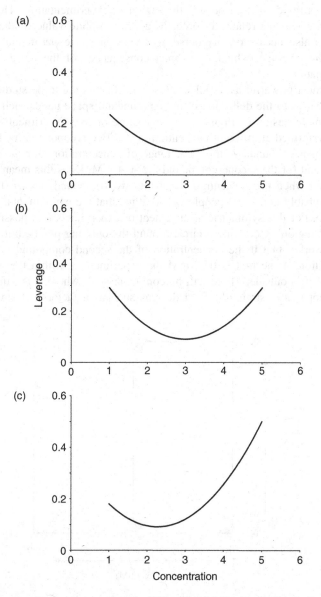

Figure 2.25 Leverage for the three designs in Table 2.24. (a) Design A; (b) design B; (c) design C

- Design A contains more replicates at the periphery of the experimentation to design B, and so results in a flatter graph of leverage. This design will provide predictions that are fairly even throughout the area of interest.
- Design C shows how the nature of replication can result in a major change in the shape of the curve for leverage. The asymmetric graph is a result of the replication regime. In fact, the best predictions are no longer in the centre of experimentation.

This approach can be employed more generally. How many experiments are necessary to produce a given degree of confidence in the prediction? How many replicates are sensible? How good is the prediction outside the region of experimentation? How do different experimental arrangements relate? In order to get an absolute value of the confidence of predictions, it is also necessary, of course, to determine the experimental error, but this together with the leverage, which is a direct consequence of the design and model, is sufficient information.

Leverage is most powerful as a tool when several factors are to be studied. There is no general agreement as to the definition of an experimental space under such circumstances. Consider the simple design of Figure 2.26, consisting of five experiments (a star design: Section 2.11) performed at different concentrations for two compounds (or factors). Where does the experimental boundary stop? The range of concentrations for the first compound is 0.5–0.9 mM and for the second compound 0.2–0.4 mM. Does this mean we can predict the response well when the concentrations of the two compounds are at 0.9 and 0.4 mM respectively? Probably not, some people would argue that the experimental region is a circle, not a square. For this symmetric arrangement of experiments it is possible to envisage an experimental region by drawing a circle around the outlying points, but imagine telling the laboratory worker that if the concentration of the second compound is 0.34 mM then if the concentration of the first is 0.77 mM the experiment is within the region, whereas if it is 0.80 mM it is outside. There will be confusion as to where it is safe to predict the response. For apparently straightforward designs such as a factorial design (Section 2.6)

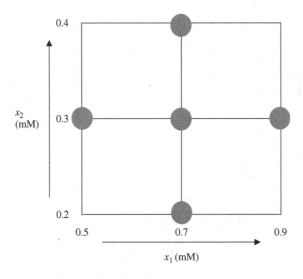

Figure 2.26 A simple two factor design

the definition of what constitutes the experimental region is, in fact, even harder. In reality information about confidence of predictions could be very important, for example, in process control – can a response (e.g. the quality of a product such as the taste of food) be predicted confidently by the control variables (e.g. manufacturing conditions)? Under what circumstances is it unsafe to predict the quality given the experimental design for the training set?

So the best solution is to produce a graph as to how confidence in the prediction varies over the experimental region. Consider the two designs of Figure 2.27. Using a linear model, the leverage for both designs is given in Figure 2.28. The consequence of the different experimental arrangement is now quite obvious, and the gradual change in confidence from the centre of the design can be visualized. Although this example is fairly straightforward, for multifactor designs (e.g. mixtures of several compounds) it is hard to produce an arrangement of samples in which there is symmetric confidence in the results over the experimental domain. Leverage can show how serious this problem is for any proposed arrangement. It

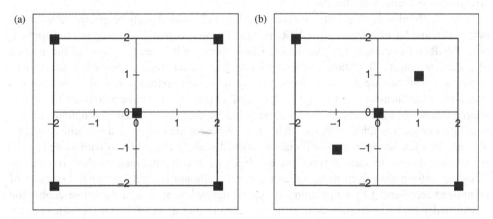

Figure 2.27 Two simple designs, used to illustrate how leverage can give guidance about the experimental region, with each axis as coded variables

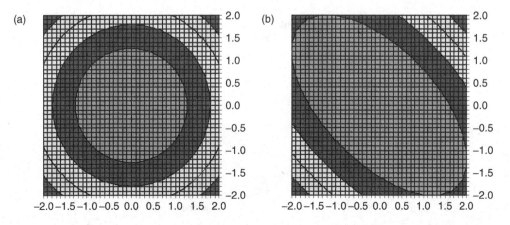

Figure 2.28 Leverage corresponding to the two designs in Figure 2.27

can also illustrate the effect of replication, and changing the nature of the mathematical model on the confidence in the prediction around the experimental region.

2.16 DESIGNS FOR MULTIVARIATE CALIBRATION

One of the biggest applications of chemometrics is multivariate calibration, as discussed in more detail in Chapter 6. Indeed, many scientists first become acquainted with chemometrics through calibration, and the associated Partial Least Squares algorithm (Section 6.6). There are a large number of packages for multivariate calibration, but the applicability and role of such methodology spawns several schools of thought, and can be a source of controversy. A difficulty with the use of many software packages is that the user may have little or no technical knowledge of the background to the methods he or she employs. Unfortunately the ability to make safe predictions depends, in part, on the design of the calibration experiment and so the nature of the training set, so it is important that the experimenter has a feel for how design influences predictions.

Many calibration experiments involve recording information such as spectra on mixtures of compounds: for example, taking NIR spectra of food (Section 13.3 discusses the application of NIR in this context in more detail). Can a spectrum be used to determine the amount of protein in wheat? To obtain a very certain answer, chromatography or other more time-consuming methods are required. The difficulty with such approaches, is that they are not suited for a fast turn-round of samples. A typical laboratory may have a contract to analyse a large number of food samples, and will be judged in part on its rate of throughput. A conventional chromatographic analysis might take 1 h per sample, whereas a spectrum could be obtained in a few seconds, especially using infrared probes. The problem with using spectra is that signals due to each component overlap, and a computational method is required to extract information from these spectra about the amount of each compound or class of compound represented by a spectrum. A typical development of a multivariate calibration model under such circumstances might involve recording a set of 25 spectra with known concentrations of one compound and then trying to form a model between these samples and the concentrations as illustrated in Figure 2.29. This calibration model can then be used in future samples to determine the concentration of an unknown. The dataset used to obtain

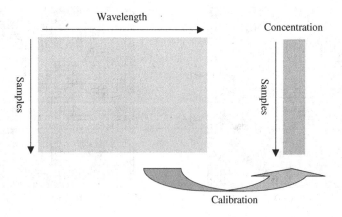

Figure 2.29   Principles of multivariate calibration – more details are discussed in Chapter 6

the model is often referred to as a 'training set', and the nature and design of the training experiments is crucial to the success of subsequent analysis. If the training set is not well designed there could be major difficulties in the use of software to predict concentrations in spectra of future unknown samples. If a training set is small, there may be correlations between the concentrations of different compounds which can invalidate calibration models. Consider, for example, a situation in which the concentrations of acetone and methanol in a training set are completely correlated. This means that if the concentration of acetone increases so does that of methanol, and similarly with a decrease. Such an experimental arrangement is shown in Figure 2.30: the concentrations of the two compounds are said to be correlated. A more satisfactory design is given in Figure 2.31: the two concentrations are

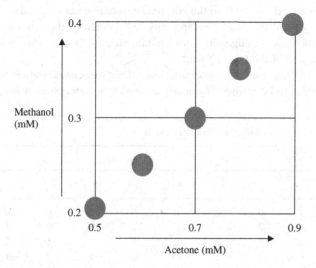

Figure 2.30   A poor calibration design

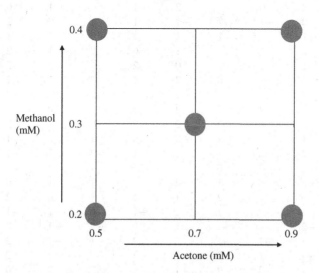

Figure 2.31   A good calibration design

now uncorrelated or orthogonal. The idea of orthogonality has been introduced previously (Sections 2.8 and 2.9). In the former design there is no way of knowing whether a change in spectral characteristic results from change in concentration of acetone or methanol. If this feature is in the training set and in all future samples, there is no problem, but if a future sample arises with a high acetone and low methanol concentration, calibration software will give a misleading answer for the concentration of each component as it has not been designed to cope with this situation. This is potentially very serious especially when the result of chemometric analysis of spectral data is used to make decisions, such as the quality of a batch of pharmaceuticals, based on the concentration of each constituent as predicted by computational analysis of spectra.

To overcome this problem, some chemometricians advocate very large training sets, often hundreds or thousands of samples so the correlations are evened out, and also to give a very representative set of spectra. However, in many real situations there is insufficient time or resources to produce such a huge amount of data, and experimental designs can be used to safely reduce the size of the training set.

As guidance, it is first useful to determine the number of concentration levels at which a particular problem is to be studied. Typically around five concentration levels are sensible.

Table 2.25 A five level, 10 factor calibration design

Experiments	Compounds									
	1	2	3	4	5	6	7	8	9	10
1	0	0	0	0	0	0	0	0	0	0
2	0	−2	−1	−2	2	2	0	−1	2	−1
3	−2	−1	−2	2	2	0	−1	2	−1	1
4	−1	−2	2	2	0	−1	2	−1	1	1
5	−2	2	2	0	−1	2	−1	1	1	0
6	2	2	0	−1	2	−1	1	1	0	2
7	2	0	−1	2	−1	1	1	0	2	1
8	0	−1	2	−1	1	1	0	2	1	2
9	−1	2	−1	1	1	0	2	1	2	−2
10	2	−1	1	1	0	2	1	2	−2	−2
11	−1	1	1	0	2	1	2	−2	−2	0
12	1	1	0	2	1	2	−2	−2	0	1
13	1	0	2	1	2	−2	−2	0	1	−2
14	0	2	1	2	−2	−2	0	1	−2	1
15	2	1	2	−2	−2	0	1	−2	1	−1
16	1	2	−2	−2	0	1	−2	1	−1	−1
17	2	−2	−2	0	1	−2	1	−1	−1	0
18	−2	−2	0	1	−2	1	−1	−1	0	−2
19	−2	0	1	−2	1	−1	−1	0	−2	−1
20	0	1	−2	1	−1	−1	0	−2	−1	−2
21	1	−2	1	−1	−1	0	−2	−1	−2	2
22	−2	1	−1	−1	0	−2	−1	−2	2	2
23	1	−1	−1	0	−2	−1	−2	2	2	0
24	−1	−1	0	−2	−1	−2	2	2	0	−1
25	−1	0	−2	−1	−2	2	2	0	−1	2

Designs exist involving kl^p mixtures, where l equals the number of concentration levels, p is an integer at least equal to 2, and k an integer at least equal to 1. Setting both k and p to their minimum values, at least 25 experiments are required to study a mixture (of more than one component) at five concentration levels.

Table 2.25 illustrates such a typical calibration design. Note various features:

- There are five concentration levels, numbered from -2 (lowest) to $+2$ (highest), analogous to factorial designs (Section 2.6), where the lowest level is -1 and the highest $+1$.
- Ten compounds are studied, each represented by a column.
- Twenty-five experiments are to be performed, each represented by a row.

The information in Table 2.25 could be presented to the experimenter as a set of concentrations, the outcome being 25 spectra.

The trick to the success of such designs is that all compounds in a mixture are 'uncorrelated' or 'orthogonal' to each other. For example, the graph of concentration levels of compound 4 versus compound 1 is given in Figure 2.32, and it is obvious that experiments at each of the five levels of compound 1 correspond to experiments at all possible five levels of compound 4. There are only a very small number of mathematical ways in which all 10 columns can be arranged to have this property. Such designs can be regarded as multilevel fractional factorials, analogous to the two level fractional factorial (Section 2.8) or Plackett–Burman and Taguchi designs (Section 2.9) and in the case of a 10 component mixture reduce the experiments from a possible 5^{10} (or almost 10 million) to 25, selecting

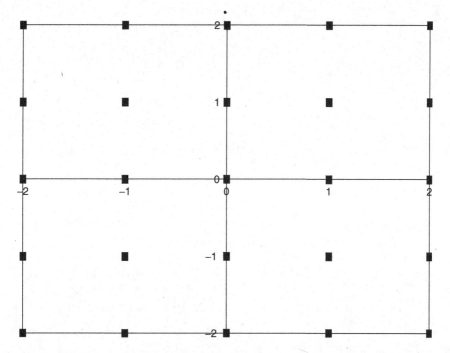

Figure 2.32 Graph of concentration of compound 4 against compound 1, in coded values, for the design in Table 2.25

these 25 out of the large number of possible combinations to maintain various important features. More detailed information of these designs are provided elsewhere [7].

REFERENCES

1. S.N. Deming and S.L. Morgan, *Experimental Design: a Chemometric Approach*, Elsevier, Amsterdam, 1993
2. B. Parodi, G. Caviglioli, A. Bachi, S. Cafaggi and G. Bignardi, Complex extemporaneous preparations: stability study of a formulation containing diethylpropion hydrochloride in hard capsules, *Acta Technologiae et Legis Medicamenti*, III, 3 (1992), 109–119
3. R.L. Plackett and J.P. Burman, The design of optimum multifactorial experiments, *Biometrika*, 33 (1946), 305–325
4. J. Cornell, *Experiments with Mixtures: Designs, Models and the Analysis of Mixture Data*, John Wiley & Sons, Inc., New York, 1990.
5. D. Vojnovic, B. Campisi, A. Mattei and L. Favretto, Experimental mixture design to ameliorate the sensory quality evaluation of extra virgin olive oils, *Chemometrics and Intelligent Laboratory Systems*, 27 (1995), 205–210
6. W. Spendley, G.R. Hext and F.R. Himsworth, Sequential application of simplex designs in optimisation and evolutionary operation, *Technometrics*, 4 (1962), 441–462
7. R.G. Brereton, Multilevel multifactor designs for multivariate calibration, *Analyst*, 122 (1997), 1521–1529

3
Statistical Concepts

3.1 STATISTICS FOR CHEMISTS

Several years ago I spoke to one of the 'grand old men' of British Scientific publishing, Ellis Horwood, whose then company marketed the text *Statistics for Analytical Chemistry* by Miller and Miller, currently in its fifth edition [1]. He told me how surprised he was that books on such an apparently 'dry old subject' like statistics in chemistry actually sold well. Many such books are oriented towards analytical chemistry, of which Gardiner [2] and Caulcutt and Boddy [3] are also excellent texts, although there is also a need in other branches of chemistry as well.

Why is there so much interest in basic statistics in chemistry, especially analytical chemistry? Early in their careers, most chemists choose not to specialize as mathematicians. Their reasons could be wide, for example a poor maths teacher, or a difficult exam, or simply a lack of inspiration. In many countries a high level of maths skills is not a requirement for achieving a good degree in chemistry, so high achieving chemists often do not necessarily require a fundamental understanding of maths (or if they do it has been long since forgotten). Yet many areas of chemistry such as analytical, physical and environmental chemistry are often based upon quantitative measurements and computations. For example, a practising chemist who may be working in a public service laboratory estimating the concentrations of chemicals in urine or blood, suddenly realizes that his or her techniques must be properly validated, both for the laboratory to maintain its accreditation and for legal reasons. So the chemist, who may have left formal training in maths 10 or even 20 years ago is suddenly faced with the need to brush up on statistics.

Some of the more fundamental uses are summarized below:

- *Descriptive statistics.* A series of measurements can be described numerically. If we record the absorbances of 100 samples, instead of presenting the user with a table of 100 numbers, it is normal to summarize the main trends. These can be done graphically, but also overall parameters such as the mean, standard deviation, skewness and so on help, and there are a variety of numbers that can be employed to give an overall picture of a dataset.
- *Distributions.* Fundamental to most statistics is the concept of a distribution. If we take a series of measurements we are sampling a population. Not all the samples will have exactly the same characteristics. Errors and noise, as discussed in Section 3.2, influence

Applied Chemometrics for Scientists   R. G. Brereton
© 2007 John Wiley & Sons, Ltd

the characteristics of each sample. In fact measurements on each sample will result in a slightly different value, even if they originate from the same source. The distribution of measurements is often approximated by a normal distribution, although in some cases this does not represent a very faithful model. When analysing samples, if enough measurements are taken, it is possible to see whether they fit into such a distribution. If there are significant problems this could be due to a different underlying error structure, but also can be due to outliers, samples that are atypical of the population or which may even have been incorrectly labelled or grouped. If we have reasonable evidence of an underlying distribution, we can then make further predictions about the population of samples. For example, if the mean concentration of a series of measurements is 4.35 mM and the standard deviation is 0.14 mM, how often would we expect to obtain a measurement greater than 4.60 mM if the measurements are normally distributed?

• *Hypothesis testing.* In many cases measurements are used to answer qualitative questions, and numbers are simply a means to an end. For example, we may be primarily interested in whether the near infrared (NIR) spectrum of an orange juice shows it is adulterated rather than whether the absorbance at a given wavelength is above a specified value. There are a wide variety of such tests. Most provide a probability that a hypothesis is either correct or not, dependent on one's aim. A common example is comparison of means. Laboratory A might have recorded 10 concentration measurements on a series of samples and obtained a mean of 7.32 mM plus a standard deviation of 0.17 mM, whereas the mean and standard deviations for laboratory B are 7.56 mM and 0.23 mM, respectively. Can we hypothesize that the methods and samples in both laboratories are identical and attach a probability to this?

• *Confidence.* An important use of statistics is to obtain confidence in predictions. For example we may predict that the concentration of a compound is 31.2 mM, but there will always be distortion in the measurement process. It may be important to be sure that this concentration is below a certain limit, for example it might be a potential pollutant. Can we be really sure it is below 35 mM, perhaps a legal limit below which our sample is acceptable? There are several methods for determining confidence limits of almost all types of parameters, to allow us to state, for example, that we are 99 % certain that the true concentration is less than 35 mM.

• *Outliers.* These are the bane of much experimental science, and are dubious samples. These can have a tremendous impact on determination of overall descriptions of data if not treated properly. Finding whether there are outliers and dealing with them is an important job.

• *Quality control.* This topic could be covered in several chapters, but is commonly regarded as part of basic statistics for chemists. In processes, often associated with manufacturing, it is important to see whether the quality of a product varies with time and if so how serious is the deviation. This is often done by taking regular measurements and comparing these to measurements from batches that are known to be acceptable. If the deviation is outside statistical limits this is an indication that there is some problem.

Below we will look at the various methods and underlying assumptions in more detail.

3.2 ERRORS

Fundamental to understanding why chemometrics and statistics is important in chemistry is to recognize the inevitability and all pervading existence of errors. Related concepts have been introduced in Sections 2.2 and 2.3 in a different context. One problem with chemistry is that at an early stage in a University education students are often given experiments that apparently 'work' first time, if correctly performed. However, in most cases the lecturer or technician has spent many weeks devising these hopefully foolproof experiments. They are of such a nature that answers such as good linear behaviour can be obtained often in a short time, as students have just a few hours in a practical class to obtain results. These are regarded as 'good textbook' experiments.

Yet in the real world it is very rare to perform a completely novel type of measurement and immediately obtain perfect behaviour. In fact there is no such thing as an error-free world. We can never observe underlying processes exactly. In real life it is necessary to accept a certain level of error or uncertainty in all measurements. How serious this is depends on the nature of an experiment. Can we tolerate 10 % error? If we primarily are after a qualitative answer – and in many cases in chemistry quantitative measurements are primarily a means to an end rather than an end in themselves (for example, to determine whether pollution is acceptable or not), it does not matter if there is some slight inaccuracy in the measurement. On the other hand, in other cases there may be legal implications, for example if the level of an additive in a food must be between certain limits, otherwise a company could be prosecuted, it may be important to measure the amount to within a high degree of accuracy. Physical modelling such as kinetics, spectroscopy or quantum mechanical predictions often assume that there is a high degree of quantitative confidence in experimental results, and interesting phenomena such as quantum tunnelling are observed by measuring small deviations from perfect behaviour.

In an industrial laboratory, where time is money, it is important to determine how cost effective it is to increase the accuracy of measurement procedures. For example, it may take a week to develop a method that results in estimates of concentrations accurate to 10 %, 1 month to reduce it to 2 % and 6 months to reduce it to 0.5 %. Is the extra time invested really worth it? Also, what parts of the system are most subject to error, and can we find the weak link and so improve our method easily?

One conceptual problem many chemists have is distinguishing between 'mistakes' and 'errors'. Statistically speaking, errors are everywhere. For example, if we use a volumetric flask for a series of dilutions, theoretically of 100 ml capacity, there will always be a distribution of volumes around the manufacturer's mean. There will never be the 'perfect' volumetric flask. This does not mean that every measurement is a mistake, we simply have to tolerate a statistical spread of results. A mistake, on the other hand, might involve mislabelling a sample or typing the wrong number into a spreadsheet. Colloquially mistakes and errors are often confused (one being an Anglo-Saxon and the other a Latin originated word; English language enthusiasts realize that there are often two equivalent sets of words in the language) but we will distinguish between the two concepts in this book and concentrate only on errors in the true statistical sense. Errors are all around us and cannot be avoided, whereas mistakes should always be avoided.

3.2.1 Sampling Errors

In most areas of applied statistics, the main errors encountered are due to sampling. Many professional statisticians have problems with chemometric data analysis, because they are often unprepared for a second and quite different type of error, as discussed in Sections 3.2.2 and 3.2.3. To illustrate the problem of sampling errors, let us say we wish to measure the concentrate of ammonia in a fish tank. If this is above a certain level, the fish may die from the toxic water, so it is important to keep this under control. However, each sample of water may have a slightly different value of ammonia, due to imperfect mixing. Without sampling the entire fish tank, something that is impracticable because we do not want to replace all the water in the tank, we will never know the exact average concentration, but we hope that we can take a representative set of water samples. The ammonia concentrations in each sample forms a distribution. The differences from the true mean concentrations can be called sampling errors. These differences are not mistakes (they are real), but they mean that the amount of ammonia is different in each sample.

In many cases these types of errors are inevitably large. For example, we might want to measure the mean concentration of a potential pollutant in an air sample in a factory. The concentration may vary according to time of day, temperature, sampling site, etc. and so we can only estimate an average. If the sampling times and sites are representative we would hope that we can build up a decent picture and obtain a meaningful value. There will however be a large spread (the concept of 'large' and how to determine this will be discussed in more detail below).

Sampling errors are real underlying errors and not a function of the measurement process. In classical statistics these are the main causes of deviation in measurements. For example, we may be interested in the size of adult fox tails in different regions of the country. There is no significant measurement error (providing the fox can be caught prior to measurement), the distribution resulting from foxes being different shapes and sizes rather than using a defective ruler. Nevertheless we could still conclude that statistically a population in one region had statistically significant differences in tail length and so perhaps did not interbreed or came from a different genetic pool to those from another region. Our conclusion hopefully has nothing to do with the type of ruler used to measure the tail length.

3.2.2 Sample Preparation Errors

In chemistry there is another and often more important source of error, that due to sample preparation. Consider a procedure involving determining the concentration of a compound in a reaction mixture by off-line high performance liquid chromatography (HPLC). The method might consist of several steps. The first could be to obtain a sample, perhaps using a syringe, of 1 ml. Then this sample might be introduced into a volumetric flask, and diluted to 100 ml. Finally 20 µl is injected onto a HPLC column. Each of these three steps is subject to error. In many cases there are even more stages, such as weighings, in a typical quantitative experiment. Sometimes one stage can be highly prone to inaccuracies. For example, if we use a 10 ml syringe to measure 1 ml, there often can be quite large percentage errors. Each part of the analysis results in its own set of errors, but there is a theorem that a combination of symmetric error distributions tends to a normal or Gaussian distribution which is generally adequate to model the overall errors. We will discuss this distribution again in Section 3.4, and some ideas will also be introduced in the context of signal processing (Section 4.7.2).

In practice it is good to have an idea of how the errors in each step combine to give an overall error. Normally we measure the standard deviation of the distribution of measurements and call this the root mean square error. Using a normal distribution, the overall error is the root mean square of each individual error, so if a four stage analysis had errors at each stage of 2 %, 4 %, 5 % and 1 % the overall error is $(2^2 + 4^2 + 5^2 + 1^2)^{1/2}$ or 6.78 %. In some modern articles the word 'error' in this context is replaced by 'uncertainty' which perhaps is a better description of the situation.

3.2.3 Instrumental Noise

The final type of error is due to the measurement process or instrument. No instrument is perfect. The size of this error can be assessed using replicate instrumental measurements. Older books tend to concentrate on this aspect because 20 years ago most instruments were not very reliable, so for example, there is a rule of thumb in atomic spectroscopy that each measurement should be repeated three times on each sample. However, as instruments improve with time, this source of error, although still important, has gradually become less significant compared with that due to sample preparation. The science of manufacturing volumetrics and measuring cylinders has not improved drastically over the years, whereas the signal to noise characteristics, for example of modern HPLC or nuclear magnetic resonance (NMR) instruments, is much better than a decade ago. This type of error contributes to measurement uncertainty in addition to sample preparation errors.

There are, of course, several factors that do influence instruments ranging from electricity supplies such as fluctuating voltages, to changes in filters and lamps. Some traditional books tend to take these problems very seriously, because many of the enthusiasts built their own instruments and were acutely aware of the effect of a defective component, and it is still very important to consider stability over time. For example, in optical spectroscopy it is well known that calibration models change with time, how important this is depends on the application. In some types of mass spectrometry, for example, the ionization mechanism is not completely reproducible so the exact relative intensities of different m/z values will differ each time a spectrum is recorded. However, in many modern instruments the problems of variability are considerably less acute to those of sample preparation.

3.2.4 Sources of Error

The experimentalist must be aware of the different sources of error, and determine whether it is worth his or her while reducing each type of error. This largely depends on the reason for measurement and often the funds available both in terms of instrumentation and development time. It is also important to understand that errors can never be completely eliminated and so statistical methods as described below will always be important when analysing numerical information from chemical measurements.

3.3 DESCRIBING DATA

Many chemists are faced with dealing with large datasets, often consisting of replicated measurements. These may, for example, be the absorbance of a reference sample taken at regular intervals, or the determination of a metabolite in urine samples of patients, or the measurement of a heavy metal pollutant in a river sample over a period of months.

A first step is to summarize the information in a variety of ways.

Table 3.1 Sixty determinations of Cu content (in ppm) in a reference sample

61.0	65.4	60.0	59.2	57.0	62.5	57.7	56.2	62.9	62.5
56.5	60.2	58.2	56.5	64.7	54.5	60.5	59.5	61.6	60.8
58.7	54.4	62.2	59.0	60.3	60.8	59.5	60.0	61.8	63.8
64.5	66.3	61.1	59.7	57.4	61.2	60.9	58.2	63.0	59.5
56.0	59.4	60.2	62.9	60.5	60.8	61.5	58.5	58.9	60.5
61.2	57.8	63.4	58.9	61.5	62.3	59.8	61.7	64.0	62.7

Data taken from Caulcutt and Boddy [3].

As an example, we will take 60 determinations of Cu content (in ppm) in a reference sample, taken from Caulcutt and Boddy's book [3] as presented in Table 3.1. From this we will illustrate a number of methods for summarizing and visualizing the data.

3.3.1 Descriptive Statistics

The simplest statistic is the mean, defined by:

$$\overline{x} = \sum_{i=1}^{I} x_i / I$$

where there are I readings in a dataset. Normally a 'bar' is placed above the symbol to denote mean. In our case this number is 60.37.

Sometimes the median is used instead. This is the middle number. To calculate this, organize all the data in order, and take the middle measurement. If there is an odd number of measurements this is straightforward, for example, if we record five numbers, the median is the third value, but for an even number of measurements, take the average of the two central numbers, for example, in our case the average of the 30th and 31st measurement. The measurements happen to be the same to one decimal place, giving a median of 60.50.

Normally the median and mean are quite close, but if they differ a lot this can be due to an asymmetric distribution or a few extreme values that will unduly influence the mean, but not the median.

The spread of readings is quite important and relates to the variability. It is normal to calculate a standard deviation (s) or variance (v), defined by:

$$v = s^2 = \sum_{i=1}^{I} (x_i - \overline{x})^2 / (I - 1)$$

For a series of samples, the divisor used is recommended to be $(I - 1)$ or 59 in our case, rather than I. This is because one degree of freedom is lost during the process of measurement (for more discussion of the idea of degrees of freedom in the context of experimental design see Section 2.2). For example, if I have only one sample in my dataset, the measurement and the average of all the measurements must be equal, this does not mean that the standard deviation of the overall distribution is zero, it is simply unknown. If I have two samples, there is only one possible value of the difference with the mean, so only one degree of freedom for such measurements, and so on. This definition of the standard deviation is called the *sample* standard deviation. The sample standard deviation and variance can also be calculated as follows:

$$v = s^2 = \sum_{i=1}^{I} x_i^2/(I-1) - \bar{x}^2 \times I/(I-1)$$

In our example, the sample standard deviation is 2.54 ppm. This implies that readings between 57.83 and 62.91 ppm are within one standard deviation of the mean. In later sections we will show how this information can be employed to provide more insight into the data.

For an overall population we often define these parameters slightly differently by:

$$v = s^2 = \sum_{i=1}^{I} (x_i - \bar{x})^2/I = \sum_{i=1}^{I} x_i^2/I - \bar{x}^2$$

and an aim of the sampling is to estimate these parameters. Notice an interesting quirk in chemometrics. When we scale matrices, for example, in preprocessing (Section 5.5) we are not estimating statistical parameters, but primarily placing variables on a similar scale, and it is usual to employ the population rather than sample statistics.

Sometimes it is useful to calculate the quartiles, the values between which a quarter of the readings lie. Three values need to be calculated, the first in our case is the average of the 15th and 16th readings, the second the median, the third the average of the 45th and 46th readings (when placed in ascending order), or in our case, 58.9, 60.5 and 62.0. This would imply, for example, that a quarter of the readings are between 60.5 and 62.0 ppm. Notice that the first and third quartiles are slightly less than one standard deviation from the mean, which would be expected from a normal distribution (as discussed in Section 3.4). Ideally, using a normal distribution, these quartiles will be 0.674 standard deviations from the mean.

3.3.2 Graphical Presentation

The simplest method of graphical presentation of the distribution is as a blob chart (Figure 3.1). Since the measurements are cited to only one decimal place, some readings overlap, and are placed above one another. This is not always necessary, but in fact is clearer, it can be hard to distinguish symbols if they are next to each other and the graph is very crowded.

The cumulative measurement frequency is often a helpful way of looking at the information graphically. The graph is presented in Figure 3.2 for our dataset. The vertical axis

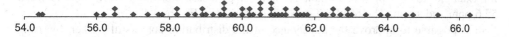

Figure 3.1   Blob chart

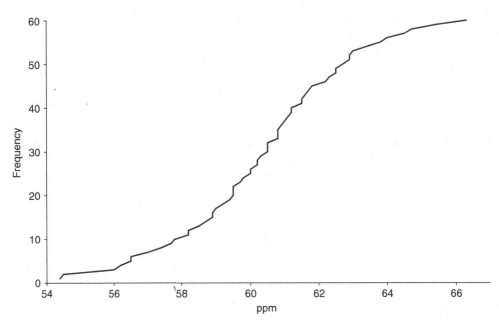

Figure 3.2 Cumulative frequency for the distribution in Table 3.1

represents the number of measurements less than or equal to the value on the horizontal axis. Real data will usually appear slightly bumpy, but the graph could be smoothed if necessary. However, such graphs can be used to detect for asymmetry or discontinuity or outliers and are easy to construct.

Often, if there are enough measurements, the data can be presented as a frequency histogram or bar chart. Figure 3.3 is of the percentage of measurements falling within ranges of 1 ppm. The vertical axis represents the percentage and the horizontal the range. Note that any range could have been chosen for the bars (sometimes called 'bins' or 'windows'), for example, 0.5 or 2 ppm and there is a balance between a small window in which case there will probably not be many measurements in each range and so the graph will not be very informative and a large window which will not easily reflect the variation and smooth away the underlying trends. Figure 3.3 has been divided into 13 windows, with an average of 4.61 measurements per bin, which is a reasonable number in this application.

It is often useful to obtain a distribution that can be presented as a continuous graph. One approach is to look at the density of measurements within a given window. The larger the window, the smoother the distribution. For example, we may choose a 5 ppm window. Centre the window on 53.0 ppm and then count the number of measurements between 50.5 ppm and 55.5 ppm (=2). Then move the centre of the window by a small amount, for example, to 53.1 ppm and count the number of measurements between 50.6 ppm and 55.6 ppm and so on. The resultant curve is presented in Figure 3.4.

It is possible to improve the appearance of the distribution curve still further, by using a smoothing function. Figure 3.5 is of the data of Figure 3.4 after a 9 point quadratic Savitzky–Golay filter (Section 4.3) has been applied to the curve. The mean and median are also indicated. Notice, firstly, that the distribution now seems quite smooth and resembles

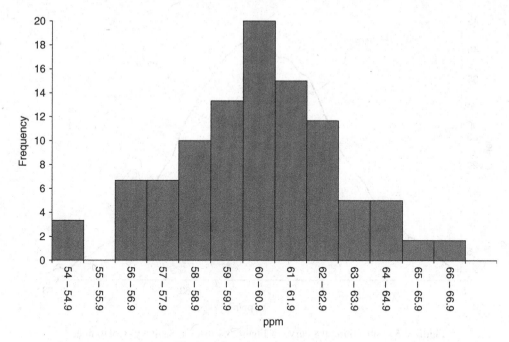

Figure 3.3 Frequency histogram

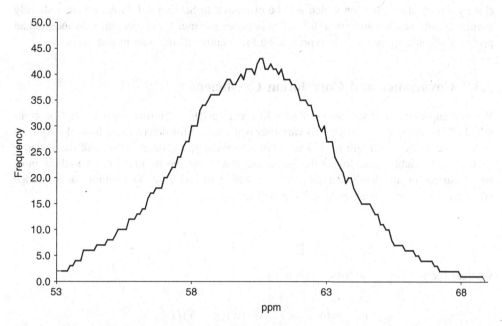

Figure 3.4 Distribution obtained from the data in Table 3.1, counting the number of measurements within a 5 ppm window, shifted by 0.1 ppm

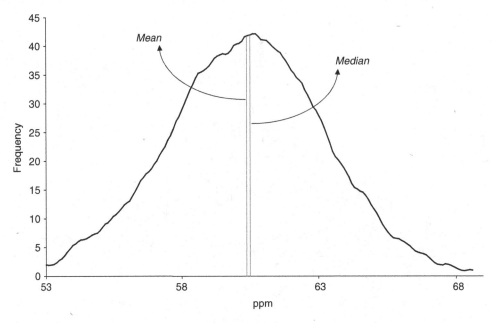

Figure 3.5 Smoothing the curve of Figure 3.4 using a Savitzky–Golay filter

closely a normal distribution which will be discussed in Section 3.4. However, it is slightly asymmetrical, which results in a difference between the mean and median, with the median probably slightly more usefully representing the 'centre' of the data in this case.

3.3.3 Covariance and Correlation Coefficient

When comparing the distribution of two different variables, further statistics can be computed. The covariance between two variables is a method for determining how closely they follow similar trends. It will never exceed in magnitude the geometric mean of the variance of the two variables, the lower the value the less close the trends. Both variables must be measured for an identical number of samples, I in this case. The sample or estimated covariance between variables x and y is defined by:

$$\text{cov}_{xy} = \sum_{i=1}^{I} (x_i - \bar{x})(y_i - \bar{y})/(I - 1)$$

whereas the population statistic is given by:

$$\text{cov}_{xy} = \sum_{i=1}^{I} (x_i - \bar{x})(y_i - \bar{y})/I$$

Unlike the variance, it is perfectly possible for a covariance to take on negative values.

Many chemometricians prefer to use the correlation coefficient, given by:

$$r_{xy} = \frac{\text{cov}_{xy}}{s_x . s_y} = \frac{\sum_{i=1}^{I} (x_i - \overline{x})(y_i - \overline{y})}{\sqrt{\sum_{i=1}^{I} (x_i - \overline{x})^2 \sum_{i=1}^{I} (y_i - \overline{y})^2}}$$

Notice that the definition of the correlation coefficient is identical for both samples and populations.

The correlation coefficient has a value between -1 and $+1$. If close to $+1$, the two variables are perfectly correlated. In many applications, correlation coefficients of -1 also indicate a perfect relationship. Under such circumstances, the value of y can be exactly predicted if we know x. The closer the correlation coefficients are to zero, the harder it is to use one variable to predict another. Some people prefer to use the square of the correlation coefficient which varies between 0 and 1.

3.4 THE NORMAL DISTRIBUTION

3.4.1 Error Distributions

Most numerically minded chemists (and biochemists, chemical engineers, biologists, geochemists, etc.) need to understand the principles behind the normal distribution. It occurs in many guises within science, but an important one in chemistry relates to the analysis of errors. If we make a number of replicated measurements on a sample then often they approximate to a normal distribution. Of course this is a simplified model, as we have seen (Section 3.2) and there are many factors that influence the spread of measurements, but most are symmetrical in nature and if there are a sufficient number of such factors they should add up to approximately a normal distribution. Some people get very concerned about the precise nature of error distributions, but we rarely have enough information to study this in detail. If we do design experiments to study error distributions in depth, we probably understand so much about a system that our original aim which might, for example, be simply to determine whether two measurement methods provide a similar answer, is often lost and our systematic design probably involves studying a process more thoroughly than is normally necessary so that different (and more controlled) source of errors take over. It is often sufficient to assume normality in the absence of any significant evidence to the contrary as a reasonable approximation, and as a means to an end; there are some simple tests which will be discussed in Section 3.5 to check roughly that our assumptions are right. It is important never to get overexcited about precise measurements of error distributions and always to remember that they are likely to originate from the measurement process rather than exist as a physically meaningful underlying property of the system (although here chemistry merges with philosophy and some people believe otherwise – but this book is not about religion, however it is interesting that people's beliefs about the nature of probability can have tremendous influence on how predictions are interpreted in many areas of science).

3.4.2 Normal Distribution Functions and Tables

The normal distribution is often presented as an equation:

$$f(x) = \frac{1}{\sigma\sqrt{2\pi}} \exp\left[-\frac{1}{2}\left(\frac{x-\mu}{\sigma}\right)^2\right]$$

This expression is often called a probability density function (pdf) and is proportional to the chance that a reading has a value of x if a dataset has a true mean of μ and standard deviation of σ. The function is scaled so that the area under the curve between plus and minus infinity is 1. The centre of the normal distribution curve is at $x = \mu$ and the standard deviation σ is reached when the height of the curve is $e^{-1/2}$ or 0.6065 of that of the centre. This is illustrated in Figure 3.6.

Usually, the normal distribution is presented as a table of numbers. Although there are quite a lot of ways in which these numbers can be reported, the cumulative normal distribution, illustrated in Table 3.2, is probably the most common, which involves determining the area under the standardized normal distribution curve: the units are standard deviations from the mean, and standardization (also discussed in Section 5.5) involves taking a dataset, subtracting the mean and dividing by the standard deviation: this transformation is useful to be able to compare data from different sources on the same scale. These areas can then be converted into probabilities. The table indicates the relative area of the curve at different numbers of standard deviations from the mean. Note, 0.5 of the data are 0 standard deviations from the mean, i.e. 50 % of the data are expected to have a value less than the average reading, and so the probability of a standardized reading being less than 0 is 0.5. At 1.32 standard deviations away from the mean, the probability becomes 0.9066. This implies that only 9.34 % (since $0.9066 + 0.0934 = 1$) of the measurements are expected to exceed 1.32 standard deviations above the mean, for a perfect normal distribution.

We can use the normal distribution tables to answer all sorts of questions. For example, what is the probability that a measurement falls between -1.0 and $+0.5$ standard deviations from the mean? Since the distribution is symmetrical, negative values are as likely as positive ones. The chances that the reading is between -1.0 and 0.0 standard deviations is given by $(0.8413 - 0.5) = 0.3413$, and between 0 and $+0.5$ standard deviations is $(0.6915 - 0.5) =$

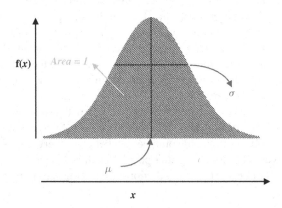

Figure 3.6 Features of a normal distribution

Table 3.2 Cumulative normal distribution

	0.00	0.01	0.02	0.03	0.04	0.05	0.06	0.07	0.08	0.09
0.0	0.5000	0.5040	0.5080	0.5120	0.5160	0.5199	0.5239	0.5279	0.5319	0.5359
0.1	0.5398	0.5438	0.5478	0.5517	0.5557	0.5596	0.5636	0.5675	0.5714	0.5753
0.2	0.5793	0.5832	0.5871	0.5910	0.5948	0.5987	0.6026	0.6064	0.6103	0.6141
0.3	0.6179	0.6217	0.6255	0.6293	0.6331	0.6368	0.6406	0.6443	0.6480	0.6517
0.4	0.6554	0.6591	0.6628	0.6664	0.6700	0.6736	0.6772	0.6808	0.6844	0.6879
0.5	0.6915	0.6950	0.6985	0.7019	0.7054	0.7088	0.7123	0.7157	0.7190	0.7224
0.6	0.7257	0.7291	0.7324	0.7357	0.7389	0.7422	0.7454	0.7486	0.7517	0.7549
0.7	0.7580	0.7611	0.7642	0.7673	0.7704	0.7734	0.7764	0.7794	0.7823	0.7852
0.8	0.7881	0.7910	0.7939	0.7967	0.7995	0.8023	0.8051	0.8078	0.8106	0.8133
0.9	0.8159	0.8186	0.8212	0.8238	0.8264	0.8289	0.8315	0.8340	0.8365	0.8389
1.0	0.8413	0.8438	0.8461	0.8485	0.8508	0.8531	0.8554	0.8577	0.8599	0.8621
1.1	0.8643	0.8665	0.8686	0.8708	0.8729	0.8749	0.8770	0.8790	0.8810	0.8830
1.2	0.8849	0.8869	0.8888	0.8907	0.8925	0.8944	0.8962	0.8980	0.8997	0.9015
1.3	0.9032	0.9049	0.9066	0.9082	0.9099	0.9115	0.9131	0.9147	0.9162	0.9177
1.4	0.9192	0.9207	0.9222	0.9236	0.9251	0.9265	0.9279	0.9292	0.9306	0.9319
1.5	0.9332	0.9345	0.9357	0.9370	0.9382	0.9394	0.9406	0.9418	0.9429	0.9441
1.6	0.9452	0.9463	0.9474	0.9484	0.9495	0.9505	0.9515	0.9525	0.9535	0.9545
1.7	0.9554	0.9564	0.9573	0.9582	0.9591	0.9599	0.9608	0.9616	0.9625	0.9633
1.8	0.9641	0.9649	0.9656	0.9664	0.9671	0.9678	0.9686	0.9693	0.9699	0.9706
1.9	0.9713	0.9719	0.9726	0.9732	0.9738	0.9744	0.9750	0.9756	0.9761	0.9767
	0.0	0.1	0.2	0.3	0.4	0.5	0.6	0.7	0.8	0.9
2.0	0.9772	0.9821	0.9861	0.9893	0.9918	0.9938	0.9953	0.9965	0.9974	0.9981
3.0	0.9987	0.9990	0.9993	0.9995	0.9997	0.9998	0.9998	0.9999	0.9999	1.0000

0.1915, so the overall chance is 53.28 % which is obtained by adding the two numbers together and multiplying by 100.

Note that most modern packages can also provide the normal distribution function, without the need to employ tables. For users of Excel, use NORMDIST.

3.4.3 Applications

We can use this reasoning to deal with practical situations. Consider a series of measurements of pH, whose average is 7.92, and standard deviation 0.24. What is the probability that a measurement exceeds 8.00? The following steps are performed:

- Standardize the measurement. The value is given by $(8.00 - 7.92)/0.24 = 0.33$, so that pH 8.00 is 0.33 standard deviations above the mean.
- Look up the value of the cumulative normal distribution 0.33 standard deviations from the mean, which is 0.6293.

• The probability that the reading exceeds this value is equal to $1 - 0.6293$ or 0.3707. Thus around 37 % of readings should exceed the value of pH 8.00.

If this prediction is not approximately obeyed by a series of measurements, either the readings really are not normally distributed or else the standard deviation/mean have not been estimated very well in the first place.

There can be real implications in making these sorts of predictions. For example, we may wish to ensure that the concentration of a contaminant or an additive is below a certain level, otherwise there could be problems with the quality of a product. If we measure the concentration in a range of products obtained by a single manufacturing process, and it appears approximately normally distributed, what concentration can we be 99 % confident will not be exceeded? If this concentration is exceeded we may be suspicious about the manufacturing process and want to check carefully what is going on, even closing down a manufacturing plant for hours or days, potentially costing a significant amount of money.

3.5 IS A DISTRIBUTION NORMAL?

It is often quite useful to see whether a series of measurements fits roughly into a normal distribution. Most series of measurements should obey normality, and most statistical tests assume this.

If, for example, we measure pH 50 times on a series of samples, are these measurements normally distributed? Sometimes they may not be because there are underlying factors that influence the measurement process that we have not yet discovered. In other cases there may be a few outliers, or rogue samples which we want to eliminate before we go further (see Sections 3.11 and 3.12). Another interesting reason is to look at residual errors after regression (Section 6.2). The majority of packages for fitting lines or curves or even multivariate approaches, assume the errors are normally distributed. If not it is sometimes possible to transform the data (e.g. if noise is known to be proportional to signal intensity, sometimes called heteroscedastic noise, there are several standard transforms). There are a number of tests which can be employed to look at the data before further analysis. Often simple problems can then be ironed out at an early stage.

3.5.1 Cumulative Frequency

One of the simplest approaches to seeing whether a distribution appears to be normal is a graphical approach involving cumulative frequency distributions. In the old days, graph paper manufacturers had a field day, and printed a huge variety of specialized and very expensive graph paper, many of which were used for statistical tests. Still many texts published even now refer in a nostalgic way to different types of graph paper, but nowadays we can do most of the work using Excel or several common numerical packages.

The trick to understanding this is to think about the cumulative normal distribution. In Section 3.4 we introduced this in tabular form. From Table 3.2 we can deduce that, for example, 50 % of the readings are expected to be greater than the mean, but only around 16 % should be more than 1 standard deviation more than the mean. If we find that 30 % of readings fall above 1 standard deviation away from the mean, we should be a bit suspicious whether the data really do fit a normal distribution, especially if we took a lot of measurements in the first place.

We will use the data in Table 3.1 to illustrate this. The first step is to arrange them in order, and then work out the cumulative frequency that a reading is below a given value: this is not too difficult to do. Because there are 60 readings, the first 50 % of the readings are less than or equal to the 30th reading which happens to be 60.5 ppm. 1/120 of the readings (0.833 %) are considered to be less than or equal to the lowest reading (54.4 ppm). 1/120 + 1/60 (2.5 %) are less than or equal to the next reading (54.5 ppm) and 1/120 + 1/60 + 1/60 (4.167 %) less than or equal to the third lowest reading and so on. These cumulative probabilities are represented in Table 3.3 (light grey shaded cells). To obtain a cumulative probability plot, simply produce a graph of these probabilities against the raw readings (Figure 3.7) similar to Figure 3.2 but with a different vertical scale.

The question we want to answer is does this graph really resemble one that would be obtained if there were a normal distribution? The way to do this is to convert the left-hand side axis to one of standard deviations away from the mean. An example is as follows.

Table 3.3 Cumulative frequencies to determine whether the data in Table 3.1 are normally distributed (see Sections 3.5.2 and 3.5.3 for description of shading)

54.4	54.5	56	56.2	56.5	56.5	57	57.4	57.7	57.8
0.833	2.500	4.167	5.833	7.500	9.167	10.833	12.500	14.167	15.833
−2.394	−1.960	−1.732	−1.569	−1.440	−1.331	−1.235	−1.150	−1.073	−1.001
−2.350	−2.311	−1.720	−1.641	−1.523	−1.523	−1.326	−1.169	−1.051	−1.012
58.2	58.2	58.5	58.7	58.9	58.9	59	59.2	59.4	59.5
17.500	19.167	20.833	22.500	24.167	25.833	27.500	29.167	30.833	32.500
−0.935	−0.872	−0.812	−0.755	−0.701	−0.648	−0.598	−0.549	−0.501	−0.454
−0.854	−0.854	−0.736	−0.657	−0.579	−0.579	−0.539	−0.461	−0.382	−0.342
59.5	59.5	59.7	59.8	60	60	60.2	60.2	60.3	60.5
34.167	35.833	37.500	39.167	40.833	42.500	44.167	45.833	47.500	49.167
−0.408	−0.363	−0.319	−0.275	−0.232	−0.189	−0.147	−0.105	−0.063	−0.021
−0.342	−0.342	−0.264	−0.224	−0.146	−0.146	−0.067	−0.067	−0.028	0.051
60.5	60.5	60.8	60.8	60.8	60.9	61	61.1	61.2	61.2
50.833	52.500	54.167	55.833	57.500	59.167	60.833	62.500	64.167	65.833
0.021	0.063	0.105	0.147	0.189	0.232	0.275	0.319	0.363	0.408
0.051	0.051	0.169	0.169	0.169	0.209	0.248	0.287	0.327	0.327
61.5	61.5	61.6	61.7	61.8	62.2	62.3	62.5	62.5	62.7
67.500	69.167	70.833	72.500	74.167	75.833	77.500	79.167	80.833	82.500
0.454	0.501	0.549	0.598	0.648	0.701	0.755	0.812	0.872	0.935
0.445	0.445	0.484	0.524	0.563	0.720	0.760	0.838	0.838	0.917
62.9	62.9	63	63.4	63.8	64	64.5	64.7	65.4	66.3
84.167	85.833	87.500	89.167	90.833	92.500	94.167	95.833	97.500	99.167
1.001	1.073	1.150	1.235	1.331	1.440	1.569	1.732	1.960	2.394
0.996	0.996	1.035	1.193	1.350	1.429	1.626	1.704	1.980	2.334

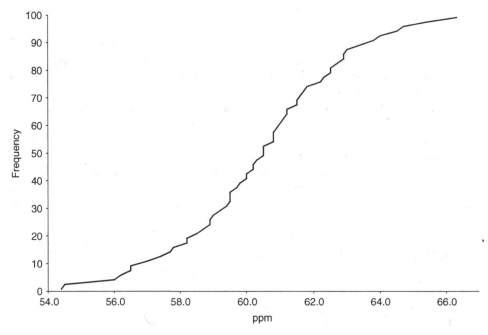

Figure 3.7 Cumulative probability graph

12.5 % of the data have a value of 57.4 or less (see Table 3.3). If the data were normally distributed, 12.5 % of the data will exceed 1.15 standard deviations above the mean, which can be seen using a normal distribution table (Table 3.2). A probability of 0.875 is 1.15 standard deviations away from the mean; hence, only 0.125 ($= 1 - 0.875$) of the data (12.5 %) would exceed this value if the underlying distribution is normal.

- The normal distribution is symmetrical, so similarly 12.5 % of the data will have a reading less than 1.15 standard deviations below the mean.
- Thus, we expect the value 57.4 to be approximately equal to the mean minus 1.15 standard deviations.

What we want to do is look at each of the 60 measurements and see whether their cumulative probability fits roughly into this pattern. Fortunately, Excel and many other statistical packages come to our rescue and the calculations are not so difficult. The NORMSINV function can be used to calculate these probabilities automatically, so NORMSINV(0.125) = -1.150. In Table 3.3, we represent in dark grey shaded cells the value of NORMSINV for all 60 probabilities. By plotting a graph of these numbers against the 60 readings, if the data are normally distributed, it should be roughly linear (Figure 3.8). It is possible to determine a probability for how well the straight-line relationship is obeyed, but graphics is often safer.

3.5.2 Kolmogorov–Smirnov Test

This complicatedly named test is often referred to as a hypothesis test. Some people like to use probabilities and want to answer 'what is the probability that the data fit the hypothesis

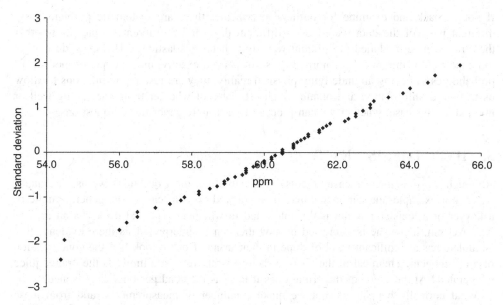

Figure 3.8 Graphical test of whether the data in Table 3.1 is normally distributed: the graph should be roughly linear if this is obeyed

that the distribution is normal?' We will be discussing the idea of hypothesis of significance tests in greater detail in Section 3.6, so only a brief introduction is given below.

What this test does is to give a numerical value of how the distribution differs from a normal distribution. The first step is to determine the standard deviation and mean for the readings (60.37 and 2.54 – see Section 3.3). Now standardize the data, to give the values in the unshaded and unbold cells in Table 3.3. This implies, for example, that a measurement of 59.5 ppm is 0.342 standard deviations less than the mean. How does this relate to what we expect from a normal distribution? Simply compare this new set of numbers in Table 3.3 to the numbers in the dark grey shaded cells. The bigger the difference, the worse the problem. In fact in most cases they are quite close, only one difference is quite large, that of the second reading (0.350 standard deviations). This reading may be an outlier, and sometimes extreme values can have a rather unpredictable behaviour, and every now and then one extreme value does not fit expectations. Some people use statistical tables or packages to then provide a probability that the distribution is normal, converting the differences to try to prove (or disprove) a hypothesis, and often use the biggest difference to determine normality: in this case the data would be rejected (i.e. not assumed to be normal), which is a little unfair, the next largest difference is 0.195 standard deviations, and the majority are really quite small. By simply removing a couple of points, the data appear to obey normal statistics quite well. Perhaps if the outliers come from individual samples we might want to check them again later.

3.5.3 Consequences

This section has shown some ways of examining data to see whether they appear to be normally distributed. If so, most conventional statistical and chemometric methods are safe.

If not go back and examine for outliers, or whether there are systematic problems (e.g. different parts of the data recorded on different days or by different people) or whether the errors can be modelled in different ways (e.g. heteroscedasticity). However, do not get too concerned if the model for normality is not obeyed exactly, most people do not regard probabilities as having an underlying physical reality, they are just approximations to allow us to look at information and obtain rough estimates of whether things are going well or not and, in this case, whether we can proceed to ask more sophisticated questions.

3.6 HYPOTHESIS TESTS

Although most chemical measurements are quantitative, most questions we ask are qualitative. For example, measurements may be employed in court cases. The general public or a lawyer or a legislator is not really interested in whether a spectral peak has an area of 22.7 AU*nm, he or she is interested in how this can be interpreted. Is there evidence that an athlete has a significant level of dope in their urine? Does it look like the concentration of a carcinogenic chemical in the air of a factory is above a safe limit? Is the orange juice adulterated? At the scene of the crime does it look as the dead person was poisoned?

What normally happens is that we make a number of measurements, and from these we try to determine some qualitative information. What we are really after is whether the measurements fit a hypothesis. As a forensic scientist we may wish to look at a blood sample and measure whether a poison is in the blood and if so whether its level is sufficiently above safe and natural limits (some poisons are quite subtle and hard to detect), and so whether the chemical evidence is consistent with a crime. Hypothesis tests are used to formulate this information statistically, and can also be called significance tests.

What we do first is to set up a hypothesis. Many people start with the null hypothesis, for example, in the absence of any specific evidence let us assume that the level of a stimulant in an athlete's urine is not significant. To disprove our null hypothesis, we require strong evidence that there really is a lot of stimulant in the sample. This is not always so easy to prove. For example, many chemicals also have a natural origin, so one has to demonstrate that the compound is at a higher level than would naturally be found. To do this one may take a set of urine samples from a series of individuals without the stimulant added, perform some measurements (which will have a spread of results due to each person's metabolism) and then see whether the level of stimulant in the contested sample differs significantly. In addition there are other factors such as breakdown of metabolites and long term storage of samples, for example, if we wish to return to perform more tests in a few months' time if the athlete takes us to court to contest the decision. A significance test usually has a probability associated with it. We may want to be more than 99 % sure that the concentration of the metabolite is so high that it could not have occurred through normal metabolic processes, so that the chance we are falsely accusing the athlete is only 1 %; in other words we are 99 % sure that the null hypothesis is wrong. Some people use different significance levels and a lot depends on whether we want to do screening where a lower level might simply be used to eliminate the samples that appear alright, or really water-tight confirmatory analysis, where a higher level might be advisable.

An understanding of the normal distribution can help us here. Let us say we knew that the concentration of a compound in normal subjects had an average of 5.32 ppm and a standard deviation of 0.91 ppm. We could reformulate the problem as follows. If we assume the errors are normally distributed, we can use a normal distribution (Section 3.4) and find that

99 % of measurements should be less than the mean plus 2.33 standard deviations [for Excel users the function NORMINV(0.99,0,1) can be used for this calculation], so if a reading is more than $5.32 + 2.33 \times 0.91 = 7.44$ ppm, it could only have arisen by chance in the background population in 1 % of the cases. Of course life is not always so simple. For example, the original set of samples might be small and so the mean and standard deviation may not have been estimated very well. In order to make these adjustments special statistical tables are used to provide probabilities corrected for sample size, as will be discussed below. We would probably use what is called a t-test if there are not many samples, however, if there are a large number of samples in the original data this tends to a normal distribution. Note that the t-test still assumes that the errors are normally distributed in a background population, but if we measure only a few samples, this perfect distribution will become slightly distorted, and there will be significant chances that the mean and standard deviation of the samples differ from the true ones.

There are a number of such hypothesis tests (most based on errors being normally distributed), the chief ones being the F- and t-test. These tests are either one- or two-tailed. A two-tailed test might provide us information about whether two measurements differ significantly, for example is a measurement significantly different from a standard sample. In this case, we do not mind whether it is less or more. For example we might be looking at a food from a given supplier. From several years we know the concentrations expected of various constituents. If we are testing an unknown sample (an example may be customs that inspect imports with a given label of authenticity), we would be surprised if the concentrations differ outside given limits, but deviations both above and below the expected range are equally interesting. A one-tailed test just looks at deviations in a single direction. The example of a potentially doped athlete is a case in question: we may be only interested if it appears that the level of dope is above a given level, we are not going to prosecute the athlete if it is too low. One-tailed tests are quite common when looking at significance of errors, and in ANOVA analysis (Section 2.3).

The use of statistical tests in a variety of applications has developed over more than 100 years, and many scientists of yesteryear would be armed with statistical tables, that had been laboriously calculated by hand. The vast majority of texts still advocate the use of such tables, and it is certainly important to be able to appreciate how to use them, but like books of logarithm tables and slide rules it is likely that their days are numbered, although it may take several years before this happens. We have already introduced the normal distribution table (Section 3.4) and will present some further tables below. It is important to recognize that common spreadsheets and almost all good scientific software will generate this information easily and more flexibly because any probability level can be used (tables are often restricted to specific probabilities). Because of the widespread availability of Excel, we will also show how this information can be obtained using spreadsheet commands.

3.7 COMPARISON OF MEANS: THE t-TEST

Consider trying to determine whether a river downstream from a factory is polluted. One way of doing this might be to measure the concentration of a heavy metal, perhaps 10 times over a working week. Let us say we choose Cd, and do this one week in January and find that the mean is 13.1 ppb. We then speak to the factory manager who says that he or she will attempt to reduce the emissions within a month. In February we return and we take eight measurements and find that their mean is now 10.4 ppb. Are we confident that this really indicates a reduction in Cd content?

There are several factors that influence the concentration estimates, such as the quality of our instrument and measurement process, and the sampling process (Section 3.2). A difference in means may not necessarily imply that we have detected a significant reduction in the amount of Cd. Can we be sure that the new mean really represents a lower concentration?

Each series of measurements has a set of errors associated with it, and the trick is to look also at the standard deviation of the measurements. If, for example, our measurements have a standard deviation of 0.1 ppb a reduction in the mean by 2.7 ppb is huge (27 standard deviations) and we can be really confident that the new value is lower. If, however, the standard deviation is 3 ppb, we may not have very much confidence.

To understand how we can determine the significance, return to the example of the normal distribution: we have already come across a simple example, of looking at the chances that a measurement can arise from population that forms a known normal distribution. We assume that the overall distribution is known very well, for example we may have recorded 100 measurements and are extremely confident about the mean, standard deviation and normality, but our new test sample consists only of a single measurement. Perhaps we performed 100 tests in January to be sure, and then come out to obtain a single sample in February just to check the situation. Then all we do is use the normal distribution. If the mean and standard deviation of the original readings were 13.1 and 1.4, February's single reading of 10.4 is $(13.1 - 10.4)/1.3 = 2.077$ standard deviations below the mean. Using the Excel function NORMDIST($-2.077, 0, 1$, FALSE) we find that only 1.9 % of the measurements are expected to fall below 2.077 standard deviations below the mean (this can also be verified from normal distribution tables). In other words, we only would expect a reading of 10.4 or less to arise about once in 50 times in the original population sampled in January, assuming the null hypothesis that there is no significant difference. This is quite low and we can, therefore, on the available evidence be more than 98 % sure that the new measurement is significantly lower than the previous measurements. We are, in fact, approximately using what is often called a one-tailed t-test for significance, as described below.

In practice we usually do not have time to perform hundreds of measurements for our reference, and often usually take several measurements of the new test population. This means that we may not be completely confident of either of the means (in January or February), both of which carry a level of uncertainty. The difference in means depends on several factors:

- Whether there is actually a true underlying difference.
- How many samples were used to obtain statistics on both the original and the new data.
- What the standard deviations of the measurements are.

Then the means of the two datasets can be compared, and a significance attached to the difference. This is the principle of the t-test.

Let us assume that we obtain the following information.

January. We take 10 measurements with a mean of 13.1, and a standard deviation of 1.8.
February. We take 8 measurements with a mean of 10.4 and a standard deviation of 2.1.
How can we use a statistical test of significance? The following steps are performed:

- First we determine what is called a 'pooled variance'. This is the average variance (the square of the standard deviation: see Section 3.3) from both groups and defined by:

$$s^2 = \frac{(n_1 - 1)s_1^2 + (n_2 - 1)s_2^2}{(n_1 + n_2 - 2)}$$

This equation takes into account the fact that different numbers of samples may be used in each set of measurements, so it would be unfair to weight a standard deviation obtained using four measurements as equally useful as one arising from 40. In our case the 'pooled variance' is 3.75, the corresponding standard deviation being 1.94.
- Next we calculate the difference between the means $(m_1 - m_2)$ or 2.7. This should be calculated as a positive number, subtracting the largest mean from the smallest.
- The third step is to calculate the t-statistic which is a ratio between the two values above, adjusted according to sample size, calculated by:

$$t = \frac{m_1 - m_2}{s\sqrt{1/n_1 + 1/n_2}} = \frac{2.7}{1.93 \times 0.474} = 2.939$$

- A final step is used to convert this to a probability. If we are interested only in whether the new value is significantly lower than the previous one, we obtain this using a one-tailed t-test, because our question is not to answer whether the new value of Cd is different to the previous one, but whether it is significantly lower: we are interested only that the pollution has decreased, as this is the claim of the factory owner. We first have to determine the number of degrees of freedom for the overall distribution, which equals $n_1 + n_2 - 2$ or 16. Then, in Excel, simply calculate TDIST(2.939,16,1) to give an answer 0.0048. This implies that the chances that there really is no reduction in the Cd content is less than 0.5 % so we are sure that there has been an achievable difference.

What happens if we are not certain that there is a difference between the means? There may, of course, really be no difference, but it also may lie in our measurement technique (too much uncertainty) or the number of samples we have recorded (take more measurements).

The t-test can be used in many different ways. A common situation is where there is a well established standard (perhaps an international reference sample) and we want to see whether our sample or analytical method results in a significantly different value from the reference method, to see whether our technique really is measuring the same thing. In this case we use the two-tailed t-test. We would be concerned if our mean is either significantly higher or lower than the reference. If you use Excel, and the TDIST function, replace the '1' by a '2' in the last parameter.

Another common reason for using a t-test is to compare analytical techniques or technicians or even laboratories. Sometimes people have what is called 'bias', that is their technique consistently over- or underestimates a value. This might involve problems with poorly calibrated apparatus, for example all balances should be carefully checked regularly, or it may involve problems with baselines. In many laboratories if a new technique is to be adopted it is good practice to first check its results are compatible with historic data.

In many books, t-test statistics are often presented in tabular form, an example being Table 3.4 of the one-tailed t-distribution. This presents the t-statistic for several probability levels. If, for example, we calculate a t-statistic of 2.650 with 13 degrees of freedom we can be 99 % certain that the means differ (using a one-tailed test – the probabilities in Table 3.4 are halved for a two-tailed test). Notice that for an infinite number of degrees of freedom the t-statistic is the same as the normal distribution: see the normal distribution table (Table 3.2), for example a 95 % probability is around 1.645 standard deviations from the mean, corresponding to the t-statistic at 5 % probability level in Table 3.4.

The t-test for significance is often very useful, but it is always important not to apply such tests blindly. Data do have to be reasonably homogeneous and fit an approximate normal distribution. Often this is so, but a major experimental problem, to be discussed in Section 3.12 involves outliers, or atypical points. These could arise from rogue measurements and can

Table 3.4   One-tailed t-test

Degrees of freedom	Probability		
	0.05	0.01	0.005
1	6.314	31.821	63.656
2	2.920	6.965	9.925
3	2.353	4.541	5.841
4	2.132	3.747	4.604
5	2.015	3.365	4.032
6	1.943	3.143	3.707
7	1.895	2.998	3.499
8	1.860	2.896	3.355
9	1.833	2.821	3.250
10	1.812	2.764	3.169
11	1.796	2.718	3.106
12	1.782	2.681	3.055
13	1.771	2.650	3.012
14	1.761	2.624	2.977
15	1.753	2.602	2.947
16	1.746	2.583	2.921
17	1.740	2.567	2.898
18	1.734	2.552	2.878
19	1.729	2.539	2.861
20	1.725	2.528	2.845
25	1.708	2.485	2.787
30	1.697	2.457	2.750
35	1.690	2.438	2.724
40	1.684	2.423	2.704
45	1.679	2.412	2.690
50	1.676	2.403	2.678
55	1.673	2.396	2.668
60	1.671	2.390	2.660
65	1.669	2.385	2.654
70	1.667	2.381	2.648
80	1.664	2.374	2.639
90	1.662	2.368	2.632
100	1.660	2.364	2.626
∞	1.645	2.327	2.576

have a very large influence on the mean and standard deviation, and it is important to eliminate these before performing most common statistical tests.

3.8 *F*-TEST FOR COMPARISON OF VARIANCES

It is very common for several laboratories to have to compare their efficiency in measuring the concentrations of compounds in samples. For example, a forensic science laboratory may be interested in determining the quantity of a toxin in a blood. There will often be several centres in a country, perhaps attached to hospitals that run regional analyses. In fact having a single institute nationally or internationally would cause arguments about who receives the funding, and also problems about transport and storage and even employment of skilled scientists who may not wish to relocate to the one place in the world or in the country that maintains a single centralized service. However, we do need to check that each laboratory is functioning in a similar fashion.

One key issue is the accuracy of the analyses, which can be assessed using the *t*-test for means (Section 3.7), the mean of several analyses on a reference sample for each laboratory are compared, and if one is very different from the others we suspect there is a problem. These difficulties will occur from time to time as new instruments and personnel are introduced, and cannot easily be controlled unless there is a single central organization that updates all laboratories (ideally in the world) simultaneously – an impossible dream. Note that many analytical chemists use the term *accuracy* to denote how a measurement (or more usually the mean of a series of measurements) differ from the true measurement.

A second issue relates to what is called the *precision* of the analyses. This relates to the spread or variance (Section 3.3) of results. We might want to compare two laboratories, A and B, and see if there is a significant difference in their variance. If a reference sample is given to laboratory A and the variance of several measurements is 2.5 ng^2/ml^2 whereas for laboratory B it is 1.1 ng^2/ml^2, is this difference significant? If there is a big difference this may suggest that one laboratory is much less reliable than the other and must improve its procedures, or else it may lose contract work in the future.

The way to check this is by calculating the *F*-ratio defined by:

$$F = \frac{s_A^2}{s_B^2} = \frac{2.5}{1.1} = 2.273$$

Is this number significantly greater than 1? It looks greater than 1 but remember this is the ratio of the square of the standard deviations. What we might want to ask is can we be 95 % certain that there genuinely is a larger spread of results in laboratory A compared with laboratory B? This depends in part on how many measurements we make. If we do not take many measurements it is possible to obtain erroneous estimates of the variance quite easily. The more measurements, the closer we expect the estimated variance to be to the true variance. Table 3.5 illustrates what might happen if we record three measurements from each laboratory. It does not look as if this apparently large difference in variances is very significant. Because we have not made many measurements we do not have many degrees of freedom (only $I - 1 = 2$) for each variance, hence there really has to be a very big difference in the variances for us to be sure.

Often people use the *F*-test to assess the significance of this ratio. This is used to compare two variances, and convert the ratio above to a probability. The way to do this is as follows:

- Calculate the F-ratio always placing the larger variance on the top of the equation (so the ratio is never less than 1), in this case it equals 2.273.
- Determine the number of degrees of freedom for each variance, in the case of Table 3.5, each variance has two degrees of freedom associated with it, one less than the number of measurements.
- Decide whether you want a two-tailed or one-tailed test. The former asks whether two variances differ significantly, but there is no problem whether one is larger than the other or vice versa, but the latter specifically asks whether one is larger than the other. In our example we want the two-tailed test.
- Then, if you use Excel, calculate the probability that there is no significant difference between the variances by using the function FDIST(*F-ratio, degrees of freedom of larger variance, degrees of freedom of smaller variance*) and multiply by 2 if two-tailed, or in our case 2*FDIST(2.273,2,2) which equals 0.611.
- Then interpret the probability. One really needs a probability that is low (perhaps 5 % or less) that the two variances are similar to be fairly sure there is a difference. A 61 % probability really suggests that there is no real reason to have doubts that the two variances are different.

Let us say we obtained these variances using a much larger dataset, say 20 measurements from laboratory A and 30 from laboratory B, then the value 2*FDIST(2.273,19,29) = 0.045 which is quite different, meaning that we are 95.5 % sure that there really is a difference.

The F-test is more often used as a one-tailed test, especially in connection with ANOVA (introduced in Section 2.3). The principle, in this case, is usually not so much to see whether one variance differs from another but whether it is significantly larger than the other. A major reason is to check whether an error such as the lack-of-fit to a model is significantly larger than the experimental error. If it is, then the lack-of-fit is significant, and the model is a bad one. The F-ratio is normally a ratio between mean error for fitting a model to the mean replicate error. Note that both errors are divided by their respective number of degrees of freedom (Section 2.2). In this case a one-tailed test is more appropriate. When looking at the significance of parameters in a multilinear model a favourite trick is to remove parameters one by one and if the model has a significant change in lack-of-fit whilst excluding one of the parameters, then this parameter is important and must be left in the model.

The F-statistic pops up in all sorts of contexts within chemistry, but mainly is used to compare the significance of two variances. It is used a lot in physical chemistry.

Many books present tables of critical F-values. As usual, these were devised in the days prior to the widespread use of spreadsheets, but are still a valuable aid. Each table is

Table 3.5 Determining the variances of two sets of three measurements

	Laboratory A	Laboratory B
Measurements	79.85	78.10
	78.89	79.57
	76.12	77.01
Variance	2.50	1.10

Table 3.6 Critical values of F-statistic: 5% one-tailed, 10% two-tailed

	1	2	3	4	5	6	7	8	9	10	11	12	13	14	15	20	30	40	50	75	100	250	500	∞
1	161.4	199.5	215.7	224.6	230.2	234.0	236.8	238.9	240.5	241.9	243.0	243.9	244.7	245.4	245.9	248.0	250.1	251.1	251.8	252.6	253.0	253.8	254.1	254.3
2	18.51	19.00	19.16	19.25	19.30	19.33	19.35	19.37	19.38	19.40	19.40	19.41	19.42	19.42	19.43	19.45	19.46	19.47	19.48	19.48	19.49	19.49	19.49	19.50
3	10.13	9.552	9.277	9.117	9.013	8.941	8.887	8.845	8.812	8.785	8.763	8.745	8.729	8.715	8.703	8.660	8.617	8.594	8.581	8.563	8.554	8.537	8.532	8.526
4	7.709	6.944	6.591	6.388	6.256	6.163	6.094	6.041	5.999	5.964	5.936	5.912	5.891	5.873	5.858	5.803	5.746	5.717	5.699	5.676	5.664	5.643	5.635	5.628
5	6.608	5.786	5.409	5.192	5.050	4.950	4.876	4.818	4.772	4.735	4.704	4.678	4.655	4.636	4.619	4.558	4.496	4.464	4.444	4.418	4.405	4.381	4.373	4.365
6	5.987	5.143	4.757	4.534	4.387	4.284	4.207	4.147	4.099	4.060	4.027	4.000	3.976	3.956	3.938	3.874	3.808	3.774	3.754	3.726	3.712	3.686	3.678	3.669
7	5.591	4.737	4.347	4.120	3.972	3.866	3.787	3.726	3.677	3.637	3.603	3.575	3.550	3.529	3.511	3.445	3.376	3.340	3.319	3.290	3.275	3.248	3.239	3.230
8	5.318	4.459	4.066	3.838	3.688	3.581	3.500	3.438	3.388	3.347	3.313	3.284	3.259	3.237	3.218	3.150	3.079	3.043	3.020	2.990	2.975	2.947	2.937	2.928
9	5.117	4.256	3.863	3.633	3.482	3.374	3.293	3.230	3.179	3.137	3.102	3.073	3.048	3.025	3.006	2.936	2.864	2.826	2.803	2.771	2.756	2.726	2.717	2.707
10	4.965	4.103	3.708	3.478	3.326	3.217	3.135	3.072	3.020	2.978	2.943	2.913	2.887	2.865	2.845	2.774	2.700	2.661	2.637	2.605	2.588	2.558	2.548	2.538
11	4.844	3.982	3.587	3.357	3.204	3.095	3.012	2.948	2.896	2.854	2.818	2.788	2.761	2.739	2.719	2.646	2.570	2.531	2.507	2.473	2.457	2.426	2.415	2.404
12	4.747	3.885	3.490	3.259	3.106	2.996	2.913	2.849	2.796	2.753	2.717	2.687	2.660	2.637	2.617	2.544	2.466	2.426	2.401	2.367	2.350	2.318	2.307	2.296
13	4.667	3.806	3.411	3.179	3.025	2.915	2.832	2.767	2.714	2.671	2.635	2.604	2.577	2.554	2.533	2.459	2.380	2.339	2.314	2.279	2.261	2.229	2.218	2.206
14	4.600	3.739	3.344	3.112	2.958	2.848	2.764	2.699	2.646	2.602	2.565	2.534	2.507	2.484	2.463	2.388	2.308	2.266	2.241	2.205	2.187	2.154	2.142	2.131
15	4.543	3.682	3.287	3.056	2.901	2.790	2.707	2.641	2.588	2.544	2.507	2.475	2.448	2.424	2.403	2.328	2.247	2.204	2.178	2.142	2.123	2.089	2.078	2.066
20	4.351	3.493	3.098	2.866	2.711	2.599	2.514	2.447	2.393	2.348	2.310	2.278	2.250	2.225	2.203	2.124	2.039	1.994	1.966	1.927	1.907	1.869	1.856	1.843
30	4.171	3.316	2.922	2.690	2.534	2.421	2.334	2.266	2.211	2.165	2.126	2.092	2.063	2.037	2.015	1.932	1.841	1.792	1.761	1.718	1.695	1.652	1.637	1.622
40	4.085	3.232	2.839	2.606	2.449	2.336	2.249	2.180	2.124	2.077	2.038	2.003	1.974	1.948	1.924	1.839	1.744	1.693	1.660	1.614	1.589	1.542	1.526	1.509
50	4.034	3.183	2.790	2.557	2.400	2.286	2.199	2.130	2.073	2.026	1.986	1.952	1.921	1.895	1.871	1.784	1.687	1.634	1.599	1.551	1.525	1.475	1.457	1.438
75	3.968	3.119	2.727	2.494	2.337	2.222	2.134	2.064	2.007	1.959	1.919	1.884	1.853	1.826	1.802	1.712	1.611	1.555	1.518	1.466	1.437	1.381	1.360	1.338
100	3.936	3.087	2.696	2.463	2.305	2.191	2.103	2.032	1.975	1.927	1.886	1.850	1.819	1.792	1.768	1.676	1.573	1.515	1.477	1.422	1.392	1.331	1.308	1.283
250	3.879	3.032	2.641	2.408	2.250	2.135	2.046	1.976	1.917	1.869	1.827	1.791	1.759	1.732	1.707	1.613	1.505	1.443	1.402	1.341	1.306	1.232	1.202	1.166
500	3.860	3.014	2.623	2.390	2.232	2.117	2.028	1.957	1.899	1.850	1.808	1.772	1.740	1.712	1.686	1.592	1.482	1.419	1.376	1.312	1.275	1.194	1.159	1.113
∞	3.841	2.996	2.605	2.372	2.214	2.099	2.010	1.938	1.880	1.831	1.789	1.752	1.720	1.692	1.666	1.571	1.459	1.394	1.350	1.283	1.243	1.152	1.106	1.000

Table 3.7 Critical values of F-statistic: 1 % one-tailed, 2 % two-tailed

	1	2	3	4	5	6	7	8	9	10	11	12	13	14	15	20	30	40	50	75	100	250	500	∞
1	4052	4999	5404	5624	5764	5859	5928	5981	6022	6056	6083	6107	6126	6143	6157	6209	6260	6286	6302	6324	6334	6353	6360	6366
2	98.50	99.00	99.16	99.25	99.30	99.33	99.36	99.38	99.39	99.40	99.41	99.42	99.42	99.43	99.43	99.45	99.47	99.48	99.48	99.48	99.49	99.50	99.50	99.499
3	34.12	30.82	29.46	28.71	28.24	27.91	27.67	27.49	27.34	27.23	27.13	27.05	26.98	26.92	26.87	26.69	26.50	26.41	26.35	26.28	26.24	26.17	26.15	26.125
4	21.20	18.00	16.69	15.98	15.52	15.21	14.98	14.80	14.66	14.55	14.45	14.37	14.31	14.25	14.20	14.02	13.84	13.75	13.69	13.61	13.58	13.51	13.49	13.463
5	16.26	13.27	12.06	11.39	10.97	10.67	10.46	10.29	10.16	10.05	9.963	9.888	9.825	9.770	9.722	9.553	9.379	9.291	9.238	9.166	9.130	9.064	9.042	9.020
6	13.75	10.92	9.780	9.148	8.746	8.466	8.260	8.102	7.976	7.874	7.790	7.718	7.657	7.605	7.559	7.396	7.229	7.143	7.091	7.022	6.987	6.923	6.901	6.880
7	12.25	9.547	8.451	7.847	7.460	7.191	6.993	6.840	6.719	6.620	6.538	6.469	6.410	6.359	6.314	6.155	5.992	5.908	5.858	5.789	5.755	5.692	5.671	5.650
8	11.26	8.649	7.591	7.006	6.632	6.371	6.178	6.029	5.911	5.814	5.734	5.667	5.609	5.559	5.515	5.359	5.198	5.116	5.065	4.998	4.963	4.901	4.880	4.859
9	10.56	8.022	6.992	6.422	6.057	5.802	5.613	5.467	5.351	5.257	5.178	5.111	5.055	5.005	4.962	4.808	4.649	4.567	4.517	4.449	4.415	4.353	4.332	4.311
10	10.04	7.559	6.552	5.994	5.636	5.386	5.200	5.057	4.942	4.849	4.772	4.706	4.650	4.601	4.558	4.405	4.247	4.165	4.115	4.048	4.014	3.951	3.930	3.909
11	9.646	7.206	6.217	5.668	5.316	5.069	4.886	4.744	4.632	4.539	4.462	4.397	4.342	4.293	4.251	4.099	3.941	3.860	3.810	3.742	3.708	3.645	3.624	3.602
12	9.330	6.927	5.953	5.412	5.064	4.821	4.640	4.499	4.388	4.296	4.220	4.155	4.100	4.052	4.010	3.858	3.701	3.619	3.569	3.501	3.467	3.404	3.382	3.361
13	9.074	6.701	5.739	5.205	4.862	4.620	4.441	4.302	4.191	4.100	4.025	3.960	3.905	3.857	3.815	3.665	3.507	3.425	3.375	3.307	3.272	3.209	3.187	3.165
14	8.862	6.515	5.564	5.035	4.695	4.456	4.278	4.140	4.030	3.939	3.864	3.800	3.745	3.698	3.656	3.505	3.348	3.266	3.215	3.147	3.112	3.048	3.026	3.004
15	8.683	6.359	5.417	4.893	4.556	4.318	4.142	4.004	3.895	3.805	3.730	3.666	3.612	3.564	3.522	3.372	3.214	3.132	3.081	3.012	2.977	2.913	2.891	2.868
20	8.096	5.849	4.938	4.431	4.103	3.871	3.699	3.564	3.457	3.368	3.294	3.231	3.177	3.130	3.088	2.938	2.778	2.695	2.643	2.572	2.535	2.468	2.445	2.421
30	7.562	5.390	4.510	4.018	3.699	3.473	3.305	3.173	3.067	2.979	2.906	2.843	2.789	2.742	2.700	2.549	2.386	2.299	2.245	2.170	2.131	2.057	2.032	2.006
40	7.314	5.178	4.313	3.828	3.514	3.291	3.124	2.993	2.888	2.801	2.727	2.665	2.611	2.563	2.522	2.369	2.203	2.114	2.058	1.980	1.938	1.860	1.833	1.805
50	7.171	5.057	4.199	3.720	3.408	3.186	3.020	2.890	2.785	2.698	2.625	2.563	2.508	2.461	2.419	2.265	2.098	2.007	1.949	1.868	1.825	1.742	1.713	1.683
75	6.985	4.900	4.054	3.580	3.272	3.052	2.887	2.758	2.653	2.567	2.494	2.431	2.377	2.329	2.287	2.132	1.960	1.866	1.806	1.720	1.674	1.583	1.551	1.516
100	6.895	4.824	3.984	3.513	3.206	2.988	2.823	2.694	2.590	2.503	2.430	2.368	2.313	2.265	2.223	2.067	1.893	1.797	1.735	1.646	1.598	1.501	1.466	1.427
250	6.737	4.691	3.861	3.395	3.091	2.875	2.711	2.583	2.479	2.392	2.319	2.256	2.202	2.154	2.111	1.953	1.774	1.674	1.608	1.511	1.457	1.343	1.297	1.244
500	6.686	4.648	3.821	3.357	3.054	2.838	2.675	2.547	2.443	2.356	2.283	2.220	2.166	2.117	2.075	1.915	1.735	1.633	1.566	1.465	1.408	1.285	1.232	1.164
∞	6.635	4.605	3.782	3.319	3.017	2.802	2.639	2.511	2.407	2.321	2.248	2.185	2.130	2.082	2.039	1.878	1.696	1.592	1.523	1.419	1.358	1.220	1.153	1.000

either for a one-tailed or two-tailed test (look carefully at the description when interpreting the table), and has a critical probability associated with it. A one-tailed F-statistic at 5 % probability gives the F-values for which one can be 95 % certain that the higher variance (e.g. a lack-of-fit error) is significantly larger than the lower variance (e.g. an experimental error). A two-tailed F-statistic at 10 % probability is identical to a one-tailed F-statistic at 5 % probability. The degrees of freedom for the larger variance are normally presented horizontally, and for the smaller variance, vertically. Tables 3.6 and 3.7 are at the 5 % and 1 % probability for a one-tailed F-statistic. Therefore, if we want to see whether one variance is significantly bigger than the other (one-tailed) at a 5 % level, and there are 12 measurements for the data with larger variance, and 8 for those with smaller variance, we look at the 5 % critical F-values for 11 and 7 degrees of freedom, respectively, and find, from Table 3.6, it is 3.603. A variance ratio exceeding this is pretty conclusive evidence that our larger variance is significantly bigger than the smaller one.

It is important to understand the motivations behind the F-test and how to use it in practice. We will use this test again below. Note that this test, like many others, depends on the errors being approximately normally distributed.

3.9 CONFIDENCE IN LINEAR REGRESSION

Newspaper headlines are rarely written by scientists. Consider a shock headline in which a poll on the eve of an election finds the underdog leading by 2 %, having trailed by margins of 10 % or more for several weeks. For British readers, a 10 % margin in a general election can translate into a huge parliamentary majority, so this apparent turnaround may suggest a substantial change in fate from a significant minority representation to forming a small majority government.

But really what the pollsters might prefer to say is that their sampling suggests that they are 95 % sure that one political party should receive between 40 % and 48 % of the votes and the other between 38 % and 46 %. If this is so, this apparent 44 % to 42 % lead between the main parties (the remaining votes going to minority parties) is not so conclusive. In fact it would still be within the bounds of possibility that the sampling is compatible with true voting intentions of 40 % to 46 % for the two parties, respectively, not far off the original 10 % gap that appeared at the beginning of the campaign. Perhaps the polling organization made it quite clear to the newspaper editor that given the short time available on the eve of the election to obtain a small sample, their polls really represent a range of possible results.

So in chemistry, we have similar problems. Consider trying to estimate the concentration of a reactant in a reaction mixture by spectroscopy or chromatography. In many cases there will be a series of steps that are used to convert a peak area into a concentration. These could involve sampling (is the reaction mixture truly homogeneous, is our pipette or syringe completely accurate?), dilution, behaviour of an HPLC instrument including baseline problems, digital resolution and stability of a calibration standard as discussed in Section 3.2. Although we may obtain an answer that the concentration of the reactant is 15 mM, there will normally be associated confidence limits, for example, being truthful we might more accurately like to state that we are '95 % confident that the concentration of the reactant is between 13 and 17 mM'. Many experimentalists are good scientists and understand the problems of predicting election results from small samples, but then do not like being told that their measurements of concentration are similarly subject to significant errors, and particularly do not enjoy saying this to their managers or sponsors. But it is rare to be able to measure concentrations very accurately and we do well to consider the nature of the measurement process and how we too can estimate the confidence of our predictions.

3.9.1 Linear Calibration

There are a huge number of uses of confidence interval calculations in chemistry. In this section we will look at one common need, that of univariate linear calibration where one parameter (e.g. concentration c) is obtained from an experimental measurement x such as a peak height, and the relationship is linear of the form $x = b_0 + b_1 c$. Note that we will illustrate this example using classical calibration, although in some cases inverse methods are more appropriate (Section 6.2).

After obtaining a set of measurements, we may want to ask a number of questions:

- What is the confidence interval of the responses x in the regression line between x and c (Section 6.2)? For example if the true concentration is 5 mM between what limits do we expect 95 % of the measurements to lie? Generally these confidence intervals are presented graphically.
- What are the confidence intervals of the parameters b_0 and b_1? For example we may estimate b_0 as 0.07 AU (for spectroscopic measurements). If the 95 % confidence intervals are ±0.10 AU, it is entirely possible that our estimate of the intercept is a little too high and there is in reality an intercept term of 0. If, however, the 95 % confidence intervals are ±0.01 AU it is very likely that there is a significant baseline or impurity, as the lower bound to b_0 is 0.06 AU.
- What are the confidence limits for the predicted concentrations? If we make a measurement of 0.78 AU, can we be 95 % certain that the concentration is between 6.23 and 6.78 mM?

3.9.2 Example

We will illustrate this with a simple example. We assume that we know that there ought to be a linear relationship between a response x and a concentration c. If we are not sure of this, we have to design experiments and use ANOVA (Section 2.3) first to discover whether our model is adequate. Table 3.8 represents the spectroscopic response (in AU) of a solution at seven different concentrations. The best fit straight line using classical calibration is $x = 0.312 + 0.177c$. We can predict $\hat{x}$ (the 'hat' means estimated) from this model and take the residual which is the difference between the observed and predicted.

Table 3.8 Example in Section 3.9.2

Concentration (mM)	Peak height (AU)	Predicted	Residual
1	0.400	0.488	−0.088
2	0.731	0.665	0.065
3	0.910	0.842	0.067
4	0.949	1.019	−0.071
5	1.308	1.196	0.112
6	1.294	1.373	−0.079
7	1.543	1.550	−0.007

There are various types of confidence intervals but the so-called Working–Hotelling limits allow us to determine confidence bands of the best fit straight line. The shape of the confidence bands depends on:

- the value of c;
- the arrangement of the experiments.

The width depends on:

- the error of regression;
- the confidence level chosen (normally 95 %).

The equation for the 95 % confidence intervals of the regression line is given by:

$$\hat{x} \pm t_{0.025, I-2} s \sqrt{\frac{1}{I} + \frac{(c - \bar{c})^2}{\sum_{i=1}^{I} (c_i - \bar{c})^2}}$$

How to interpret this?

- $\hat{x}$ is the predicted value of x. For a value of $c = 6.2$ mM this is 1.408 in our example.
- $t_{0.025, I-2}$ refers to a one-tailed t-test at 2.5 % level, for $I-2$ or five degrees of freedom, in this case. We have introduced the t-statistic previously (Section 3.7) and the critical value of t can be obtained used the TINV function in Excel; it equals 2.57 in this case. It is possible to determine confidence bands for any specific probability level, the 0.025 referring to the level.
- s is simply the root mean square error between predicted and observed and equals 0.0898 in this case: note that it is normal to divide the total square error by the number of degrees of freedom which is 5, and not 7, as the equation has two terms in it. The bigger this number, the worse the experimental error.
- The expression under the square root sign effectively determines the shape of the confidence bands and is determined by the experimental design of the original calibration experiment, i.e. the arrangement of the observations, and in our case becomes:

$$\sqrt{\frac{1}{7} + \frac{(c - \bar{c})^2}{28}}$$

Hence at 6.2 mM, we can say that at a 95 % confidence level, $x = 1.408 \pm 0.130$. That is we expect 95 % of the measurements to fall between 1.278 and 1.538 AU. These bands can be represented graphically as in Figure 3.9. In general we would expect only one in 20 points to lies outside these confidence intervals, although one experimental point is just outside in this example.

There are, in fact, quite a few other ways of calculating the confidence intervals of a straight line, but these are beyond the scope of this text. In particular we may sometimes want to predict c from x using inverse calibration and slightly different equations come into play.

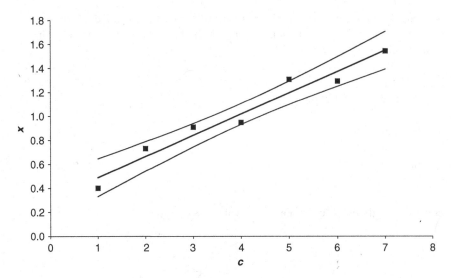

Figure 3.9   95 % confidence bands for the best fit regression line for the data in Table 3.8

The equation for the confidence for prediction of a single point, x differs slightly and is given by:

$$\hat{x} \pm t_{0.025, I-2} s \sqrt{1 + \frac{1}{I} + \frac{(c - \bar{c})^2}{\displaystyle\sum_{i=1}^{I} (c_i - \bar{c})^2}}$$

3.9.3 Confidence of Prediction of Parameters

Sometimes we are not so much interested in the predictions of the points but in the actual parameters, for example, our experiment might be to measure an extinction coefficient. We might be concerned about how well we can determine a physical parameter. The formula for the 95 % bounds for the intercept is given by:

$$t_{0.025, I-2} s \sqrt{\frac{\displaystyle\sum_{i=1}^{I} c_i^2}{I \displaystyle\sum_{i=1}^{I} (c_i - \bar{c})^2}} = 2.57 \times 0.0898 \times \sqrt{\frac{140}{7 \times 28}} = 0.195$$

and for the slope by:

$$t_{0.025, I-2} s \sqrt{\frac{1}{\displaystyle\sum_{i=1}^{I} (c_i - \bar{c})^2}} = 2.57 \times 0.0898 \times \sqrt{\frac{1}{28}} = 0.0437$$

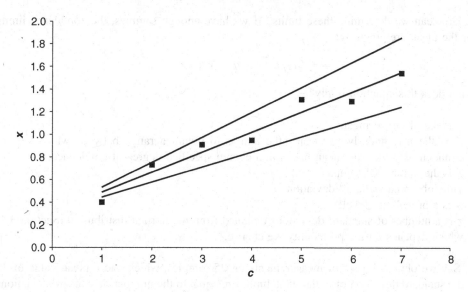

Figure 3.10 Range of possible best fit regression lines for the data in Table 3.8

These equations state that the 95 % confidence limits of the intercept are between 0.117 and 0.506 and of the slope between 0.133 and 0.220. In real terms we can be pretty sure that there really is an intercept term, so, for example, that there is likely to be an impurity or background or baseline problem with our sample, but there is a wide range of possible slopes consistent with the data. This latter problem may be of some concern if the measured gradient is then used to predict a concentration for an unknown sample, for example, since the 95 % limits represent almost a two fold uncertainty in the slope in this case. Using the average intercept term, Figure 3.10 represents the extreme best fit straight lines using the two extreme values of the slope and the average predicted intercept.

3.10 MORE ABOUT CONFIDENCE

In Section 3.9 we discussed primarily the problem of confidence in linear regression. There are many other reasons for determining confidence limits in chemistry.

3.10.1 Confidence in the Mean

Confidence limits for means provide quite useful information. Consider taking 10 analyses of river samples and determining the amount of a potential pollutant, the average value being 20 ppm. How confident can we be of this value? Maybe we are 90 % confident that the true value lies between 15 ppm and 25 ppm? The legal limit for prosecution might be 18 ppm. Although the average is above this, we are not really very sure of this. The confidence limits depend on the standard deviation of the measurements and the number of samples. If we really are not too confident, either we have to drop the case and move onto something else, or analyse more samples.

How can we determine these limits? If we have enough samples, the confidence limits for the mean are given by:

$$\bar{x} - z_{\alpha/2}(s/\sqrt{I}) < \mu < \bar{x} + z_{a/2}(s/\sqrt{I})$$

What does this equation imply?

- $\bar{x}$ is the observed mean;
- μ is the true underlying mean – the aim is to obtain a range between which we are confident that the true mean lies with a certain specified degree of confidence;
- I is the number of samples;
- s the observed standard deviation;
- α is a probability level;
- z is a number of standard deviations obtained from the normal distribution (Section 3.4) which depends on the probability we choose.

Say we obtained a set of measurements on 25 samples, which had a mean value of 30 and a standard deviation of 3, the 95 % limits are equal to the number of standard deviations for which 97.5 % of the data either side of the mean falls. Figure 3.11 illustrates this. We find the value is 1.96 standard deviations from the mean [which can be obtained either from a table or using the function NORMINV(0.975,0.1) in Excel], so the 95 % confidence limits of the mean are given by:

$$30 - 1.96(3/5) < 30 < 30 + 1.96(3/5)$$

or

$$28.82 < 30 < 31.18$$

Some people prefer to use the t-statistic (Section 3.7), so that the equation becomes:

$$\bar{x} - t(s/\sqrt{I}) < \bar{x} < \bar{x} + t(s/\sqrt{I})$$

where t is now the t-statistic at the desired probability level. This is useful when sample size is small (remember that for a large number of samples, the t-distribution converges to the normal distribution). For 95 % probabilities we want to use either the two-tailed t-test for $I - 1$ (=24 in this case) degrees of freedom at 95 % probability, or the one-tailed t-test

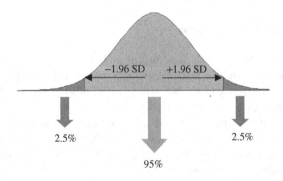

Figure 3.11 Illustration of area under a normal distribution curve for which $\alpha = 95\%$

at 97.5 % probability. The default Excel function is for a two-tailed test and in this case is given by TINV(0.05,24) or 2.06, slightly greater than using the normal distribution, but the discrepancy is so small that it is hardly worth the effort of using a different equation. If sample sizes are quite small (15 or less) the correction of using the t-distribution rather than the normal distribution can be important, and perhaps should be taken into account.

3.10.2 Confidence in the Standard Deviation

Sometimes we want to look at confidence in the standard deviation. For example we may have obtained 10 samples from laboratory A and find that the standard deviation of concentration measurements appears to be 5 mg/ml, but how confident can we be in the standard deviation? When sample sizes are small, the predicted standard deviation can differ considerably from the true standard deviation. Perhaps the 95 % confidence limits for the true standard deviation are between 4 mg/ml and 6 mg/ml. This can be quite important information if we want to compare precision of different laboratories and methods. If we state that one particular laboratory really is less precise than another, we could get into trouble unless we check our facts carefully beforehand.

The equation for the confidence limits of the variance (the square of the standard deviation) is given by:

$$\frac{(I-1)s^2}{\chi^2_{\alpha/2}} < \sigma^2 < \frac{(I-1)s^2}{\chi^2_{1-\alpha/2}}$$

What does this imply?

- s is the observed standard deviation;
- σ is the true underlying standard deviation;
- I is the number of samples;
- α is a probability level;
- χ^2 is a statistic dependent on the number of degrees of freedom ($= I - 1$) and probability level.

The χ^2 distribution is given in Table 3.9, but it is possible to use the CHIINV function in Excel to generate the values for any probability level and any number of degrees of freedom. We can illustrate this by determining the 95 % confidence limit of the true standard deviation if we use 10 samples to determine an experimental standard deviation of 8. The value of $\chi^2_{0.975}$ for 9 degrees of freedom is 2.700 [=CHIINV(0.975,9) in Excel] and the corresponding value of $\chi^2_{0.025}$ is 19.023, so using our equation we have:

$$\frac{9 \times 64}{19.023} < \sigma^2 < \frac{9 \times 64}{2.700} \text{ or } 30.28 < \sigma^2 < 213.33$$

These limits look large, but that is because we are dealing with a variance rather than standard deviation, and the limits for the standard deviation are between 5.50 and 14.61.

One requires many more measurements to be able to estimate a standard deviation accurately than a mean, so beware statements about relative precision. If your favourite student or technician is audited and thought to be imprecise, you can normally question the results of the audit unless a very large number of repeat measurements are taken. Accusations of lack of precision are quite hard to prove. It may require 40 or 50 replicates before one can

Table 3.9 The χ^2 distribution

Degrees of freedom	Probability							
	0.99	**0.975**	**0.95**	**0.9**	**0.1**	**0.05**	**0.025**	**0.01**
1	0.00016	0.00098	0.00393	0.01579	2.706	3.841	5.024	6.635
2	0.0201	0.0506	0.1026	0.2107	4.605	5.991	7.378	9.210
3	0.115	0.216	0.352	0.584	6.251	7.815	9.348	11.345
4	0.297	0.484	0.711	1.064	7.779	9.488	11.143	13.277
5	0.554	0.831	1.145	1.610	9.236	11.070	12.832	15.086
6	0.872	1.237	1.635	2.204	10.645	12.592	14.449	16.812
7	1.239	1.690	2.167	2.833	12.017	14.067	16.013	18.475
8	1.647	2.180	2.733	3.490	13.362	15.507	17.535	20.090
9	2.088	2.700	3.325	4.168	14.684	16.919	19.023	21.666
10	2.558	3.247	3.940	4.865	15.987	18.307	20.483	23.209
11	3.053	3.816	4.575	5.578	17.275	19.675	21.920	24.725
12	3.571	4.404	5.226	6.304	18.549	21.026	23.337	26.217
13	4.107	5.009	5.892	7.041	19.812	22.362	24.736	27.688
14	4.660	5.629	6.571	7.790	21.064	23.685	26.119	29.141
15	5.229	6.262	7.261	8.547	22.307	24.996	27.488	30.578
20	8.260	9.591	10.851	12.443	28.412	31.410	34.170	37.566
25	11.524	13.120	14.611	16.473	34.382	37.652	40.646	44.314
30	14.953	16.791	18.493	20.599	40.256	43.773	46.979	50.892
40	22.164	24.433	26.509	29.051	51.805	55.758	59.342	63.691
50	29.707	32.357	34.764	37.689	63.167	67.505	71.420	76.154
100	70.065	74.222	77.929	82.358	118.498	124.342	129.561	135.807

obtain a reliable estimate of the standard deviation. For confidence in the means, however, many less samples are required.

It is possible to determine confidence in almost anything, such as correlation coefficients, slopes, covariances and so on, the list is endless but beyond the scope of this text. However some of the more common calculations have been indicated in this section and Section 3.9.

3.11 CONSEQUENCES OF OUTLIERS AND HOW TO DEAL WITH THEM

Outliers are the bane of chemometricians. Many a paper, thesis and official report has been written about outliers. Outliers can arise for many reasons.

One of the simplest problems arises from mislabelling samples. More sophisticated is a misdiagnosis, for example let us say we want to look at the chemical basis of cancer, and one sample is falsely diagnosed as cancerous. An outlier can arise from a sample that simply does not fit into a given group, e.g. a sample of rat urine inserted into a series of samples of horse urine. Or perhaps the balance in the laboratory was not properly calibrated: the technician might have had a bad day? Sometimes the existence of a so-called outlier does not imply a mistake but just a sample that should not be part of a training set.

If outliers are in a series of measurements, what is the effect?

The simplest case involves a single measurement on a series of samples. We might want to see whether the measurements form a normal distribution (Section 3.5). Outliers are simply very large or very small measurements. If they are retained in the dataset, they will be represented by extreme values and we may conclude, falsely, that the data do not fall into a normal distribution. In Section 3.12 we will look at a number of ways to spot outliers, but the influence of one or more outlying observations can be quite extreme on our conclusions about whether measurements are normally distributed, unless they are spotted and removed first.

Important consequences of outliers also involve their influence on the best fit line in linear regression. Table 3.10 represents a series of measurements as a function of concentration. In Figure 3.12(a) we obtain the best fit straight line graph using a least squares criterion, with an equation (using classical calibration, see Section 6.2) of $x = 0.832 + 0.281c$. The problem of using least squares for regression is that the deviation from the outlier results in a large sum of square error, so this influential point pushes the straight line in its direction and has far greater effect than the majority of other points.

Removing this point (at concentration $= 2$), results in a best fit straight line of $x = 0.657 + 0.313c$, quite a difference as illustrated in Figure 3.12(b). To show that it really is the outlier that changes the gradient, look at Figure 3.12(c) where we remove a different point (at concentration $= 2.5$) and the line becomes $x = 0.842 + 0.280c$, almost identical to the original model. So one point, which itself is an outlier and so perhaps should be ignored or at least not taken very seriously, in fact has an enormous effect on the result. The three best fit straight lines are superimposed in Figure 3.13. It can easily be seen that the line when the proposed outlier is removed is quite different from the other two lines but probably most faithfully reflects the underlying trend. Notice that the effect is not particularly predictable. In this example, the outlier in fact disproportionately influences the intercept term and, in fact, underestimates the gradient.

There are various ways of overcoming this problem, the simplest being to remove the outlier. We can also use different criteria for fitting a straight line. A well known method involves using fuzzy membership functions. Instead of trying to minimize the sum of squares deviation from the straight line we can maximize a so-called membership function, between the observed and estimated points along the line. The closer these values, the higher the membership function, often people use a Gaussian so that the value of the function is given by $\exp[-(x_i - \hat{x}_i)^2 / s^2]$, s relating to how strongly a deviation from the best fit straight line

Table 3.10 A series of measurements obtained as a function of concentration

Concentration	Measurement
0.5	0.892
1	0.944
1.5	1.040
2	2.160
2.5	1.468
3	1.501
3.5	1.859
4	1.929
4.5	2.046
5	2.214

influences the value of this function. A large deviation will have a small influence so the main aim is to maximize the fit to the model of the best points rather than minimize the error of the worst points.

There are also several methods based on using the median of regression parameters. A simple (but rather tedious one although this can be automated quite easily) involves calculating the slope between all possible pairs of points in the dataset. In our case there are 10 points which results in 45 possible pairs of points (pair 1:2, 1:3, 1:4 .... 1:10, 2 : 3, 2:4 .... 2:10, 3:4, 3:5, etc.). Simply take the median slope. To find the intercept use

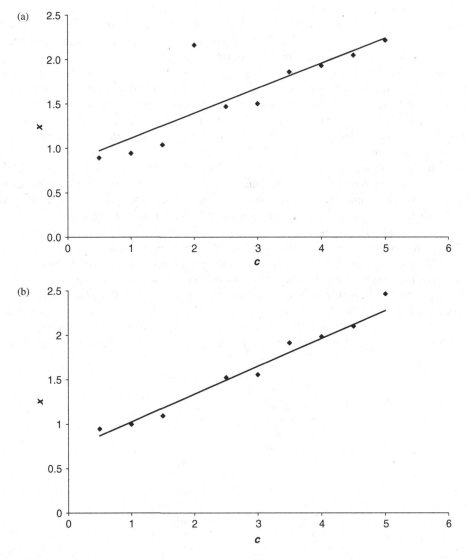

Figure 3.12 Three best fit straight lines calculated for the data in Table 3.10, removing different samples in (b) and (c)

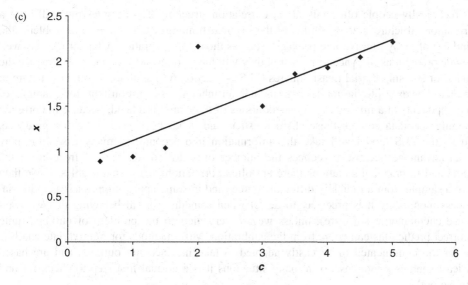

Figure 3.12 (*continued*)

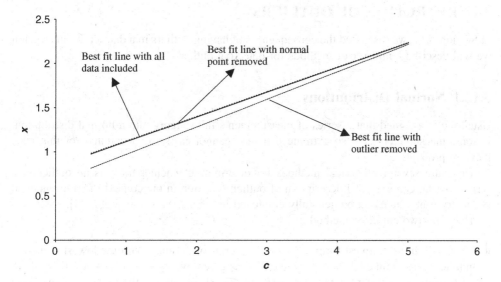

Figure 3.13 Three best fit straight lines in Figure 3.12 superimposed

this median slope and calculate the intercept with this slope, between all pairs of points, and use the median intercept as the estimate of the intercept. The advantage of median based approaches is that they are not unduly influenced by large values.

The influence of outliers in multivariate data analysis is an important one, and several commercial software packages devote specialist facilities for the detection and elimination of outliers. One problem is that an outlier may simply be a sample that, whereas entirely legitimate, does not fit into the training set. When a series of spectra is recorded, it is usually assumed that there are certain features common to the entire dataset that we take advantage of. Technically people often talk about correlation structure. The samples must all have a correlation structure that is similar to the original training set. If there is a problem, the blind use of many multivariate packages such as those for calibration (Chapter 6), may well provide predictions of concentrations but they will have no physical meaning. In practice the model [often using Partial Least Squares (PLS) – Section 6.6] assumes a certain structure in the data. For example the model may include an inherent assumption that, for example, of 10 components in a mixture, the concentrations of three are correlated, because the original (training set) data contain these characteristics, and it has been given no other guidance, and so the PLS model will take this information into account. In many cases this may be an advantage because it reduces the number of factors or degrees of freedom in the model and in practical situations there are often correlations between variables. If we then have a sample from a radically different source (and in some applications such as industrial process monitoring it is precisely to detect when something is misbehaving that we want to use chemometric software), unless we are very alert to the problem of outliers, quite incorrect predictions can come from the calibration. Most methods for multivariate analysis are far too complicated to be easily adapted to take into account outliers, and are based on least squares principles, so in many situations it is a normal first step to check for and remove outliers.

3.12 DETECTION OF OUTLIERS

In Section 3.11, we discussed the consequences of having outliers in a dataset. In this section we will describe a number of methods for detecting outliers.

3.12.1 Normal Distributions

Usually it is assumed that a series of measurements fall roughly into a normal distribution. If some measurements are so extreme that they cannot easily be accounted for this way, they are probably outliers.

There are several numerical methods for determining whether there is an outlier in a series of data. Quite a good description of outlier detection in the context of environmental chemistry (but which can be generally employed in other areas) is available [4].

There are two common methods:

- *Dixon's Q*. In its simplest form if we order a series of I values with the lowest ranked 1, and the highest ranked I then the value of Q is given by $(x_2 - x_1)/(x_I - x_1)$ if we think the lowest is an outlier, or $(x_I - x_{I-1})/(x_I - x_1)$ if we suspect the highest value. There are a few variations on the theme for large sample sizes, but this equation is the most straightforward. A table of 90 % critical values is given in Table 3.11. This test normally is only valuable if the aim is to determine whether a single extreme reading is an outlier.
- *Grubb's test*. This involves calculating $G = |(x_i - \bar{x})|/s$, where x_i is the suspect outlier, and the other statistics are the mean and sample standard deviation (Section 3.3). The higher this is the greater the likelihood of outliers, and critical values of G are presented in Table 3.12.

These two tests are best used simply to find one outlier in a dataset and remove it, but can be used repeatedly on the same data, each time stripping outliers until none are left.

Table 3.11 95 % confidence values for Dixon's Q-statistic

N	4	5	6	7	8	9	10
Q	0.83	0.71	0.62	0.57	0.52	0.49	0.46

N, number of samples.

Table 3.12 95 % confidence values for Grubb's test

N	4	5	6	7	8	9	10
G	1.48	1.71	1.89	2.02	2.13	2.21	2.29

N, number of samples.

3.12.2 Linear Regression

Another instance for finding out whether there are outliers is prior to performing linear regression. In Section 3.11 we discussed how one can use methods such as fuzzy regression or mean methods to cope with outliers, but an alternative approach is to detect and remove the outliers first. We introduced a dataset in Table 3.10 where a likely outlier influences the regression line. One way of detecting outliers is to calculate the residuals of the best fit data, as in Table 3.13. Often we standardize the residuals, that is subtract their mean (which equals zero in this case) and divide by their standard deviation. A graph of standardized residuals against c can be produced (Figure 3.14), and obvious outliers visualized. These standardized residuals can also be interpreted numerically and there are a number of statistical criteria as to how many standard deviations from the mean residual are indicative of an outlier, based on a normal distribution of errors, although graphical methods are in many cases adequate.

In fact quite a few different types of graphs of residuals can be obtained, and if there are large datasets it can be valuable, for example, to scan several hundred measurements and see if there is one or more problem measurement that could ultimately influence the

Table 3.13 Standardized residuals for best fit straight line (using all the data) for the dataset in Table 3.10

Conc.	Measurement	Prediction	Residual	Standardized residual
0.5	0.892	0.973	−0.081	−0.289
1	0.944	1.113	−0.169	−0.604
1.5	1.040	1.254	−0.214	−0.764
2	2.160	1.394	0.765	2.729
2.5	1.468	1.535	−0.067	−0.239
3	1.501	1.676	−0.175	−0.623
3.5	1.859	1.816	0.043	0.154
4	1.929	1.957	−0.028	−0.100
4.5	2.046	2.097	−0.051	−0.182
5	2.214	2.238	−0.024	−0.084

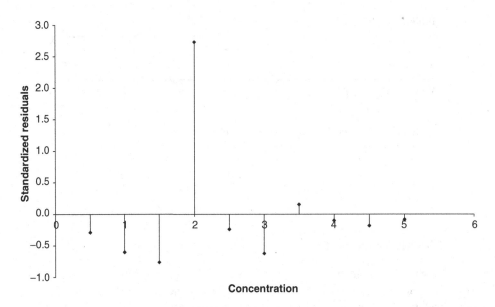

Figure 3.14   Standardized residuals

entire regression line. Some people use alternative methods for standardization, a common approach being so-called 'studentized' residuals, where the error in measurement is assumed to vary according to how far the experiment is from the centre, so the standard deviation differs according to which datapoint we are looking at.

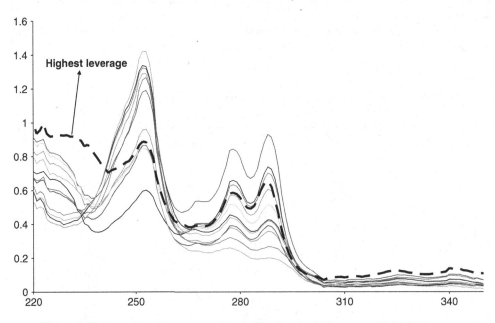

Figure 3.15   A series of spectra used for calibration, including one with high leverage

3.12.3 Multivariate Calibration

Detecting outliers in multivariate calibration (Chapter 6) is another common task. There are several approaches, usually to identify inappropriate samples in a training set or after calibration whether a sample we are trying to predict really does fit into the structure of the original training set. A simple statistic is that of leverage, which we have already introduced in another context (Section 2.15). PLS1 is probably the most widespread multivariate method for calibration, and since the PLS scores are orthogonal (see Section 6.6), for sample i in a training set, leverage can be defined by:

$$h_i = 1/I + \sum_{a=1}^{A} \left(t_{ia}^2 / \left[\sum_{i=1}^{I} t_{ia}^2 \right] \right)$$

Table 3.14 Calculation of leverage

	PLS scores				
	Comp. 1	Comp. 2	Comp. 3	Comp. 4	Comp. 5
	−0.333	−0.374	0.270	−0.427	−0.112
	−1.071	0.406	0.003	−0.146	0.198
	0.568	−0.134	0.083	0.073	−0.057
	1.140	0.318	0.758	−0.301	0.070
	0.242	0.902	0.025	0.018	0.130
	−0.744	0.522	−0.101	0.120	−0.034
	−0.635	−1.269	0.214	0.014	−0.176
	1.402	0.778	−0.465	0.045	−0.273
	0.543	−0.512	0.376	0.658	0.104
	−1.189	−0.112	−0.164	0.072	−0.091
	−0.863	0.591	−0.238	0.066	0.028
	0.939	−1.114	−0.762	−0.193	0.213
$\sum_{i=1}^{I} t_{ia}^2$	9.198	5.593	1.732	0.796	0.248

Leverage	$\left(t_{ia}^2 / \left[\sum_{i=1}^{I} t_{ia}^2 \right] \right)$				
0.442	0.012	0.025	0.042	0.229	0.050
0.422	0.125	0.030	0.000	0.027	0.157
0.145	0.035	0.003	0.004	0.007	0.013
0.708	0.141	0.018	0.332	0.113	0.020
0.304	0.006	0.145	0.000	0.000	0.068
0.221	0.060	0.049	0.006	0.018	0.005
0.567	0.044	0.288	0.026	0.000	0.125
0.832	0.214	0.108	0.125	0.002	0.299
0.832	0.032	0.047	0.082	0.544	0.043
0.295	0.154	0.002	0.015	0.006	0.033
0.268	0.081	0.062	0.033	0.005	0.003
0.965	0.096	0.222	0.335	0.047	0.182

This has the property that the sum of leverage for all the objects in a training set equals the number of PLS components $(A) + 1$ (if the data have been centred). Figure 3.15 is of a series of spectra. There are some suspicious spectra but it would not be obvious at first glance whether there is any single spectrum which is an outlier; a spectrum found to have large leverage is indicated but there appear, without further analysis, to be several candidates. Performing a calibration against the known concentration of one compound gives the results of Table 3.14 in which we see that the leverage of the 12th sample is high (almost equal to 1), and so is most likely to be an outlier. There are other potential outliers also, but if we calibrate against known concentrations of different compounds in the mixture we actually find that the 12th sample has consistently high leverage whereas others reduce in value. In fact, from other sources, it is known that this latter sample contains an extra compound not present in the remaining samples, so we hope it would be flagged as an outlier. Many multivariate calibration packages produce a facility for using leverage to determine outliers.

3.13 SHEWHART CHARTS

Another important topic relates to quality control, especially of industrial processes. This involves studying how one or more parameters, measured using with time, vary. Many of the statistical tests discussed above can be employed in this context, but are used to look at changes in time.

Often there are limits to the acceptable quality of a product. Some of these may be established for legal reasons, for example, if one markets a pharmaceutical drug it is important to certify that the composition of active ingredients are within certain limits. If outside these limits, there could be unknown physiological consequences as clinical trials are performed only on a range of products with certain well defined composition: and in the extreme a company can be sued billions of dollars and go out of business or have such a bad press that nobody wants to buy drugs from them anymore, if a product was found to be outside the limits demonstrated to be safe and effective in clinical trials. Normally batches are tested before going on the market, but rigorous checks can take days or weeks, and there can be major economic consequences if a large amount of product has to be destroyed. The profit margins in many manufacturing processes are surprisingly tight, especially since the price must take into account huge development costs not only for the products that reach the market but for the vast majority of lead compounds that do not get to the manufacturing plant, as discussed in Section 9.1. So a failure to detect problems with processes as they take place can have big consequences.

However, a neurotic factory manager who is always shutting down a plant as soon as he or she suspects there may be a problem can also cost the company money, as employees must be paid even if there is no product, and factories cost money even if nothing is happening. So it is important to have some means of monitoring what is going on and to only close down production when it appears there is a real difficulty with the product.

The Shewhart chart is commonly used for this purpose in quality control, and allows the visualization of the change in value of a parameter in time as a process is proceeding. Typically one may record a measurement, such as the concentration of a compound or a pH or a temperature or a spectroscopic peak height, at regular intervals during a process. In fact in many applications, lots of different measurements are recorded with time. From previous work it is normally known what the typical properties of an acceptable batch or process are; for example, we may have measured the mean and standard deviation

(Section 3.3) of parameters associated with acceptable products and processes. We may know, for example, that if the process is going well we expect a mean concentration of an ingredient of 16.73 mg/g, with a standard deviation of 0.26 mg/g. What happens if we find that the ingredient in our product is 17.45 mg/g? Does this mean that we should shut the factory down? Or could it simply be accounted for by small fluctuations in the process which are probably quite acceptable? Should we have a sleepless night, and call the engineer out at 2 a.m., often at great cost, or just go home and keep a watchful eye the next day? These decisions can have huge economic consequences because of the cost of idle factories or destroyed batches balanced against what may often be quite tightly defined profit margins.

The principle of Shewhart charts is simply to plot each variable with time and then if it exceeds a limit, do something. Normally two types of limit are calculated: the warning limit, maybe something is going wrong, but we do not really know but need to check the system more thoroughly; and the action limit, where we push the panic button and have to do something. This is illustrated in Figure 3.16.

How are these limits determined? There are quite a few rules. The simplest is to use 2 standard deviations away from the mean of the batch measurement for the warning limit and 3 standard deviations for the action limit, so in our example the upper action limit is 17.51 mg/g. These statistics are often obtained from historic measurements on batches that past muster and are regarded as marketable. The chances that these are breached by an acceptable process are around 5 % and 0.2 %, respectively, if we assume the errors are normally distributed (using a two-tailed test).

There are, however, huge numbers of variations on the theme. Sometimes, instead of measuring one parameter per batch, we might take several measurements of the same parameter and calculate the mean at each successive point in time. In such a case our mean batch measurement is less likely to deviate from the true mean and the limits are reduced to $s/(I)^{1/2}$, where I is the number of measurements. So in our example above, if, instead of taking a

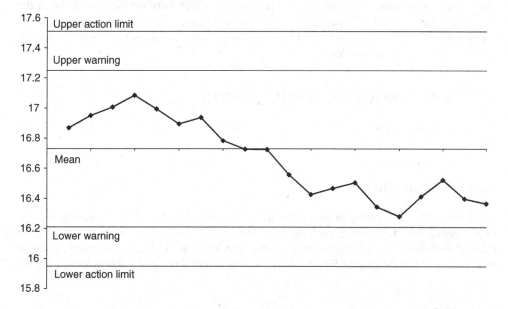

Figure 3.16 Shewhart chart

single measurement every time we test the product, we take the mean of 10 measurements, then the upper action limit becomes much tighter, and equals $16.73 + 3 \times 0.26/(10)^{1/2}$ or 16.98, with all the other limits adjusted as appropriate.

There are many other ways of interpreting Shewhart charts and several rules for making deductions about a process, for example, is a process deviating randomly or in a systematic way? If a measurement increases systematically, this would suggest that the process is moving in a certain direction rather than just randomly fluctuating, and, for example if 6 points decrease or increase successively this may be cause for concern, even if the limits are not reached. Or if several successive points are above the mean, even if within the limits, does this imply a problem? A point above the warning limit may not be a big deal, we expect it to happen occasionally, but several successive points even if just above the warning limits are much less likely to arise by chance. These sophistications to interpretation of Shewhart charts can be used to give the user a bigger feel for whether there may be a problem. In many typical situations there could be a large number of parameters that can be monitored, continuously, and so an equally important need for several criteria to alert the operator to difficulties.

Note that errors in many processes do not necessarily behave in a completely uncorrelated manner. Consider a typical temperature controller, often this may switch on once the temperature drops below a certain level, and off once it has reached a maximum. If the temperature is set to 30 °C, what normally happens is that the temperature oscillates around certain limits, for example between 29.5 °C and 30.5 °C, and some systems will wait until the bottom temperature is reached before switching on the heater. This depends on the quality of engineering, obviously a temperature controller used for a domestic fish tank is going to be of lower quality than one used for an industrial plant, but even so, there may be dozens of such sensors in a plant, some of variable quality simply because they are old and need replacing or because components wear out, or because they have not been serviced recently. The effect may only be indirectly noticed, very small oscillations in temperature could have an influence on other parts of the system, which may not necessarily be at the exact time the temperature changes, but later, so the factors influencing deviation for the mean can be quite complex and hard to track down. Control charts have a major role to play for detecting these problems.

3.14 MORE ABOUT CONTROL CHARTS

In addition to the Shewhart chart, there are a variety of other control charts available. Some of the more widespread are listed below.

3.14.1 Cusum Chart

The Cusum chart is a common alternative to the Shewhart chart. Instead of plotting a graph of the actual measurement at each point in time, plot the cumulative difference of each reading with the mean. The calculation is illustrated in Table 3.15, and the graph presented in Figure 3.17 for the data of Section 3.13. We can see that the process is starting to move

Table 3.15 Cusum calculation

Measurement	Difference with mean of 6.73	Cusum
16.869	0.139	0.139
16.950	0.220	0.359
17.006	0.276	0.635
17.084	0.354	0.989
16.994	0.264	1.253
16.895	0.165	1.418
16.937	0.207	1.626
16.783	0.053	1.679
16.727	−0.003	1.676
16.725	−0.005	1.671
16.558	−0.172	1.499
16.427	−0.303	1.196
16.468	−0.262	0.933
16.506	−0.224	0.709
16.345	−0.385	0.324
16.280	−0.450	−0.125
16.413	−0.317	−0.442
16.524	−0.206	−0.648
16.399	−0.331	−0.980
16.367	−0.363	−1.342

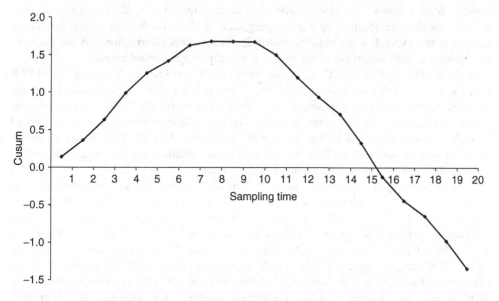

Figure 3.17 Cusum chart

in one direction and then moves in the opposite way. There are various ways of interpreting the axes, discussed in the specialist statistical literature, but this method of presenting the data represents a good graphical approach and it is easy to monitor both the Shewhart and Cusum charts simultaneously.

3.14.2 Range Chart

The range chart is another popular approach when each batch is sampled several times. The range is simply the difference between the maximum and minimum measurements at each point in time. It can be compared with the expected range of the data. One can assume that the ranges are normally distributed and so set up action and warning limits, just as for the Shewhart chart.

3.14.3 Multivariate Statistical Process Control

A major trend in chemometrics over the past decade has been in multivariate process control. This is a huge subject often the basis of numerous conferences and PhD theses. In reality Multivariate Statistical Process Control (MSPC) has been quite slow to get off the ground and most chemical engineers still use simple univariate measures. A problem with the conventional univariate approach is that there may be an enormous number of parameters measured in time, at regular intervals, so in a typical day hundreds of thousands of pieces of information are available. Many of the measurements are related so it makes sense to form a multivariate model rather than several univariate models. In addition some modern plants using spectroscopic monitoring such as NIR, which in itself, can generate a new spectrum every few seconds (Section 8.2); include this information and the database becomes unmanageable, there could be hundreds of possible control charts.

Methods such as Principal Components Analysis (PCA) (Sections 5.2 and 5.3) and PLS (Section 6.6) come to our rescue. Why plot a single parameter when we could plot the Principal Component (suitably scaled) of several parameters? And why not calibrate our spectroscopic data to the concentration of a known additive, and use PLS predictions of concentrations as our variable that we monitor with time? Instead of following a number of physical measurements, follow the concentration, obtained by real time multivariate calibration, of one or more components in the product.

More sophisticated multivariate measurements can be obtained and are regularly used by chemical engineers, who often first define regions of a process or of a product as Normal Operating Conditions (NOC) samples; in the case of a reaction or a product these are the samples arising from an acceptable process. An aim of MSPC is to look for samples whose characteristics (normally spectroscopic) deviate significantly from the characteristics of the NOC samples, by calculating a statistic for this deviation. There are two common ones. The D-statistic is a measurement of the distance a process is to the centre of the NOC samples and determines whether a specific sample has a systematic deviation from the steady state region. Usually the Mahalanobis distance (Sections 5.8 and 10.4) from the NOC class of spectra is computed. The Q-statistic calculates the residuals from the new samples when trying to fit them to a Principal Components model (Sections 5.2 and 5.3) which is based on the NOC samples. The Q-statistic measures nonsystematic variations which are not explained by the NOC model. If both statistics are assumed to follow normal distributions, the control limit for these statistics can be computed, usually 95 % and 99 % limits are calculated. If samples appear to be exceeding these limits, it is probably that there is some difficulty in the process.

The use of chemometrics in on-line process control is highly developed theoretically and slowly being implemented in certain high tech industries such as the pharmaceutical and food industries for front line products, but there is a long way to go. The majority of processes are still monitored using fairly straightforward univariate control charts. Nevertheless this is a major potential growth area, especially in terms of applications.

REFERENCES

1. J.N. Miller and J.C. Miller, *Statistics and Chemometrics for Analytical Chemistry*, Fifth Edition, Pearson Prentice Hall, Hemel Hempstead, 2005
2. W.P. Gardiner, *Statistical Analysis Methods for Chemists: a Software-based Approach*, Royal Society of Chemistry, Cambridge, 1997
3. R. Caulcutt and R. Boddy, *Statistics for Analytical Chemists*, Chapman and Hall, London, 1983
4. A Fallon and C. Spada, Detection and Accommodation of Outliers in Normally Distributed Data Sets (http://www.ce.vt.edu/program_areas/environmental/teach/smprimer/outlier/outlier.html)

4

Sequential Methods

4.1 SEQUENTIAL DATA

In this chapter we will be exploring a variety of methods for the handling of sequential series of data in chemistry. Such information is surprisingly common and requires a large battery of methods for interpretation.

- *Environmental and geological processes.* An important source of data involves recording samples regularly with time. Classically such time series occur in environmental chemistry and geochemistry. A river might be sampled for the presence of pollutants such as polyaromatic hydrocarbons or heavy metals at different times of the year. Is there a trend and can this be related to seasonal factors? Different and fascinating processes occur in rocks, where depth in the sediment relates to burial time. For example, isotope ratios are a function of climate, as relative evaporation rates of different isotopes are temperature dependent: certain specific cyclical changes in the Earth's rotation have resulted in the ice ages and so climate changes, leaving a systematic chemical record. A whole series of methods for time series analysis can be applied to explore such types of cyclicity, which are often quite hard to elucidate. Many of these data analytical approaches were first used by economists and geologists who also encounter related problems.
- *Industrial process control.* In industry, time series occur in the manufacturing process of a product. It may be crucial that a drug has a certain well defined composition, otherwise an entire batch is unmarketable. Sampling the product frequently in time is essential, for two reasons. The first is monitoring, simply to determine whether the quality is within acceptable limits. The second is for control, to predict the future and check whether the process is getting out of control. It is costly to destroy a batch, and not economically satisfactory to obtain information about acceptability several days after the event. As soon as the process begins to go wrong it is often advisable to stop the plant and investigate. However, too many false alarms can be equally inefficient. A whole series of methods have been developed for the control of manufacturing processes, an area where chemometrics can often play a key and crucial role. Some techniques are also introduced in Section 3.14 as well as in this chapter.

- *Time series analysis.* Many methods have been developed to improve the quality of sequential datasets. One of the difficulties is that long term and interesting trends are often buried within short term random fluctuations. Statisticians distinguish between two types of noise, a deterministic one that may well be important for future predictions and a nondeterministic one that simply involves day to day (or second to second) changes in the environment. Interestingly, the statistician Hermann Wold, who is known among many chemometricians for the early development of the Partial Least Squares (PLS) algorithm, is probably more famous for his work on time series, studying this precise problem. A large battery of methods are used to smooth time series, many based on so-called 'windows', whereby data are smoothed over a number of points in time. A simple method is to take the average reading over five points in time, but sometimes this could miss out important information about cyclicity especially for a process that is sampled slowly compared with the rate of oscillation. Another problem that frequently occurs is that it is not always possible to sample evenly in time yet many methods of time series analysis expect regular measurements: several approaches to interpolation are available under such circumstances.
- *Chromatograms and spectra.* Chemists frequently come across sequential series in chromatography and spectroscopy. An important aim might be to smooth a chromatogram. A number of methods have been developed here such as the Savitzky–Golay filter. A problem is that if a chromatogram is smoothed too much the peaks are blurred and lose resolution, negating the benefits, so optimal filters have been developed. Another common need is for derivatives, sometimes spectra are routinely displayed in the derivative mode (e.g. electron spin resonance spectroscopy) and there are a number of rapid computational methods for such calculations that do not emphasize noise too much.
- *Fourier transform techniques.* The Fourier transform has revolutionized spectroscopy such as nuclear magnetic resonance (NMR) and infrared (IR), over the past three decades. The raw data are not obtained as a directly interpretable spectrum but as a time series, where all spectroscopic information is muddled up and a mathematical transformation is required to obtain a comprehensible spectrum. There are a number of advantages of using Fourier methods for data acquisition, an important one is that information can be obtained much faster, often a hundred times more rapidly than using so-called continuous wave methods. This has allowed the development, for example of ^{13}C NMR, as a routine analytical tool, because the low abundance of ^{13}C is compensated by faster data acquisition. However, special methods are required to convert this 'time domain' information to a 'frequency domain' spectrum. Parallel with Fourier transform spectroscopy have arisen a large number of approaches for enhancement of the quality of such data, often called Fourier deconvolution, involving manipulating the time series prior to Fourier transformation. These are different to the classical methods for time series analysis used in economics or geology. Sometimes it is even possible to take non-Fourier data, and Fourier transform it back to a time series, then use deconvolution methods and Fourier transform back again.

In this chapter we will be discussing approaches for handling such types of data.

4.2 CORRELOGRAMS

Correlograms are often obtained from time series analysis, and are useful for determining cyclicity in a series of observations recorded in time. These could arise from a geological process, a manufacturing plant, or environmental monitoring, the cyclic changes being due to season of the year, time of day, or even hourly events. An alternative method for analysing these data involves Fourier transformation, but the latter is usually applied to quite well behaved data such as NMR spectra. Time series analysis which is based, in part, on correlograms, has wide applications throughout science ranging from economics to geology to engineering. The principles are useful in chemistry where there is an underlying cyclicity buried within other trends. Approaches for improving the appearance and useful-ness of correlograms, such as smoothing functions, will be described in Sections 4.3, 4.5 and 4.6.

4.2.1 Auto-correlograms

The simplest method involves calculate an auto-correlogram. Consider the information depicted in Figure 4.1, which represents a process changing with time. It appears that there is some cyclicity, but this is buried within the noise. The principle of correlograms is to calculate correlation coefficients between two time series shifted by different amounts, the difference in time often called a 'lag'. The closer the correlation coefficient is to 1, the more similar the two series. For auto-correlograms, both series are identical, and simply involve calculating the correlation coefficient between the same time series shifted by a given amount. If there are 30 readings in the original data, the correlation coefficient for a lag of 3 points will result in an overlap of 27 data points, comparing the first reading with the fourth, the second with the fifth, the third with the sixth and so on. Two such series are given in Table 4.1, corresponding to the data of Figure 4.1, the resultant correlogram being presented in Figure 4.2. The cyclic pattern is clearer than in the original data. Note that the graph is symmetric about the origin as expected.

An auto-correlogram emphasizes primarily cyclical features. Sometimes there are noncyclical trends superimposed over the time series. Such situations regularly occur in economics. Consider trying to determine the factors relating to expenditure in a seaside resort. A cyclical factor will undoubtedly be seasonal, there being more business in the

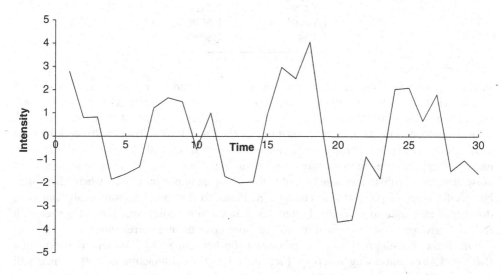

Figure 4.1   A time series: are there cyclical trends?

Table 4.1 Auto-correlogram calculation

Raw data	Lag of 3
2.77692	
0.79418	
0.81984	
−1.8478	2.77692
−1.6141	0.79418
−1.3071	0.81984
1.21964	−1.8478
1.65351	−1.6141
1.48435	−1.3071
−0.5485	1.21964
0.99693	1.65351
−1.7163	1.48435
−1.9951	−0.5485
−1.9621	0.99693
0.93556	−1.7163
2.96313	−1.9951
2.48101	−1.9621
4.04153	0.93556
−0.0231	2.96313
−3.6696	2.48101
−3.5823	4.04153
−0.8557	−0.0231
−1.8099	−3.6696
2.03629	−3.5823
2.09243	−0.8557
0.66787	−1.8099
1.80988	2.03629
−1.4799	2.09243
−1.0194	0.66787
−1.5955	1.80988

summer (except for countries near the equator where there is no summer or winter, of course). However, other factors such as interest rates, and exchange rates, will also come into play and the information will be mixed up in the resultant statistics. Expenditure can also be divided into food, accommodation, clothes and so on. Each will be influenced to different extents by seasonality. Correlograms will specifically emphasize the cyclical causes of expenditure. In chemistry, correlograms are most useful when time dependent noise interferes with uncorrelated (white) noise, for example in a river where there may be specific types of pollutants or changes in chemicals that occur spasmodically but once discharged take time to dissipate. Figure 4.3 shows a more noisy time series together with its auto-correlogram. The cyclical trends are more obvious in a correlogram.

Often the correlogram can be processed further either by Fourier transformation (Section 4.4) or smoothing functions (Section 4.3), or a combination of both, which will

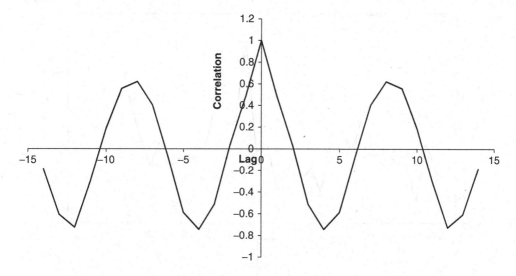

Figure 4.2 Auto-correlogram

be discussed below. Sometimes the results can be represented in the form of probabilities, for example the chance that there really is a genuine underlying cyclical trend of a given frequency. Such calculations, though, make certain assumptions about the underlying noise distributions and experimental error and cannot always be generalized.

4.2.2 Cross-correlograms

It is possible to extend these principles to the comparison of two independent time series. Consider measuring the levels of Ag and Ni in a river with time. Although each may show a cyclical trend, are there trends common to both metals? Very similar approaches are employed as described above, but two different time series are now compared. The cross-correlogram is of one set of data correlated against the other at various different lag times. Note that the cross-correlogram is no longer symmetrical.

4.2.3 Multivariate Correlograms

In many real world situations there may be a large number of variables that change with time, for example the composition of a manufactured product. In a chemical plant the resultant material could depend on a huge number of factors such as the quality of the raw material, the performance of the apparatus, even the time of day, which could, for example, relate to who is on shift or small changes in power supplies. Instead of monitoring each factor individually, it is common to obtain an overall statistical indicator, for example a Principal Component (Sections 5.2 and 5.3). The correlogram is computed from this mathematical summary of the raw data rather than the concentration of an individual constituent.

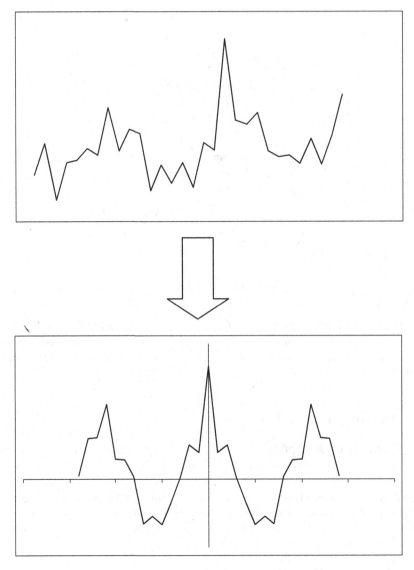

Figure 4.3   Auto-correlogram of a very noisy time series

4.3 LINEAR SMOOTHING FUNCTIONS AND FILTERS

Many types of chemical data consist of noisy sequential series. Buried within the noise are signals such as chromatographic or spectroscopic peaks, information on the quality of a manufactured product or cyclic changes in concentration of a compound. A key role of chemometrics is to obtain as informative a signal as possible after removing the noise. An important technique involves smoothing the data. Too much smoothing, however, and the signal is reduced in intensity and resolution. Too little smoothing, and noise remains. There is a huge literature on optimum filters that are used to find the best smoothing function without unduly distorting peakshapes.

The simplest methods involve linear filters whereby the resulting smoothed data are a linear function of the raw data. One of the most basic is a 3-point moving average (MA). Each point is replaced by the average of itself and the points before and after. The filter can be extended to a 5-, 7-, etc. point MA, the more the points, the smoother the signal becomes. The more the points in the filter, the greater the reduction in noise, but the higher the chance of blurring the signal. The number of points in the filter is often called a 'window'. The optimal filter depends on noise distribution and signal width.

MA filters have the disadvantage in that they require data to be approximated by a straight line, but peaks in spectra, for example, are often best approximated by curves e.g. a polynomial. As the window is moved along the data, a new best fit curve should be calculated each time. The value of the best fit polynomial in the centre of the window is taken as the new filtered datapoint. Because this procedure is computationally intensive a simple method for calculation was devised by Savitzky and Golay in 1964 [1] involving simple summations which are shown to be equivalent to curve fitting. These Savitzky–Golay filters are widely employed in analytical chemistry especially chromatography. Each filter is characterized by a window size and an order of a polynomial (e.g. quadratic) for the fitting. For small window sizes the coefficients are presented in Table 4.2. If there are, for example, 5 points in a window and one chooses a quadratic model, the intensity of the smoothed point in the centre is the sum of five original intensities in the window, the intensity at $j = -2$ from the centre being multiplied by $-3/35$, that of the intensity at $j = -1$ from the centre by $12/35$, and so on.

Figure 4.4 illustrates a typical portion of a spectrum or a noisy analytical signal. It looks as if there are two peaks, but it is not entirely clear due to the presence of noise. In Figure 4.5, 3-, 5- and 7-point MAs are calculated. The 3-point MA clearly suggests there may be two peaks, whereas increasing the window size loses resolution whilst only slightly reducing the noise. In this case a 5-point MA is overkill. Applying 5-, 7- and 9-point quadratic/cubic Savitzky–Golay filters in Figure 4.6 (note that a quadratic approximation

Table 4.2 Savitzky–Golay smoothing functions. The coefficients are given in the table: multiply each data point in a window by the coefficient divided by the normalization constant

	\multicolumn{5}{c}{Window size}				
j	5	7	9	7	9
	\multicolumn{3}{c}{Quadratic/cubic}		\multicolumn{2}{c}{Quartic/quintic}	
−4			−21		15
−3		−2	14	5	−55
−2	−3	3	39	−30	30
−1	12	6	54	75	135
0	17	7	59	131	179
1	12	6	54	75	135
2	−3	3	39	−30	30
3		−2	14	5	−55
4			−21		15
Normalization constant	35	21	231	231	429

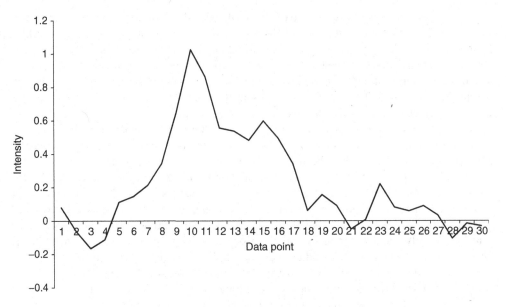

Figure 4.4 Noisy analytical signal

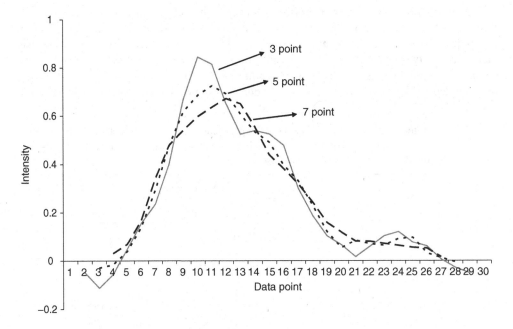

Figure 4.5 Moving average filters applied to the data in Figure 4.4

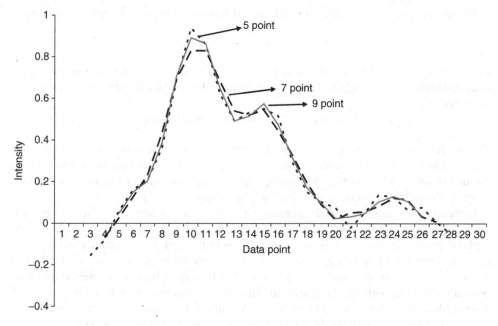

Figure 4.6 Savitzky–Golay filters applied to the data in Figure 4.4

gives the same results as a cubic in this case) preserves the resolution even when using a 9-point window.

Several other MA methods have been proposed in the literature, two of the best known being the Hanning window (named after Julius Von Hann) and the Hamming window (named after R.W. Hamming). These involve weighted averages. For a simple 3-point MA all three points in the window are equally important, whereas for a 3-point Hanning window the weight of the central point is 0.5, and the others 0.25. Often these windows are represented diagrammatically, the shape of the filter relating to the relative weights given to each datapoint in the window.

Although most conventional spectroscopic and chromatographic filters involve calculating means or a linear combination of the original data, in certain areas such as process analysis, there can be spikes (or outliers) in the data which often should be best ignored. Running Median Smoothing (RMS) functions calculate the median rather than the mean over a window. Different types of smoothing remove different features in the data and often a combination of several approaches is recommended especially for real world problems. Dealing with outliers is an important issue: sometimes these points are due to measurement errors. We discuss some of the methods for detecting outliers and how to deal with them in retrospect in Sections 3.12 and 3.13; in this section we describe how to deal with outliers 'on the fly'. Many processes take time to deviate from the expected value, a sudden glitch in the system unlikely to be a real underlying effect. Often a combination of filters is used, for example a 5-point median smoothing followed by a 3-point Hanning window. These methods are very easy to implement computationally and it is possible to view the results of different filters.

Reroughing is a another technique that should be mentioned. The 'rough' is given by:

$$Rough = Original - Smooth$$

where the smoothed data are obtained by a method described above. The rough represents residuals, or, hopefully noise, and can in itself be smoothed. New original data are calculated by:

$$Reroughed = Smooth + Smoothed\ rough$$

This is useful if there is suspected to be a number of sources of noise. One type of noise may genuinely reflect difficulties in the underlying data, the other may be due to outliers that do not reflect a long term trend. Smoothing the rough can remove one of these sources. More details of these procedures are available in Grznar *et al.* [2].

In the case of spectroscopy and chromatography the choice of smoothing methods is often quite straightforward. However, when analysing data from monitoring a process it is often harder to determine, and depends on the overall aim of the analysis. Sometimes the aim is an alert whenever the value of a parameter (e.g. the concentration of a constituent of a fabric) is outside a predetermined range: false alerts could be caused by measurement error or an anomalous sample, and may be costly. In other cases it may be more interesting to see if there is a background process building up which could in due course lead to an unacceptable product. In many cases slight deviations from an 'average' product can be tolerated and destroying a batch or closing a manufacturing plant is expensive, in other circumstances exceeding predefined limits could be dangerous and even illegal (e.g. in pharmaceuticals). Some industrial processes are monitored by techniques which in themselves are not very reliable, and so a temporary glitch could be ignored, whereas others are monitored by highly reliable but perhaps more expensive assays. Such considerations are important when deciding on the nature of the filter to be employed when monitoring a time series. More on process monitoring is discussed in Sections 3.13 and 3.14.

4.4 FOURIER TRANSFORMS

Fourier transform techniques are widely used throughout chemistry, examples being in NMR, IR and X-ray analysis. The mathematics of Fourier transformation has been well established for two centuries. However, the first reports involved transforming a continuous distribution of data, but most modern instrumental data are obtained as discrete digitized series, i.e. sampled at distinct intervals in time. In the case of instrumental data, these intervals are normally regular, and a computational method called the Discrete Fourier Transform (DFT) is available.

DFTs involve mathematical transformation between two types of data. In Fourier transform NMR (FTNMR) the raw data are acquired as a function of time, often called the 'time domain' or more specifically a Free Induction Decay (FID). FTNMR has been developed over the years because it is much quicker to obtain data than using conventional (continuous wave) methods. An entire spectrum can be sampled in a few seconds, rather than minutes, speeding up the procedure of data acquisition by one or two orders of magnitude. This has meant that it is possible to record spectra of small quantities of compounds or of natural abundance of isotopes such as ^{13}C, now routine in modern chemical laboratories. The trouble with this is that the spectroscopist cannot easily interpret time domain data,

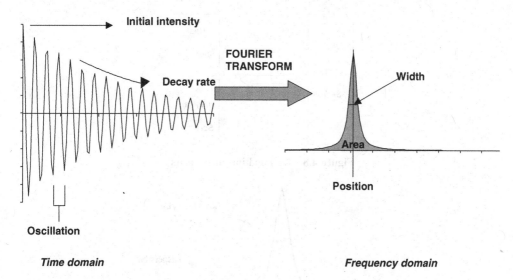

Figure 4.7 Transformation from time to frequency domain

Table 4.3 Equivalence of time and frequency domain parameters

Time domain	Frequency domain
Initial intensity	Peak area
Oscillation frequency	Peak position
Decay rate	Peak width

and here arises the need to use DFTs to transform this information into something that makes better sense. Each spectroscopic peak in the time domain is characterized by three parameters, namely, intensity, oscillation rate and decay rate. It is transformed into a recognizable spectral peak in the frequency domain characterized by an area, position and width, as illustrated in Figure 4.7. Each parameter in the time domain corresponds to a parameter in the frequency domain as indicated in Table 4.3.

- The faster the rate of oscillation in the time series the farther away the peak is from the origin in the spectrum.
- The faster the rate of decay in the time series, the broader the peak in the spectrum.
- The higher the initial intensity in the time series, the greater the area of the transformed peak.

In real spectra, there will be several peaks, and the time series generally appear much more complex than that in Figure 4.7, consisting of several superimposed curves, each corresponding to one peak in the spectrum. The beauty of Fourier transform spectroscopy, however, is that all the peaks can be observed simultaneously, so allowing rapid acquisition of data, but a mathematical transform is required to make the data comprehensible.

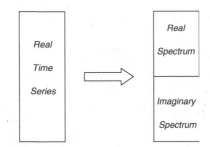

Figure 4.8   Real and imaginary pairs

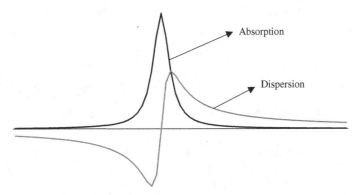

Figure 4.9   Absorption and dispersion line shapes

Fourier transforms are mathematically quite sophisticated procedures and what usually happens is that the raw data (e.g. a time series) is transformed into two frequency domain spectra, one which is real and the other imaginary (this terminology comes from complex numbers). The true spectrum is represented by only half the transformed data as indicated in Figure 4.8. Ideally the real half of the spectrum corresponds to *absorption* peak shapes and the imaginary half to *dispersion* peakshapes (Figure 4.9). Absorption peaks are easiest to interpret and are similar to the normal (non-Fourier) peak shapes.

However, often these two peak shapes are mixed together in the real spectrum, due to small imperfections in acquiring the data, often called phase errors. There are a variety of solutions to this problem, a common one being to correct this by adding together proportions of the real and imaginary data until an absorption peak shape is achieved. An alternative is to take the absolute value or magnitude spectrum which is the square root of the sum of squares of the real and imaginary parts. Although easy to calculate and always positive, it is important to realize it is not quantitative: the peak area of a two component mixture is not equal to the sum of peak areas of each individual component. Because sometimes spectroscopic peak areas (or heights) are used for chemometric pattern recognition studies this limitation is important to appreciate.

There are two fundamental ways data can be Fourier transformed.

- A *forward transform* converts real and imaginary pairs (e.g. a spectrum) to a real series (e.g. a FID). Note that the conventional notation is the opposite to what a spectroscopist might expect.
- An *inverse transform* converts a real series to real and imaginary pairs.

There is some inconsistency in the rather diverse literature on Fourier transforms, and, when reading literature from different authors using different methods for acquiring data, it is always important to check precisely which method has been applied, and to check carefully notation.

An important property of DFTs depends on the rate which data are sampled. Consider the time series of Figure 4.10, each cross indicating a sampling point. If it is sampled half as fast, it will appear that there is no oscillation, as every alternative datapoint will be eliminated. Therefore, there is no way of distinguishing such a series from a zero frequency series. The oscillation frequency in Figure 4.10 is called the *Nyquist* frequency. Anything that oscillates faster than this frequency will appear to be at a lower frequency that it really is, because the sampling rate cannot cope with faster oscillations, therefore the rate of sampling establishes the range of observable frequencies: the higher the rate, the greater the range. In order to increase the range of frequencies, a higher sampling rate is required, and so more data points must be collected per unit time. The equation:

$$N = 2 \, S \, T$$

links the number of data points acquired (e.g. $N = 4000$), the range of observable frequencies (e.g. $S = 500$ Hz) and the acquisition time (which would have to be $T = 4$ s in this example). Higher frequencies are 'folded over' or 'aliased'.

The Nyquist frequency is not only important in instrumental analysis. Consider sampling a geological core where depth relates to time, to determine whether the change in concentrations of a compound, or isotopic ratios, display cyclicity. A finite amount of core is needed to obtain adequate quality samples, meaning that there is a limitation in samples per length of core. This, in turn, limits the maximum frequency that can be observed. More frequent sampling, however, may require a more sensitive analytical technique.

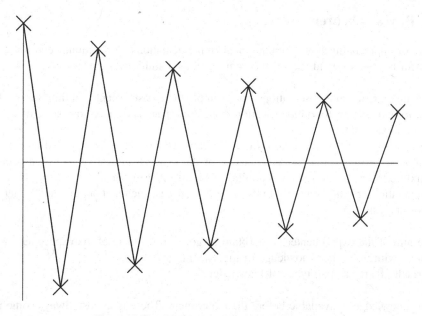

Figure 4.10 Discrete sampling of a time series

A final consideration relates to algorithms used for Fourier transforms. DFT methods became widespread in the 1960s partly because Cooley and Tukey developed a rapid computational method, the Fast Fourier Transform (FFT). This method required the number of sampling points to be a power of 2, e.g. 1024, 2048, etc. and many people still associate powers of 2 with Fourier transformation. However, there is no special restriction on the number of data points in a time series, the only consideration relating to the speed of computation.

More computationally oriented readers will find excellent descriptions of Fourier methods, among others, in various editions of the *Numerical Recipes* books [3]. Even if the book is not based around your favourite programming language, the detailed descriptions of algorithms are outstanding in these classic texts.

4.5 MAXIMUM ENTROPY AND BAYESIAN METHODS

Over the past two decades there has been substantial scientific interest in the application of maximum entropy techniques with notable successes, for the chemist, in areas such as NMR spectroscopy and crystallography. Maxent has had a long statistical vintage, one of the modern pioneers being Ed Jaynes, but the first significant scientific applications were in the area of deblurring of IR images of the sky, involving the development of the first modern computational algorithm, in the early 1980s. Since then, there has been an explosion of interest and several implementations available within commercial instrumentation. The most spectacular successes have been in the area of image analysis, for example NMR tomography, as well as forensic applications such as obtaining clear car number plates from hazy police photos. However, there has been a very solid and large amount of literature in the area of analytical chemistry.

4.5.1 Bayes' Theorem

A basis of appreciating how computational implementations of maximum entropy can be successful is to understand Bayes' theorem. Various definitions are necessary:

- Data are experimental observations, for example the measurement of a time series of FID prior to Fourier transformation (Section 4.4). Data space is a space consisting of a dataset for each experiment.
- The desired result or information that is required from the data is called a 'map', for example a clean and noise free spectrum, or the concentration of several compounds in a mixture. Map space exists in a similar fashion to data space.
- An operation or transformation links these two spaces, such as Fourier transformation or factor analysis.

The aim of the experimenter is to obtain as good an estimate of map space as possible, consistent with his or her knowledge of the system.

Normally there are two types of knowledge:

- Prior knowledge is available before the experiment. There is almost always some information available about chemical data. An example is that a true spectrum will always be

positive: we can reject statistical solutions that result in negative intensities. Sometimes much more detailed information such as lineshapes or compound concentrations is known.
- Experimental information. This refines the prior knowledge to give a posterior model of the system.

Thomas Bayes was an 18th century clergyman, who is attributed to proposing a theorem, often presented in the following simple form:

Probability (answer given new information) $\propto$

Probability (answer given old information) $\times$ Probability (new information given answer)

or

$$p \text{ (map | experiment)} = p \text{ (map | prior info)} \times p \text{ (experiment | map)}$$

The '|' symbol stands for 'given by'.

Many scientists do not use this prior information, and for cases where data are quite good, this can be perfectly acceptable. However, chemical data analysis is most useful where the answer is not so obvious, and the data difficult to analyse. The Bayesian method allows prior information or measurements to be taken into account. It also allows continuing experimentation, improving a model all the time. Each time an experiment is performed, we can refine our assessment of the situation and so iterate towards a probability that weighs up all the evidence. Bayesian statistics is all the rage in areas such as forensic science, where a specific case will often be based on several bits of evidence, no one piece in itself being completely conclusive and more examples in the context of biological pattern recognition are described in Section 10.5.

4.5.2 Maximum Entropy

Maxent can be used as a method for determining which of many possible underlying distributions is most likely. A simple example is that of the toss of a six sided unbiased die. What is the most likely underlying probability distribution, and how can each possible distribution be measured? Figure 4.11 illustrates a flat distribution and Figure 4.12 a skew distribution. The concept of entropy was introduced in its simplest form defined by:

$$S = -p_i \sum_{i=1}^{I} p_i \, \log(p_i)$$

where p_i is the probability of outcome i. In the case of our die, there are six outcomes, and in Figure 4.11 each outcome has a probability of 1/6. The distribution with maximum entropy is the most likely underlying distribution. The entropy of the distribution pictured in Figure 4.11 is 1.7917 and that pictured in Figure 4.12 is 1.7358, demonstrating that the even distribution of probabilities is best, in the absence of other evidence – we have no reason to suspect bias. These distributions can be likened to spectra sampled at 6 data points – if there is no other information the spectrum with maximum entropy is a flat distribution.

However, if some experiments have been performed, constraints can be added. For example, it might be suspected that the die is actually a biased die as a mean of 4.5 instead of 3.5 is obtained after a very large number of tosses of the die. What distribution is expected now? Entropy allows the probability of various models to be calculated.

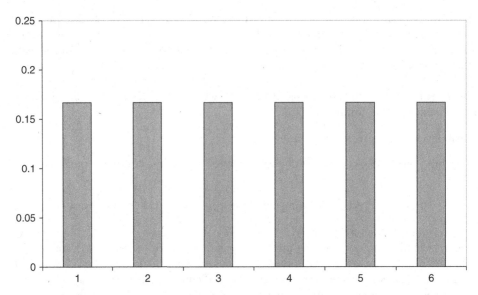

Figure 4.11   Probabilities of obtaining a specific result from an unbiased die

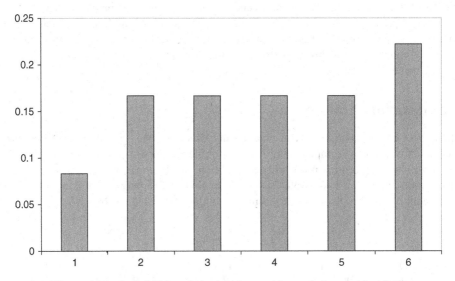

Figure 4.12   Probabilities of obtaining a specific result from a biased die

In addition, there are a number of other definitions of entropy that can take various forms of prior information into account.

4.5.3 Maximum Entropy and Modelling

In practice there is an infinite, or at least very large, number of statistically identical models that can be obtained from a system. If we know a chromatographic peak consists of two components, we can come up any number of ways of fitting the chromatogram. Figure 4.13

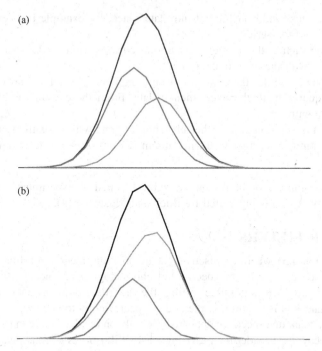

Figure 4.13 Two ways of fitting a peak to two underlying components

illustrates two such solutions. Which is more likely? Both correspond to identical least squares fit to the data. On the whole, in the absence of further information, the second is less preferable, as one peak is somewhat asymmetrical, and both peaks have different shapes. Most definitions of entropy will pick the answer of Figure 4.13(a). It is this entropy that encodes prior knowledge of the system, Figure 4.13(a) being more probable (in the absence of other information) than Figure 4.13(b).

In the absence of any information at all, of course, a flat chromatogram is the best answer, with equal absorbance at every elution time sampled, and, therefore, the experimental observations have changed the solutions considerably, and will pick two underlying peaks that fit the data well consistent with maximum entropy. Into the entropic model information can be built relating to knowledge of the system. Normally a parameter defined by:

Entropy function − Statistical fit function

is obtained. High entropy is good, but not at the cost of a numerically poor fit to the data. However, a model that fits the original (and possibly very noisy) experiment well is not a good model if entropy is too high. Therefore there is a balance between low least squares error and maximizing entropy, and these two criteria often need adjusting in an iterative way.

The basis of a computational approach is as follows, illustrated as applied to Fourier transform data (Section 4.4).

• Guess the solution, e.g. a spectrum using whatever knowledge is available. In NMR it is possible to start with a flat spectrum.

- Then take this guess and transform it into data space, for example Fourier transforming the spectrum to a time series.
- Using a statistic such as the χ^2 statistic (introduced in Section 3.10.2), see how well this guess compares with the experimental data.
- Improve the guess a little: there will be a large number of solutions all of similar (and better) least squares fit to the experimental data. From these solutions choose the one with maximum entropy.
- Then improve the solution still further choosing the one with maximum entropy, and so on, until an optimum solution (with maximum entropy and low least squares error) is obtained.

There are a large number of books on the subject aimed at experimental scientists, one introductory set of essays being edited by Buck and Macauley [4].

4.6 FOURIER FILTERS

Previously (Section 4.3) we have discussed a number of linear filter functions that can be used to enhance the quality of spectra and chromatograms. When performing Fourier transforms (Section 4.4), it is possible to filter the raw (time domain) data prior to Fourier transformation, and this is a common method in spectroscopy to enhance resolution.

The width of a peak in a spectrum depends primarily on the decay rate in the time domain. The faster the decay, the broader the peak. Figure 4.14 illustrates a broad peak together with

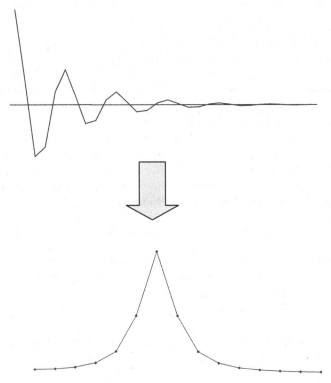

Figure 4.14 A broad peak together with its Fourier transform (the phased real part of the frequency domain is illustrated in this and subsequent diagrams)

its corresponding time domain. If it is desired to increase resolution, a simple approach is to change the shape of the time domain function so that the decay is slower. In some forms of spectroscopy (such as NMR) the time series contains a term due to exponential decay and can be characterized by:

$$f(t) = e^{-at} w(t)$$

Multiplying the time series by a positive exponential of the form:

$$g(t) = e^{+bt}$$

reduces the decay rate to give a new time series, since the exponential term is reduced in magnitude, providing b is less than a.

Figure 4.15 shows the change in appearance of the time domain as it is multiplied by a positive exponential. The transform is presented superimposed over the original peak shape. It is clear that the peak is now sharper, and so, if there were a cluster of neighbouring partially overlapping peaks, it would be better resolved.

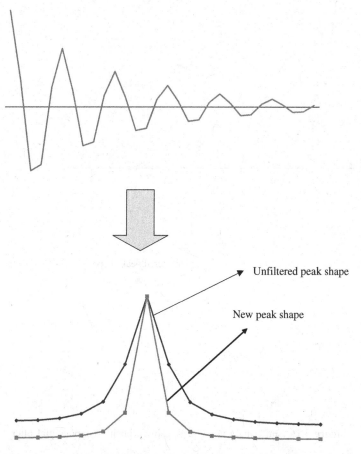

Unfiltered peak shape

New peak shape

Figure 4.15 Effect of multiplying the time series in Figure 4.14 by a positive exponential. The new (sharper) peak shape is compared with the old broader one

Theoretically it could be possible to conceive multiplying the original time series by increasingly positive exponentials until peaks are one datapoint wide and so infinitely sharp, within the limitations of the sampling frequency. Clearly there is a flaw in our argument, as otherwise it would be possible to obtain indefinitely narrow peaks and so achieve any desired resolution.

The difficulty is that real spectra always contain noise. Figure 4.16 represents a noisy time series, together with the exponentially filtered data. The filtered time series amplifies noise substantially which interferes with signals. The result of filtering in the transform is compared with that of the raw data. Although the peak width has indeed decreased, the

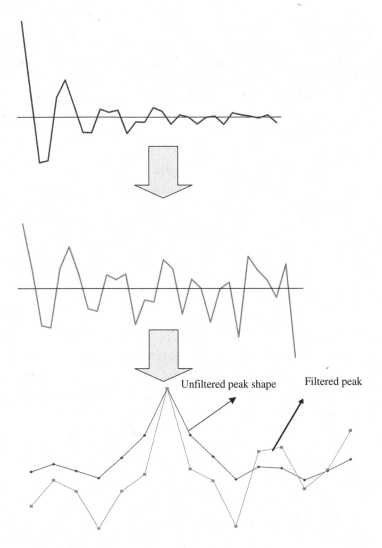

Figure 4.16   Filtering a noisy peak using a positive exponential filter

noise has increased. In addition to making peaks hard to identify, noise also reduces the ability to determine integrals and so concentrations and sometimes to accurately pinpoint peak positions.

How can this be solved? Clearly there are limits to the amount of peak sharpening that is practicable, but the filter function can be improved so that noise reduction and resolution enhancement are applied simultaneously. One common method is to multiply the time series by a double exponential filter of the form:

$$g(t) = e^{+bt-ct^2}$$

where the first term of the exponential increases with time and the second decreases. Providing the values of b and c are chosen correctly, the first term will result in increased resolution and the second term in decreased noise. These two terms can be optimized theoretically if peak shapes and noise levels are known in advance, but, in most practical cases, they are chosen empirically.

The effect on the noisy data of Figure 4.16 is illustrated in Figure 4.17, for a typical double exponential filter. The filtered time series decays more slowly than the original, but there is not much increase in noise at the end, because the second (negative) exponential term reduces the intensity at the end of the time series. The peakshape in the transform is almost as narrow as that obtained using a single exponential, but noise is dramatically reduced.

There are a large number of related filter functions, but the overall principles are to reduce noise whilst improving resolution.

A major motivation of filters is to enhance resolution (see Figure 4.18), but they can be employed in other cases, for example to reduce noise. Each type of filter and each application is specifically tailored. Powerful applications are in image analysis and X-ray processing, where data are often acquired using Fourier methods.

In most forms of NMR and IR spectroscopy raw data are acquired as a time series, and must be Fourier transformed to obtain an interpretable spectrum. However, it is not necessary to use time series data and any spectrum or chromatogram can be enhanced by the methods using Fourier filters. This process is often called *Fourier self deconvolution.*

Normally, three steps are employed, as illustrated in Figure 4.19:

• Transform the spectrum into a time series. This time series does not exist in reality but can be handled by a computer.
• Then apply a Fourier filter to the time series.
• Finally transform the spectrum back, resulting in improved quality.

Identical principles are applicable to chromatograms.

Some people get confused by the difference between Fourier filters and linear smoothing and resolution functions (Section 4.3). In fact both methods are related by the convolution theorem, and both have similar aims, to improve the quality of spectroscopic or chromatographic data, and both are linear methods.

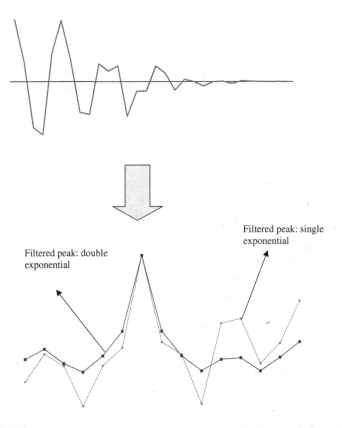

Figure 4.17 Double exponential filtering, combined smoothing and resolution enhancement

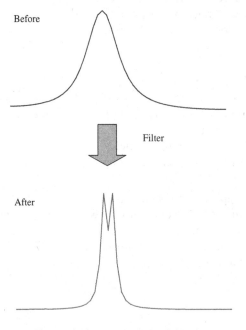

Figure 4.18 Ideal result of filtering

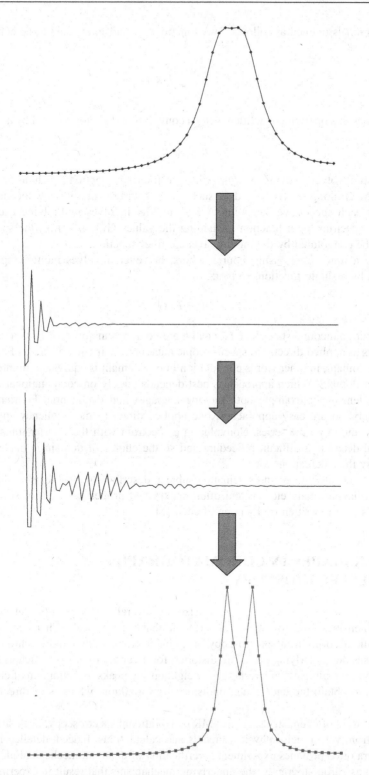

Figure 4.19 Illustration of Fourier self deconvolution

Convolution is defined as follows. Two functions, f and g, are said to be convoluted to give h, if:

$$h_i = \sum_{j=-p}^{j=p} f_j g_{i+j}$$

Sometimes this operation is written using a convolution operator denoted by a '*', so that:

$$h = f^* g$$

Convolution involves moving a windowing or filtering function (such as a Savitzky–Golay or MA) along a series of data such as a spectrum, multiplying the data by that function at each successive datapoint. A 3-point MA involves multiplying each set of 3 points in a spectrum by a function containing the values (1/3, 1/3, 1/3), the spectrum can be said to be convoluted by the moving average filter function.

Filtering a time series, using Fourier filters, however, involves multiplying the *entire* time series by a single function, so that:

$$H_i = F_i . G_i$$

The convolution theorem states that f, g and h are Fourier transforms of F, G and H. Hence linear filters as applied directly to spectroscopic data have their equivalence as Fourier filters in the time domain, in other words convolution in one domain is equivalent to multiplication in the other domain. Which approach is best depends largely on computational complexity and convenience. For example both moving averages and exponential Fourier filters are easy to apply, so are easy approaches, one applied direct to the frequency spectrum and the other to the raw time series. Convoluting a spectrum with the Fourier transform of an exponential decay is a difficult procedure and so the choice of domain is made according to how easy the calculations are.

Fourier filters are a large and fascinating subject, described in most texts on digital signal processing, having many elegant and often unexpected properties. One, of several, useful texts in this area is written by Lynn and Fuerst [5].

4.7 PEAKSHAPES IN CHROMATOGRAPHY AND SPECTROSCOPY

A typical chromatogram or spectrum consists of several peaks, at different positions, of different intensities, and sometimes of different shape. Each peak either corresponds to a characteristic absorption in spectroscopy or a characteristic compound in chromatography. In most cases the underlying peaks are distorted for a variety of reasons such as noise, poor digital resolution, blurring, or overlap with neighbouring peaks. A major aim of chemometric methods is to obtain the underlying, undistorted information, which is of direct interest to the chemist.

In some areas of chemistry such as NMR or vibrational spectroscopy, peak shapes can be predicted from very precise physical models and calculations. Indeed, detailed information from spectra often provides experimental verification of quantum mechanical ideas. In other areas, such as chromatography, the underlying mechanisms that result in experimental peak shapes are less well understood, being the result of complex mechanisms for partitioning between mobile and stationary phases, so the peaks are often modelled very empirically.

4.7.1 Principal Features

Peaks can be characterized in a number of ways, but a common approach, for symmetrical peakshapes, as illustrated in Figure 4.20, is to characterize each peak by:

1. a position at the centre (e.g. the elution time or spectral frequency);
2. a width, normally at half height;
3. an area.

The relationship between area and height is dependent on the peak shape, although heights are often easiest to measure experimentally. If all peaks have the same shape, the relative heights of different peaks are directly related to their relative areas. However, area is usually a better measure of chemical properties such as concentration so it is important to have some idea about peak shapes before using heights as indicators of concentration. In chromatography, widths of peaks may differ from column to column or even from day to day using a similar column according to instrumental performance or even across a chromatogram, but the relationship between overall area and concentration should remain stable, providing the detector sensitivity remains constant. In spectroscopy peak shapes are normally fairly similar on different instruments, although it is important to realize that filtering can distort the shape and in some cases filtering is performed automatically by the instrumental hardware, occasionally without even telling the user. Only if peak shapes are known to be

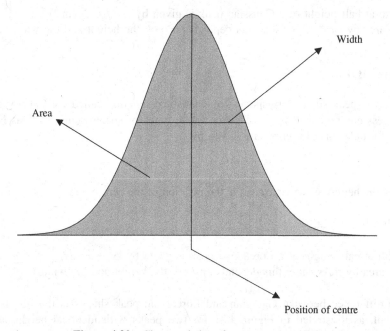

Figure 4.20 Characteristics of a symmetrical peak

fairly constant in a series of measurements can heights be safely substituted for areas as measurements of chemical properties.

Sometimes the width at a different proportion of the peak height is reported rather than the half width. Another common measure is when the peak has decayed to a small percentage of the overall height (for example 1 %), which is often taken as the total width of the peak, or using triangulation, which fits the peak (roughly) to the shape of a triangle and looks at the width of the base of the triangle.

Three common peak shapes are important, although in some specialized cases it is probably best to consult the literature on the particular measurement technique, and for certain types of instrumentation, more elaborate models are known and can safely be used.

4.7.2 Gaussians

These peak shapes are common in most types of chromatography and spectroscopy. A simplified formula for a Gaussian is given by:

$$x_i = A \exp[-(x_i - x_0)^2/s^2]$$

where A is the height at the centre, x_0 is the position of the centre and s relates to the peak width.

Gaussians crop up in many different situations and have also been discussed in the context of the normal distribution in statistical analysis of data (Section 3.4). However, in this chapter we introduce these in a different context: note that the pre-exponential factor has been simplified in this equation to represent shape rather than an area equal to 1.

It can be shown that:

- the width at half height of a Gaussian peak is given by $\Delta_{1/2} = 2s(\ln 2)^{1/2}$;
- and the area by $(\pi)^{1/2}A\,s$ (note this depends on *both* the height and the width).

4.7.3 Lorentzians

The Lorentzian peak shape corresponds to a statistical function called the Cauchy distribution. It is less common but often arises in certain types of spectroscopy such as NMR. A simplified formula for a Lorentzian is given by:

$$x_i = A/[1 + (x_i - x_0)^2/s^2]$$

where A is the height at the centre, x_0 is the position of the centre and s relates to the peak width.

It can be shown that:

- the width at half height of a Lorentzian peak is given by $\Delta_{1/2} = 2s$;
- and the area by $\pi A s$ (note this depends on *both* the height and the width).

The main difference between Gaussian and Lorentzian peak shapes is that the latter has a bigger tail, as illustrated in Figure 4.21 for two peaks with identical heights and half widths.

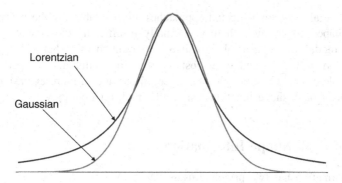

Figure 4.21   Gaussian and Lorentzian peak shapes

4.7.4 Asymmetric Peak Shapes

In many forms of chemical analysis, especially chromatography, it is hard to obtain symmetrical peak shapes. Although there are a number of quite sophisticated models that can be proposed, a very simple first approximation is that of a Lorentzian/Gaussian peak shape. Figure 4.22(a) represents a tailing peak shape. A fronting peak is illustrated in Figure 4.22(b): such peaks are much rarer. The way to handle such peaks is to model the sharper half by a Gaussian (less of a tail) and the broader half by a Lorentzian, so, for example, the right-hand half of the peak in Figure 4.22(a) is fitted as a 'half' Lorentzian peak shape. From each half a different width is obtained, and these are used to describe the data.

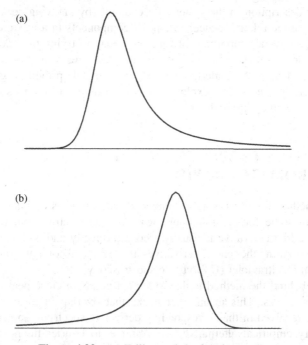

Figure 4.22   (a) Tailing and (b) fronting peaks

Although several very sophisticated approaches are available for the analysis of asymmetric peak shapes, in practice, there is normally insufficient experimental evidence for more detailed modelling, and, as always, it is important to ask what further insights this extra information will give, and what cost in obtaining better data or performing more sophisticated calculations. The book by Felinger [6] provides a more extensive insight into a large number of peak shape functions especially in chromatography.

4.7.5 Use of Peak Shape Information

Peakshape information has two principal uses.

- *Curve fitting* is quite common. There are a variety of computational algorithms, most involving some type of least squares minimization. If peak shapes are known in advance this can substantially reduce the number of parameters required for the curve fitting algorithms. For example, if there are suspected (or known) to be three peaks in a cluster, of Gaussian peak shape, then nine parameters need to be found, namely the three peak positions, peak widths and peak heights for each of the peaks. In any curve fitting routine it is important to determine whether there is certain knowledge of the peak shape, and sometimes of certain features, for example, the positions of each component. It is also important to appreciate that many chemical data are not of sufficient quality for very sophisticated models, so the three peak shape models described in Sections 4.7.2–4.7.4, whilst very much approximations, are in many cases adequate. Over-fitting can be dangerous unless very detailed information is available about the system (and of course the digital resolution and signal to noise ratio must be adequate). The result of curve fitting can be a better description of the system, for example, by knowing peak areas it may be possible to determine relative concentrations of components in a mixture.
- *Simulations* also have an important role and are a way of trying to understand a system. If the application of a chemometric method provides a result that is close to a similar application to real data, the underlying peak shape models provide a good description. Simulations can also be used to explore how well different techniques work, and under what circumstances they break down.

4.8 DERIVATIVES IN SPECTROSCOPY AND CHROMATOGRAPHY

A well known method for processing instrumental data involves derivatives (of the mathematical kind). In some forms of measurement such as electron spin resonance (ESR), by convention, the first derivative is usually recorded directly and is the normal method of obtaining information. In other cases, derivatives are calculated computationally. Derivatives are quite common in ultraviolet (UV)/visible spectroscopy.

The principle behind the method is that inflection points in close peaks become turning points in the derivatives. This means that peaks that overlap in a normal spectrum are more likely to be resolved in the corresponding derivative spectrum, and so derivatives are a simple and very empirical alternative, for example, to Fourier filters (Section 4.6), for resolution enhancement.

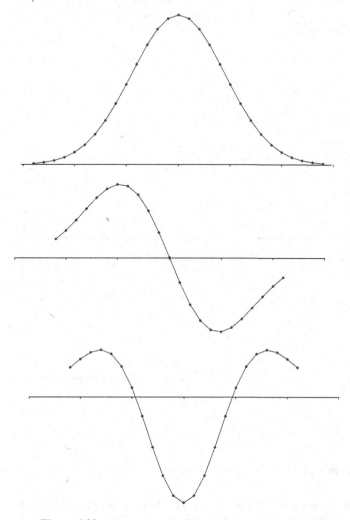

Figure 4.23 First and second derivatives of a Gaussian

The first and second derivatives of a pure Gaussian are presented in Figure 4.23.

- The first derivative equals zero at the centre of the peak, and is a good way of accurately pinpointing the position of a broad peak. It exhibits two turning points.
- The second derivative is a minimum at the centre of the peak, crosses zero at the positions of the turning points for the first derivative, and exhibits two further turning points farther apart than in the first derivative.
- The apparent peak width is reduced using derivatives.

These properties are most useful when there are several closely overlapping peaks.

Figure 4.24 illustrates the effect of calculating the first and second derivatives of two closely overlapping peaks. The second derivative clearly indicates that there are two peaks and helps pinpoint their positions. The appearance of the first derivative suggests that the

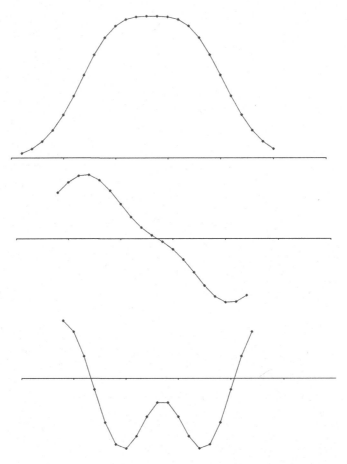

Figure 4.24 First and second derivatives of two closely overlapping peaks

peak cluster is not pure but, in this case, probably does not provide definitive evidence of how many peaks there are and what their positions are. It is, of course, possible to continue and calculate the third, fourth, etc. derivatives.

There are, however, two disadvantages of calculating derivatives. First, they are computationally intense, as a fresh calculation is required for each datapoint in a spectrum or chromatogram. Second, and most importantly, they amplify noise substantially, and, therefore, perform best when there are low signal to noise ratios. These limitations can be overcome by using Savitzky–Golay coefficients similar in principle to those described previously (Section 4.3), which involve rapid calculation of higher derivatives. Tables of coefficients have been published [1] according to window size (how many data points are used to compute the derivative?) and the type of approximation required (quadratic, cubic, etc.). The coefficients for small window sizes are presented in Table 4.4. The principles are to simultaneously smooth the data as well as determining derivatives by finding a best fit model to the new, improved, peak shape. Similar to double exponential filters in Fourier spectroscopy (Section 4.6) noise reduction is balanced against peak sharpening. The deconvolution theorem shows that there are equivalent methods for filtering in both the frequency and time domain, and the trick is to balance resolution against noise.

Table 4.4 Savitzky–Golay filters for derivatives

	Window size					
	5	7	9	5	7	9
j	Quadratic			Cubic/quartic		
First derivatives						
−4			−4			86
−3		−3	−3		22	−142
−2	−2	−2	−2	1	−67	−193
−1	−1	−1	−1	−8	−58	−126
0	0	0	0	0	0	0
1	1	1	1	8	58	126
2	2	2	2	1	67	193
3		3	3		−22	142
4			4			−86
Normalization constant	10	28	60	12	252	1188
Second derivatives						
	Quadratic/cubic			Quartic/quintic		
−4			28			−4158
−3		5	7		−117	12243
−2	2	0	−8	−3	603	4983
−1	−1	−3	−17	48	−171	−6963
0	−2	−4	−20	−90	−630	−12210
1	−1	−3	−17	48	−171	−6963
2	2	0	−8	−3	603	4983
3		5	7		−117	12243
4			28			−4158
Normalization constant	7	42	462	36	1188	56628

Another common, but very different, application of derivatives, is in chromatographic peak purity determinations. It is very common to use hyphenated chromatography (such as diode array high performance liquid chromatography) to determine whether an individual compound is pure or not. This can have important consequences in process control. For example, many commercial synthetic reactions contain small traces of isomers, with similar chromatographic properties. Sometimes, in pharmaceutical chemistry, these impurities can have a major physiological effect on the performance of a drug. Using hyphenated chromatography, spectra are recorded regularly across a chromatographic peak. In Chapter 7 we will discuss these multivariate data matrices in greater detail. However, if there is an impurity it is likely that the spectral characteristics across a chromatographic peak will change. There are several ways of measuring this change, but one is to calculate the derivative of the spectra, scaled to a constant amount, at one or more wavelengths. If the nature of the spectrum is unchanged, the derivative will be close to 0. As the spectrum changes, indicative of an impurity, the derivative deviates from 0. The application is described in greater detail in Section 7.7. Variations on this theme are often employed in spectroscopic data processing software.

Derivatives are a widespread and conceptually simple method for enhancing the appearance of chromatograms and spectra. Using the Savitzky–Golay method, they are also easy to calculate. The main requirement is that peaks are characterized by several data points. In high field NMR, however, a peak width may be only 3 or 4 data points, so there is insufficient information for meaningful calculations of derivatives. In UV/visible spectroscopy, a peak may be 20 or more data points in width and so these methods are quite useful for revealing underlying trends. There is no single universal solution to all the problems a spectroscopist or chromatographer faces, and a good appreciation of the wide battery of potential methods available is important.

4.9 WAVELETS

Wavelet transforms are increasingly finding their way into chemometric software packages and this section provides a general discussion of their potential.

Wavelet transforms are normally applied to datasets whose size is a power of two, for example consisting of 512 or 1024 data points. If the dataset is longer, it is conventional to simply clip the data to a conveniently sized window.

A wavelet is a general function, usually, but by no means exclusively of time which can be modified by *translation* (b) or *dilation* (expansion/contraction) (a). It is also an antisymmetric function: the first half of each function is always equal to but opposite in sign to the second half. A very simple example is a function, the first half consisting of -1s and the second of $+1$s. Consider a small spectrum 8 data points in width. A basic wavelet function consists of four -1s and four $+1$s ($a = 1$). This covers the entire spectrum and is said to be a wavelet of level 0. It is completely expanded and there is no room to translate this function. The function can be halved in size ($a = 2$), to give a wavelet of level 1. This can now be translated (changing b), so there are two possible wavelets of level 1 each consisting of two -1s and two $+1$s covering half the spectral width.

Wavelets are often denoted by $\{n, m\}$ where n is the level and m the translation. The lower the level, the wider the wavelet.

Seven possible wavelets for an 8-point series are presented in Table 4.5. The smallest is a 2-point wavelet. It can be seen that for a series consisting of 2^N points:

- there will be N levels numbered from 0 to $N - 1$;
- there will be 2^n wavelets at level n;
- and $2^N - 1$ wavelets in total if all levels are used.

Table 4.5 Some very simple wavelets

Level 0	-1	-1	-1	-1	1	1	1	1	$\{0, 1\}$
Level 1	-1	-1	1	1					$\{1, 1\}$
					-1	-1	1	1	$\{1, 2\}$
Level 2	-1	1							$\{2, 1\}$
			-1	1					$\{2, 2\}$
					-1	1			$\{2, 3\}$
							-1	1	$\{2, 4\}$

Of course, the function of Table 4.5 is not very useful, and there have been a number of more sophisticated wavelets proposed in the literature, often named after their originators. There is much interest in determining the optimal wavelet for any given purpose.

The key to wavelet transforms is that it is possible to express a dataset in terms of a sum of wavelets. For a spectrum 512 data points long, this includes up to 511 possible wavelets as described above, plus a scaling factor. This result of the transform is sometimes expressed by:

$$\boldsymbol{h} = \boldsymbol{W} \boldsymbol{f}$$

where $\boldsymbol{f}$ is the raw data (e.g. a spectrum of 512 points in time), $\boldsymbol{W}$ is a square matrix (with dimensions 512×512 in our example) and $\boldsymbol{h}$ are the coefficients of the wavelets, the calculation determining the best fit coefficients.

For more details about matrices see Section 2.4. It is beyond the scope of this book to provide details as to how to obtain $\boldsymbol{W}$ but many excellent papers exist on this topic.

There are two principal uses for wavelets in the analysis of chemical spectroscopic or chromatographic data.

- The first involves smoothing. If the original data consists of 512 data points, and is exactly fitted by 512 wavelets, choose the most significant wavelets (those with the highest coefficients), e.g. the top 50. In fact if the nature of the wavelet function is selected with care only a small number of such wavelets may be necessary to model a spectrum which, in itself, consists of only a small number of peaks. Replace the spectrum simply with that obtained using the most significant wavelets, or the wavelet coefficients.
- The second involves data compression. Instead of storing all the raw data, store simply the coefficients of the most significant wavelets. This is equivalent to saying that if a spectrum is recorded over 1024 data points but consists of only five overlapping Gaussians, it is more economical (and, in fact useful to the chemist) to store the parameters for the Gaussians rather than the raw data. In certain areas, such as liquid chromatography-mass spectrometry (LC-MS), there is a huge redundancy of data, most mass numbers having no significance, and many data matrices being extremely sparse. Hence it is quite useful to reduce the amount of information, and why not do this using the most important wavelets?

Wavelets are a computationally sophisticated method for achieving these two aims and are an area of active research within the data analytical community.

An excellent review published by Walczak and Massart [7] provides more information about the properties and uses of wavelets.

REFERENCES

1. A. Savitzky, and M.J.E. Golay, Smoothing + differentiation of data by simplified least squares procedures, *Analytical Chemistry*, 36 (1964), 1627–1639
2. J. Grznar, D.E. Booth, P. Sebastian and A. Robust, Smoothing approach to statistical process control, *Journal of Chemical Information and Computer Science*, 37 (1997), 241–248
3. W.H. Press, B.P. Flannery, S.A. Teukolsky and W.T. Vetterling, *Numerical Recipes in C, Second Edition*, Cambridge University Press, Cambridge, 1993; W.H. Press, B.P. Flannery, S.A. Teukolsky

and W.T. Vetterling, *Numerical Recipes in Fortran 77*, *Second Edition*, Cambridge University Press, Cambridge, 1993

4. B. Buck and V.A. Macauley (editors), *Maximum Entropy in Action*, Clarendon Press, Oxford, 1991
5. P.A. Lynn and W. Fuerst, *Introductory Digital Signal Processing with Computer Applications*, John Wiley & Sons, Ltd, Chichester, 1989
6. A. Felinger, *Data Analysis and Signal Processing in Chromatography*, Elsevier, Amsterdam, 1998
7. B. Walczak and D.L. Massart, Noise suppression and signal compression using the wavelet packet transform, *Chemometrics Intelligent Laboratory Systems*, 36 (1997), 81–94

5

Pattern Recognition

5.1 INTRODUCTION

One of the first and most publicized success stories in chemometrics is pattern recognition. Much chemistry involves using data to determine patterns. For example, can infrared (IR) spectra be used to classify compounds into ketones and esters? Is there a pattern in the spectra allowing physical information to be related to chemical knowledge? There have been many spectacular successes of chemical pattern recognition. Can a spectrum be used in forensic science, for example to determine the cause of a fire? Can a chromatogram be used to decide on the origin of a wine and, if so, what main features in the chromatogram distinguish different wines? And is it possible to determine the time of year the vine was grown? Is it possible to use measurements of heavy metals to discover the source of pollution in a river?

There are several groups of methods for chemical pattern recognition.

5.1.1 Exploratory Data Analysis

Exploratory Data Analysis (EDA) consists mainly of the techniques of Principal Components Analysis (PCA) and Factor Analysis (FA). The statistical origins are in biology and psychology. Psychometricians have for many years had the need to translate numbers such as answers to questions in tests into relationships between individuals. How can verbal ability, numeracy and the ability to think in three dimensions be predicted from a test? Can different people be grouped by these abilities? And does this grouping reflect the backgrounds of the people taking the test? Are there differences according to educational background, age, sex, or even linguistic group?

In chemistry, we, too, need to ask similar questions, but the raw data are normally chromatographic or spectroscopic. An example involves chemical communication between animals: animals recognize each other more by smell than by sight, and different animals often lay scent trails, sometimes in their urine. The chromatogram of a urine sample may contain several hundred compounds, and it is often not obvious to the untrained observer which are most significant. Sometimes the most potent compounds are only present in small quantities. Yet animals can often detect through scent marking whether there is an in-heat member of the opposite sex looking for a mate, or whether there is a dangerous intruder

Applied Chemometrics for Scientists   R. G. Brereton
© 2007 John Wiley & Sons, Ltd

entering their territory. EDA of chromatograms of urine samples can highlight differences in chromatograms of different social groups or different sexes, and give a simple visual idea as to the main relationships between these samples.

5.1.2 Unsupervised Pattern Recognition

A more formal method of treating samples is unsupervised pattern recognition, often called cluster analysis. Many methods have their origins in numerical taxonomy. Biologists measure features in different organisms, for example various body length parameters. Using a couple of dozen features, it is possible to see which species are most similar and draw a picture of these similarities, such as a dendrogram, phylogram or cladogram, in which more closely related species are closer to each other. The main branches can represent bigger divisions, such as subspecies, species, genera and families.

These principles can be directly applied to chemistry. It is possible to determine similarities in amino acid sequences in myoglobin in a variety of species, for example. The more similar the species, the closer the relationship: chemical similarity mirrors biological similarity. Sometimes the amount of information is huge, for example in large genomic or crystallographic databases such that cluster analysis is the only practicable way of searching for similarities.

Unsupervised pattern recognition differs from exploratory data analysis in that the aim of the methods are to detect similarities, whereas using EDA there is no particular prejudice as to whether or how many groups will be found. This chapter will introduce these approaches which will be expanded in the context of biology in Chapter 11.

5.1.3 Supervised Pattern Recognition

There are many reasons for supervised pattern recognition, mostly aimed at classification. Multivariate statisticians have developed a large number of discriminant functions, many of direct interest to chemists. A classic example is the detection of forgery in banknotes. Can physical measurements such as width and height of a series of banknotes be used to identify forgeries? Often one measurement is not enough, so several parameters are required before an adequate mathematical model is available.

Equivalently in chemistry, similar problems occur. Consider using a chemical method such as IR spectroscopy to determine whether a sample of brain tissue is cancerous or not. A method can be set up in which the spectra of two groups, cancerous and noncancerous tissues, are recorded: then some form of mathematical model is set up and finally the diagnosis of an unknown sample can be predicted.

Supervised techniques require a training set of known groupings to be available in advance, and try to answer a precise question as to the class of an unknown sample. It is, of course, always first necessary to establish whether chemical measurements are actually good enough to fit into the predetermined groups. However, spectroscopic or chromatographic methods for diagnosis are often much cheaper than expensive medical tests, and provide a valuable first diagnosis. In many cases chemical pattern recognition can be performed as a form of screening, with doubtful samples being subjected to more sophisticated tests. In areas such as industrial process control, where batches of compounds might be produced at hourly intervals, a simple on-line spectroscopic test together with chemical data analysis is often an essential first step to determine the possible acceptability of a batch. The methods in this chapter are expanded in Chapter 10 in the context of biology and medicine, together with several additional techniques.

5.2 PRINCIPAL COMPONENTS ANALYSIS

5.2.1 Basic Ideas

PCA is probably the most widespread multivariate statistical technique used in chemometrics, and because of the importance of multivariate measurements in chemistry, it is regarded by many as the technique that most significantly changed the chemist's view of data analysis.

There are numerous claims to the first use of PCA in the literature. Probably the most famous early paper was by Pearson in 1901 [1]. However, the fundamental ideas are based on approaches well known to physicists and mathematicians for much longer, namely those of eigen-analysis. In fact, some school mathematics syllabuses teach ideas about matrices which are relevant to modern chemistry. An early description of the method in physics was by Cauchy in 1829 [2]. It has been claimed that the earliest nonspecific reference to PCA in the chemical literature was in 1878 [3], although the author of the paper almost certainly did not realize the potential, and was dealing mainly with a simple problem of linear calibration. It is generally accepted that the revolution in the use of multivariate methods took place in psychometrics in the 1930s and 1940s of which Hotelling's paper is regarded as a classic [4]. An excellent more recent review of the area with a historical perspective, available in the chemical literature has been published by the Emeritus Professor of Psychology from the University of Washington, Paul Horst [5].

Psychometrics is well understood to most students of psychology and one important area involves relating answers in tests to underlying factors, for example, verbal and numerical ability as illustrated in Figure 5.1. PCA relates a data matrix consisting of these answers to a number of psychological 'factors'. In certain areas of statistics, ideas of factor analysis and PCA are intertwined, but in chemistry both approaches have a different meaning.

Natural scientists of all disciplines, from biologists, geologists and chemists have caught on to these approaches over the past few decades. Within the chemical community the first major applications of PCA were reported in the 1970s, and form the foundation of many modern chemometric methods.

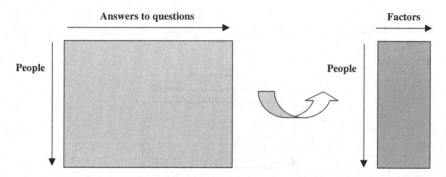

Figure 5.1 Typical psychometric problems.

A key idea is that most chemical measurements are inherently *multivariate*. This means that more than one measurement can be made on a single sample. An obvious example is spectroscopy: we can record a spectrum at hundreds of wavelengths on a single sample. Traditional approaches are *univariate* in which only one wavelength (or measurement) is used per sample, but this misses much information. Another common area is quantitative structure – property relationships, in which many physical measurements are available on a number of candidate compounds (bond lengths, dipole moments, bond angles, etc.); can we predict, *statistically*, the biological activity of a compound? Can this assist in pharmaceutical drug development? There are several pieces of information available. PCA is one of several multivariate methods that allows us to explore patterns in these data, similar to exploring patterns in psychometric data. Which compounds behave similarly? Which people belong to a similar group? How can this behaviour be predicted from available information?

As an example, Figure 5.2 represents a chromatogram in which a number of compounds are detected with different elution times, at the same time as their spectra [such as an ultraviolet (UV)/visible or mass spectrum] are recorded. Coupled chromatography, such as diode array high performance chromatography or liquid chromatography mass spectrometry, is increasingly common in modern laboratories, and represents a rich source of multivariate data. The chromatogram can be represented as a data matrix.

What do we want to find out about the data? How many compounds are in the chromatogram would be useful information. Partially overlapping peaks and minor impurities are the bugbears of modern chromatography. What are the spectra of these compounds? Figure 5.3 represents some coeluting peaks. Can we reliably determine their spectra? By looking at changes in spectral information across a coupled chromatogram, multivariate methods can be employed to resolve these peaks and so find their spectra. Finally, what are the quantities of each component? Some of this information could undoubtedly be obtained by better chromatography, but there is a limit, especially with modern trends to recording more and more data, more and more rapidly. In many cases the identities and amounts of unknowns may not be available in advance. PCA is one tool from multivariate statistics that can help sort out these data. Chapter 7 expands on some of these methods in the context of coupled chromatography, whereas the discussion in this chapter is restricted primarily to exploratory approaches.

The aims of PCA are to determine underlying information from multivariate raw data. There are two principal needs in chemistry.

The first is to interpret the Principal Components (PCs) often in a quantitative manner.

- The number of significant PCs. In the case of coupled chromatography this could relate to the number of compounds in a chromatogram, although there are many other requirements.

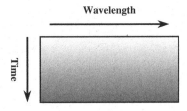

Figure 5.2 Typical multivariate chromatographic information

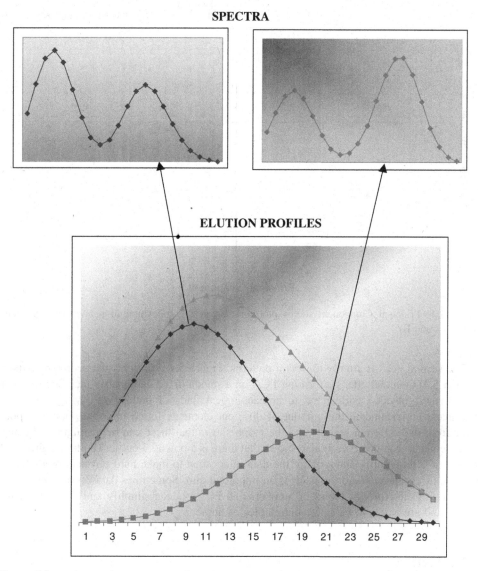

Figure 5.3 Using a chromatogram of partially overlapping peaks to obtain spectra of individual compounds

- The characteristics of each PC, usually the *scores* relating to the objects or samples (in the example of coupled chromatography, the elution profiles) and the *loadings* relating to the variables or measurements (in coupled chromatography, the spectra).

In the next section we will look in more detail how this information is obtained. Often this information is then related to physically interpretable parameters of direct interest to the chemist, or is used to set up models for example to classify or group samples. The numerical information is interesting and can be used to make predictions often of the origin or nature of samples.

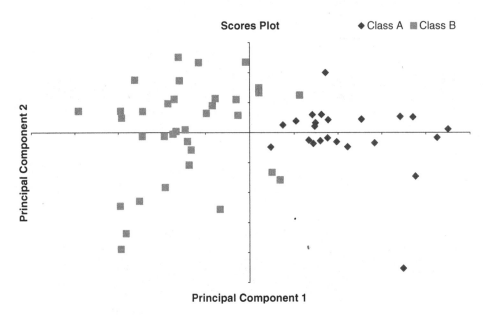

Figure 5.4 Principal Component scores plot for elemental composition of pots in two different groups (A and B)

The second need is simply to obtain patterns. Figure 5.4 represents the scores plots (see Section 5.3) obtained after performing PCA on a standardized data matrix (see Section 5.5) whose rows (objects) correspond to archaeological finds of pottery and whose columns (variables) correspond to the amount of different elements found in these pots. The pots come from two different regions and the graph shows that these can be distinguished using their elemental composition. It also shows that there is a potential outlier (bottom right). The main aim is to simplify and explore the data rather than to make hard physical predictions, and graphical representation in itself is an important aim. Sometimes datasets are very large or difficult to interpret as tables of numbers and PC plots can simplify and show the main trends, and are easier to visualize than tables of numbers.

5.2.2 Method

In order to become familiar with the method it is important to appreciate the main ideas behind PCA. Although chemists have developed their own terminology, it is essential to recognize that similar principles occur throughout scientific data analysis, whether in physics, quantum mechanics or psychology.

As an illustration, we will use the case of coupled chromatography, such as diode array high performance liquid chromatography (HPLC). For a simple chromatogram, the underlying dataset can be described as a sum of responses for each significant compound in the data, which are characterized by (a) an elution profile and (b) a spectrum, plus noise or instrumental error. In matrix terms, this can be written as:

$$X = C.S + E \tag{5.1}$$

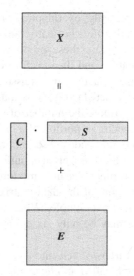

Figure 5.5 Multivariate data such as occurs in diode array high performance liquid chromatography

where X is the original data matrix or coupled chromatogram, C is a matrix consisting of the elution profiles of each compound, S is a matrix consisting of the spectra of each compound and E is an error matrix.

This is illustrated in Figure 5.5. For those not expert in coupled chromatography, this example is of a data matrix each of whose dimensions correspond to variables related sequentially, one in time (chromatography) and one in frequency (spectroscopy), and each compound in the mixture has a characteristic time/frequency profile. The data matrix consists of a combination of signals from each constituent, mixed together, plus noise.

Consider a two way chromatogram recorded over 10 min at 1 s intervals (600 points in time), and over 200 nm at 2 nm intervals (100 spectroscopic points), containing three underlying compounds:

- X is a matrix of 600 rows and 100 columns;
- C is a matrix of 600 rows and 3 columns, each column corresponding to the elution profile of a single compound;
- S is a matrix of 3 rows and 100 columns, each row corresponding to the spectrum of a single column;
- E is a matrix of the same size as X.

For more on matrices see Section 2.4.

If we observe X, can we then predict C and S? Many chemometricians use a 'hat' notation to indicate a *prediction* so it is also possible to write Equation (5.1) as:

$$X \approx \hat{C}.\hat{S}$$

Ideally the predicted spectra and chromatographic elution profiles are close to the true ones, but it is important to realize that we can *never directly or perfectly* observe the underlying data. There will always be measurement error even in practical spectroscopy.

Chromatographic peaks may be partially overlapping or even embedded meaning that chemometric methods will help resolve the chromatogram into individual components.

One aim of chemometrics is to obtain these predictions after first treating the chromatogram as a multivariate data matrix, and then performing PCA. Each compound in the mixture can be considered a 'chemical' factor with its associated spectra and elution profile, which can be related to PCs, or 'abstract' factors, by a mathematical transformation.

A fundamental first step is to determine the number of significant factors or components in a matrix. In a series of mixture spectra or portion of a chromatogram, this should, ideally, correspond to the number of compounds under observation.

The *rank* of a matrix relates to the number of significant components in the data, in chemical terms to the number of compounds in a mixture. For example, if there are six components in a chromatogram the rank of the data matrix from the chromatogram should ideally equal 6. However, life is never so simple. What happens is that noise distorts this ideal picture, so even though there may be only six compounds, it may sometimes appear that the rank is 10 or more.

Normally the data matrix is first transformed into a number of PCs and the *size* of each component is measured. This is often called an *eigenvalue*: the earlier (or more significant) the components, the larger their size. It is possible to express eigenvalues as a percentage of the entire data matrix, by a simple technique.

- Determine the sum of squares of the entire data, S_{total}.
- For each PC determine its own sum of squares (which is usually equal to the sum of squares of the scores vector as discussed below), S_k for the kth component. This is a common definition the *eigenvalue* although there is other terminology in the literature.
- Determine the percentage contribution of each PC to the data matrix ($100 S_k / S_{total}$). Sometimes the *cumulative* contribution is calculated.

Note that there are several definitions of eigenvalues, and many chemometricians have adopted a rather loose definition, that of the sum of squares of the scores, this differs from the original formal definitions in the mathematical literature. However, there is no universally agreed set of chemometrics definitions, every group or school of thought has their own views.

One simple way of determining the number of significant components is simply by the looking at the size of each successive eigenvalue. Table 5.1 illustrates this. The total sum of squares for the entire dataset happens to be 670, so since the first three PCs account for around 95 % of the data (or 639/670), so it is a fair bet that there are only three components in the data. There are, of course, more elaborate approaches to estimating the number of significant components, to be discussed in more detail in Section 5.10.

The number of nonzero components will never be more than the smaller of the number of rows and columns in the original data matrix X. Hence if this matrix consists of 600 rows

Table 5.1   Illustration of size of eigenvalues in Principal Component Analysis

Total		PC1	PC2	PC3	PC4	PC5
670	Eigenvalue	300	230	109	20	8
	%	44.78	34.34	16.27	2.99	1.19
	Cumulative %	44.78	79.11	95.37	98.36	99.55

(e.g. chromatographic elution times) and 100 columns (e.g. spectral wavelengths), there will never be more than 100 nonzero eigenvalues, but, hopefully, the true answer will be very much smaller, reflecting the number of compounds in the chromatogram.

PCA results in an abstract mathematical transformation of the original data matrix, which, for the case of a coupled chromatogram, may take the form:

$$\mathbf{X} \approx \hat{C}.\hat{S} = T.P$$

where T are called the scores, and ideally have the same dimensions as C, and P the loadings ideally having the same dimensions as S. A big interest is how to relate the abstract factors (scores and loadings) to the chemical factors, and Sections 7.8, 7.9, 8.1, 8.3 and 8.4 will introduce a number of techniques in various applications. Note that the product and number of abstract factors should ideally equal the product and number of chemical factors. Purely numerical techniques can be use to obtain the abstract factors.

Each scores matrix consists of a series of column vectors, and each loadings matrix a series of row vectors, the number of such vectors equalling the rank of the original data matrix, so if the rank of the original data matrix is 8 and spectra are recorded at 100 wavelengths, the loadings matrix consists of 8 row vectors 100 data points in length. Many authors denote these vectors by t_a and p_a where a is the number of the PC (1, 2, 3 up to the matrix rank). The scores matrices T and P are composed of several such vectors, one for each PC. If we want to interpret these vectors, these can be related to the true spectra and elution profiles by the transformations discussed in Chapter 7 in greater detail. In many areas of chemical pattern recognition, of course, the scores and loadings are an end in themselves and no further transformation to physical factors is required.

Scores and loadings have important properties, the main one being called *orthogonality* (introduced also in Sections 2.6, 2.8 and 2.9). This is often expressed in a number of ways:

- The product between any two loadings or scores vectors is 0.
- The correlation coefficient between any two loadings or scores vectors is 0 providing they are centred.

The original variables (e.g. 100 wavelengths) are reduced to a number of significant PCs (e.g. 3 or 4) each of which is orthogonal to each other. In practice, PCA has acted as a form

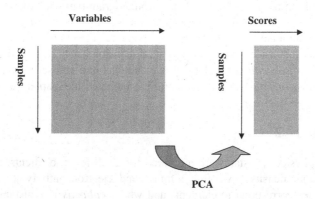

Figure 5.6 Overview of data simplification by PCA in chemistry

of variable reduction, reducing the large original dataset (e.g. recorded at 100 wavelengths) to a much smaller more manageable dataset (e.g. consisting of three PCs) which can be interpreted more easily, as illustrated in Figure 5.6. The loadings represent the means to this end.

The loadings vectors for each component are also generally *normalized*, meaning that their sum of squares equals one, whereas the sum of squares of the scores vectors are often equal to the corresponding eigenvalue. There are of course several different PC algorithms and not everyone uses the same scaling methods, however orthogonality is always obeyed.

5.3 GRAPHICAL REPRESENTATION OF SCORES AND LOADINGS

Many revolutions in chemistry relate to the graphical presentation of information. For example, fundamental to the modern chemist's way of thinking is the ability to draw structures on paper in a convenient and meaningful manner. Years of debate preceded the general acceptance of the Kekulé structure for benzene: today's organic chemist can write down and understand complex structures of natural products without the need to plough through pages of numbers of orbital densities and bond lengths. Yet, underlying these representations are quantum mechanical probabilities, so the ability to convert from numbers to a simple diagram has allowed a large community to think clearly about chemical reactions.

So with statistical data, and modern computers, it is easy to convert from numbers to graphs. Many modern multivariate statisticians think geometrically as much as numerically, and concepts such as PCs are often treated as much as objects in an imaginary space than mathematical entities. The algebra of multidimensional space is the same as that of multivariate statistics. Older texts, of course, were written before the days of modern computing, so the ability to produce graphs was more limited. However, now it is possible to obtain a large number of graphs rapidly using simple software. There are many ways of visualizing PCs and this section will illustrate some of the most common.

We will introduce two case studies.

5.3.1 Case Study 1

The first relates to the resolution of two compounds (I=2-hydroxypyridine and II=3-hydroxypyridine) by diode array HPLC. The chromatogram (summed over all wavelengths) is illustrated in Figure 5.7. More details are given in Dunkerley *et al.* [6]. The aim is to try to obtain the individual profiles of each compound in the chromatogram, and also their spectra. Remember that a second, spectroscopic, dimension has been recorded also. The raw data are a matrix whose *columns* relate to wavelengths and whose *rows* relate to elution time. Further discussions of data of this nature are included in Chapter 7.

5.3.2 Case Study 2

In this case five physical constants are measured for 27 different elements, namely melting point, boiling point, density, oxidation number and electronegativity, to form a 27×5 matrix, whose *rows* correspond to elements and whose *columns* to constants. The data are presented in Table 5.2. The aims are to see which elements group together and also which

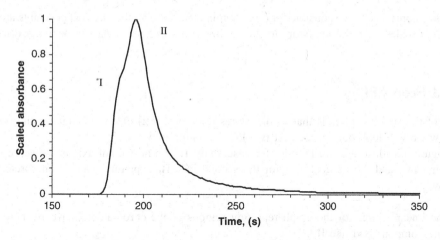

Figure 5.7 Chromatographic profile for case study 1

Table 5.2 Case study 2

Element	Group	Melting point (K)	Boiling) point K	Density (mg/cm³)	Oxidation number	Electronegativity
Li	1	453.69	1615	534	1	0.98
Na	1	371	1156	970	1	0.93
K	1	336.5	1032	860	1	0.82
Rb	1	312.5	961	1530	1	0.82
Cs	1	301.6	944	1870	1	0.79
Be	2	1550	3243	1800	2	1.57
Mg	2	924	1380	1741	2	1.31
Ca	2	1120	1760	1540	2	1
Sr	2	1042	1657	2600	2	0.95
F	3	53.5	85	1.7	−1	3.98
Cl	3	172.1	238.5	3.2	−1	3.16
Br	3	265.9	331.9	3100	−1	2.96
I	3	386.6	457.4	4940	−1	2.66
He	4	0.9	4.2	0.2	0	0
Ne	4	24.5	27.2	0.8	0	0
Ar	4	83.7	87.4	1.7	0	0
Kr	4	116.5	120.8	3.5	0	0
Xe	4	161.2	166	5.5	0	0
Zn	5	692.6	1180	7140	2	1.6
Co	5	1765	3170	8900	3	1.8
Cu	5	1356	2868	8930	2	1.9
Fe	5	1808	3300	7870	2	1.8
Mn	5	1517	2370	7440	2	1.5
Ni	5	1726	3005	8900	2	1.8
Bi	6	544.4	1837	9780	3	2.02
Pb	6	600.61	2022	11340	2	1.8
Tl	6	577	1746	11850	3	1.62

physical constants are responsible for this grouping. Because all the physical constants are on different scales, it is first necessary to standardize (Section 5.5) the data prior to performing PCA.

5.3.3 Scores Plots

One of the simplest plots is that of the scores (Section 5.2.2) of one PC against the other. Below we will look only at the first two PCs, for simplicity.

Figure 5.8 illustrates the PC plot for case study 1. The horizontal axis is the scores for the first PC, and the vertical axis for the second PC. This 'picture' can be interpreted as follows:

- The linear regions of the graph represent regions of the chromatogram where there are pure compounds, I and II.
- The curve portion represents a region of coelution.
- The closer to the origin, the lower the intensity.

Hence the PC plot suggests that the region between 187 and 198 s (approximately) is one of coelution. The reason why this method works is that the spectrum over the chromatogram changes with elution time. During coelution the spectral appearance changes most, and PCA uses this information.

How can these graphs help?

- The pure regions can inform us of the spectra of the pure compounds.
- The shape of the PC plot informs us of the amount of overlap and quality of chromatography.
- The number of bends in a PC plot can provide information about the number of different compounds in a complex multipeak cluster.

Figure 5.9 illustrates the scores plot for case study 2. We are not in this case trying to determine specific factors or pull out spectra, but rather to determine where the main

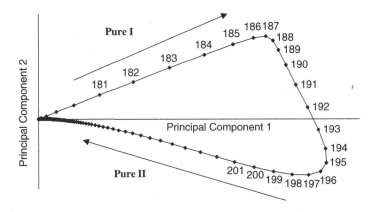

Figure 5.8 Scores plot for the chromatographic data of case study 1: the numbers refer to elution times (in s)

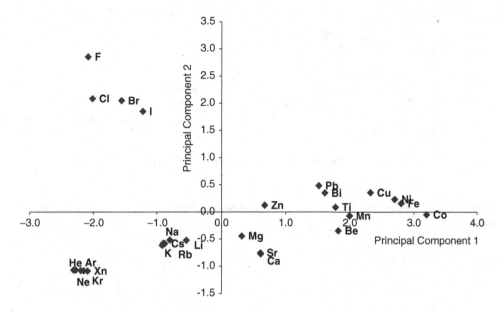

Figure 5.9 Scores plot of the first two PCs for case study 2

groupings are. We can see that the halides cluster together at the top left, and the inert gases in the bottom left. The metals are primarily clustered according to their groups in the periodic table. This suggests that there are definitive patterns in the data which can be summarized graphically using PCs. Many more statistically based chemometricians often do not particularly like these sort of graphical representations which cannot very easily be related to physical factors, but they are nevertheless an extremely common way of summarizing complex data, which we will use in several contexts later in this book.

5.3.4 Loadings Plots

It is not, however, only the scores that are of interest but sometimes the loadings. Exactly the same principles apply in that the value of the loadings at one PC can be plotted against that at the other PC. The result for case study 1 is shown in Figure 5.10. This figure looks quite complicated, this is because both spectra overlap and absorb at similar wavelengths, and should be compared with the scores plot of Figure 5.8, the pure compounds lie in the same directions. The pure spectra are presented in Figure 5.11. Now we can understand these graphs a little more:

- High wavelengths, above 325 nm belong mainly to compound I and are so along the direction of pure I.
- 246 nm is a wavelength where the ratio of absorbance of compound I to II is a maximum, whereas for 301 nm, the reverse is true.

More interpretation is possible, but it can easily be seen that the loadings plots provide detailed information about which wavelengths are most associated with which compound.

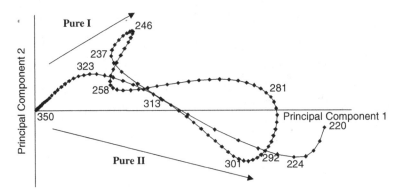

Figure 5.10 Loadings plot for case study 1 (compare with Figure 5.8): wavelengths (in nm) are indicated

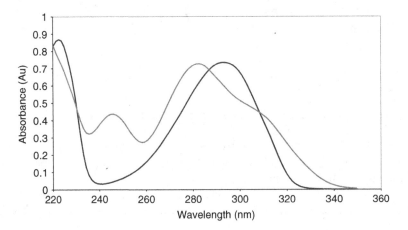

Figure 5.11 Spectra of the two pure compounds in case study 1

The loadings plot for case study 2 is illustrated in Figure 5.12. We can see several features. The first is that melting point, boiling point and density seem closely clustered. This suggests that these three parameters measure something very similar, which is unsurprising, as the higher the melting point, the higher the boiling point in most cases. The density (at room temperature) should have some relationship to melting/boiling point also particularly whether an element is in gaseous, liquid or solid state. We can see that electronegativity is in quite a different place, almost at right angles to the density/boiling/melting point axis, and this suggests it follows entirely different trends.

We can see also that there are relationships between scores and loadings in case study 2. The more dense, high melting point, elements are on the right in the scores plot, and the more electronegative elements at the top end, so we can look at which variable influences which object by looking at both plots together, as discussed.

Loadings plots can be used to answer a lot of questions about the data, and are a very flexible facility available in almost all chemometrics software.

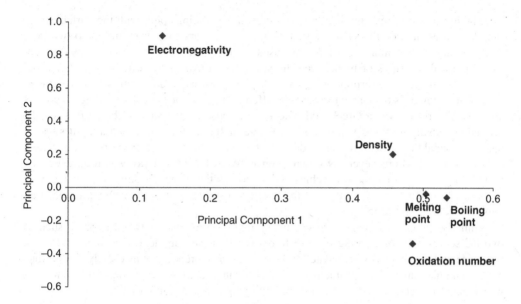

Figure 5.12   Loadings plot for case study 2

5.3.5 Extensions

In many cases, more than two significant PCs are necessary to adequately characterize the data, but the same concepts can be employed, except there are many more possible graphs. For example, if four significant components are calculated, we can produce six possible graphs, of each possible combination of PCs, for example, PC 4 versus 2, or PC 1 versus 3 and so on. Each graph could reveal interesting trends. It is also possible to produce three-dimensional PC plots, whose axes consist of the scores or loadings of three PCs (normally the first three) and so visualize relationships between and clusters of variables in three-dimensional space.

5.4 COMPARING MULTIVARIATE PATTERNS

PC plots are often introduced only by reference to the independent loadings or scores plot of a single dataset. Yet there are common patterns within different graphs. Consider taking measurements of the concentration of a mineral in a geochemical deposit. This information could be presented as a table of sampling sites and observed concentrations. A much more informative approach would be to produce a picture in which physical location and mineral concentration are superimposed, such as a coloured map, each different colour corresponding to a concentration range of the mineral. Two pieces of information are connected, namely geography and concentration. So in many areas of multivariate analysis, one aim may be to connect the samples (e.g. geographical location/sampling site) represented by scores, to the variables (e.g. chemical measurements), represented by loadings. Graphically this requires the superimposition of two types of information.

A biplot involves superimposition of a scores and a loadings plot, with the variables and samples represented on the same diagram. It is not necessary to restrict biplots to two PCs, but, of course, when more than three are used, graphical representation becomes difficult, and numerical measures of fit between the scores and loadings are often employed, using statistical software, to determine which variables are best associated with which samples.

A different need is to be able to compare different types of measurements using procrustes analysis. Procrustes was a Greek god who kept a house by the side of the road where he offered hospitality to passing strangers, who were invited in for a meal and a night's rest in his very special bed which Procrustes described as having the unique property that its length exactly matched whomsoever lay down upon it. What he did not say was the method by which this 'one-size-fits-all' was achieved: as soon as the guest lay down Procrustes went to work upon them, stretching them if they were too short for the bed or chopping off their legs if they were too long.

Similarly, procrustes analysis in chemistry involves comparing two diagrams, such as two PC scores plots originating from different types of measurement. One such plot is the reference and the other is manipulated to resemble the reference plot as closely as possible. This manipulation is done mathematically, involving rotating, stretching and sometimes translating the second scores plot, until the two graphs are as similar as possible.

It is not necessary to restrict data from each type of measurement technique to two PCs, indeed in many practical cases four or five PCs are employed. Computer software is available to compare scores plots and provide a numeric indicator of the closeness of the fit. Procrustes analysis can be used to answer quite sophisticated questions. For example, in sensory research, are the results of a taste panel comparable with chemical measurements? If so, can the rather expensive and time-consuming taste panel be replaced by chromatography? A second use of procrustes analysis is to reduce the number of tests: an example being of clinical trials. Sometimes 50 or more bacteriological tests are performed but can these be reduced to 10 or less? A way to check this is by performing PCA on the results of all 50 tests, and compare the scores plot when using a subset of 10 tests. If the two scores plots provide comparable information, the 10 selected tests are just as good as the full set of tests. This can be of significant economic benefit. A final and important application is when several analytical techniques are employed to study a process, an example being the study of a reaction by IR, UV/visible and Raman spectroscopy, does each type of spectrum give similar answers? A consensus can be obtained using procrustes analysis.

5.5 PREPROCESSING

Many users of chemometric software simply accept without much insight the results of PCA: yet interpretation depends critically on how the original data have been handled. Data preprocessing or scaling can have a significant influence on the outcome, and also relate to the chemical or physical aim of the analysis. In fact in many modern areas such as metabolomics (Section 10.10), it is primarily the method for preprocessing that is difficult and influences the end result.

As an example, consider a data matrix consisting of 10 rows (labelled from 1 to 10) and eight columns (labelled from A to H), illustrated in Table 5.3(a). This could represent a portion of a two way diode array HPLC data matrix, whose elution profile in given in Figure 5.13, but similar principles apply to other multivariate data matrices, although the chromatographic example is especially useful for illustrative purposes as both dimensions

Table 5.3 Simple example for Section 5.5. (a) Raw data; (b) column mean centred data;(c) Column standardized data (d) row scaled

(a)

	A	B	C	D	E	F	G	H
1	0.318	0.413	0.335	0.196	0.161	0.237	0.290	0.226
2	0.527	0.689	0.569	0.346	0.283	0.400	0.485	0.379
3	0.718	0.951	0.811	0.521	0.426	0.566	0.671	0.526
4	0.805	1.091	0.982	0.687	0.559	0.676	0.775	0.611
5	0.747	1.054	1.030	0.804	0.652	0.695	0.756	0.601
6	0.579	0.871	0.954	0.841	0.680	0.627	0.633	0.511
7	0.380	0.628	0.789	0.782	0.631	0.505	0.465	0.383
8	0.214	0.402	0.583	0.635	0.510	0.363	0.305	0.256
9	0.106	0.230	0.378	0.440	0.354	0.231	0.178	0.153
10	0.047	0.117	0.212	0.257	0.206	0.128	0.092	0.080

(b)

	A	B	C	D	E	F	G	H
1	−0.126	−0.231	−0.330	−0.355	−0.285	−0.206	−0.175	−0.146
2	0.083	0.045	−0.095	−0.205	−0.163	−0.042	0.020	0.006
3	0.273	0.306	0.146	−0.030	−0.020	0.123	0.206	0.153
4	0.360	0.446	0.318	0.136	0.113	0.233	0.310	0.238
5	0.303	0.409	0.366	0.253	0.206	0.252	0.291	0.229
6	0.135	0.226	0.290	0.291	0.234	0.185	0.168	0.139
7	−0.064	−0.017	0.125	0.231	0.184	0.062	0.000	0.010
8	−0.230	−0.243	−0.081	0.084	0.064	−0.079	−0.161	−0.117
9	−0.338	−0.414	−0.286	−0.111	−0.093	−0.212	−0.287	−0.220
10	−0.397	−0.528	−0.452	−0.294	−0.240	−0.315	−0.373	−0.292

(c)

	A	B	C	D	E	F	G	H
1	−0.487	−0.705	−1.191	−1.595	−1.589	−1.078	−0.760	−0.818
2	0.322	0.136	−0.344	−0.923	−0.909	−0.222	0.087	0.035
3	1.059	0.933	0.529	−0.133	−0.113	0.642	0.896	0.856
4	1.396	1.361	1.147	0.611	0.629	1.218	1.347	1.330
5	1.174	1.248	1.321	1.136	1.146	1.318	1.263	1.277
6	0.524	0.690	1.046	1.306	1.303	0.966	0.731	0.774
7	−0.249	−0.051	0.452	1.040	1.026	0.326	0.001	0.057
8	−0.890	−0.740	−0.294	0.376	0.357	−0.415	−0.698	−0.652
9	−1.309	−1.263	−1.033	−0.497	−0.516	−1.107	−1.247	−1.228
10	−1.539	−1.608	−1.635	−1.321	−1.335	−1.649	−1.620	−1.631

(d)

	A	B	C	D	E	F	G	H
1	0.146	0.190	0.154	0.090	0.074	0.109	0.133	0.104
2	0.143	0.187	0.155	0.094	0.077	0.109	0.132	0.103
3	0.138	0.183	0.156	0.100	0.082	0.109	0.129	0.101
4	0.130	0.176	0.159	0.111	0.090	0.109	0.125	0.099
5	0.118	0.166	0.162	0.127	0.103	0.110	0.119	0.095
6	0.102	0.153	0.167	0.148	0.119	0.110	0.111	0.090
7	0.083	0.138	0.173	0.171	0.138	0.111	0.102	0.084
8	0.066	0.123	0.178	0.194	0.156	0.111	0.093	0.078
9	0.051	0.111	0.183	0.213	0.171	0.112	0.086	0.074
10	0.041	0.103	0.186	0.226	0.181	0.112	0.081	0.071

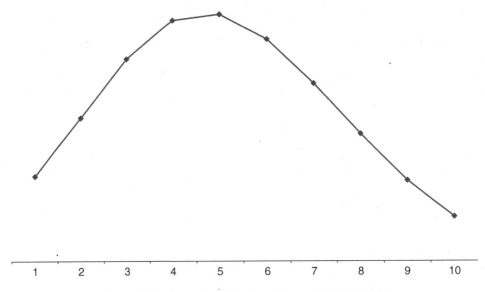

Figure 5.13   Summed profile formed from data in Table 5.3

have an interpretable sequential meaning (which is not necessarily so in most other types of data analysis) and provides a situation that illustrates several different consequences of data preprocessing.

The resultant PC scores and loadings plots are given in Figure 5.14 for the first two PCs. Several deductions are possible, for example:

- There are probably two main compounds, one which has a region of purity between points 1 and 3, and the other between points 8 and 10.
- Measurements (e.g. spectral wavelengths) A, B, G and H correspond mainly to the first (e.g. fastest eluting) chemical component, whereas measurements D and E to the second chemical component.

PCA has been performed directly on the raw data, something statisticians in other disciplines very rarely do. It is important to be very careful when using packages that have been designed primarily by statisticians, on chemical data. Traditionally, what is mainly interesting to statisticians is deviation around a mean, for example, how do the mean characteristics of a forged banknote vary? What is an 'average' banknote? In chemistry we are often (but by no means exclusively) interested in deviation above a baseline, such as in spectroscopy.

It is, though, possible to mean centre the columns. The result of this is presented in Table 5.3(b). Notice now that the sum of each column is now 0. Almost all traditional statistical packages perform this operation prior to PCA whether desired or not. The PC plots are presented in Figure 5.15. The most obvious difference is that the scores plot is now centred around the origin. However, the relative positions of the points in both graphs change slightly, the biggest effect being on the loadings. In practice, mean centring can have quite a large influence in some cases, for example if there are baseline problems or only a small region of the data is recorded.

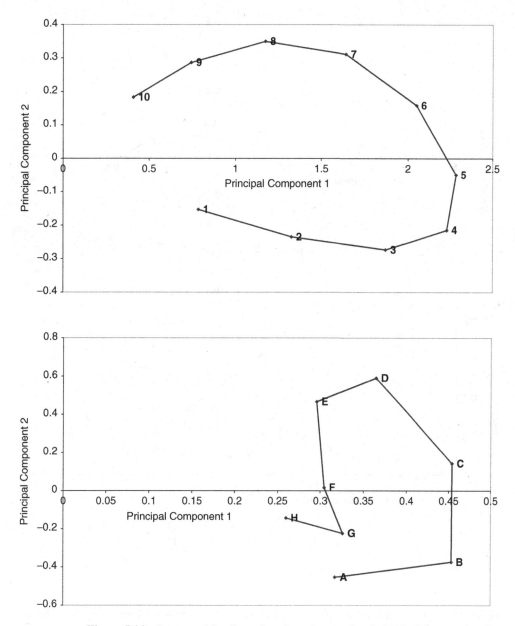

Figure 5.14 Scores and loadings plots from the raw data in Table 5.3

Note that it is also possible to mean centre the rows, and also double mean centre data both simultaneously down the columns and along the rows (see Section 7.2.1 for more details), however, this is rarely done in chemometrics.

Standardization is another common method for data scaling and first requires mean centring: in addition, each variable is also divided by its standard deviation, Table 5.3(c) for our example. This procedure has been discussed in the context of the normal distribution in Section 3.4. Note an interesting feature that the sum of squares of each column equals

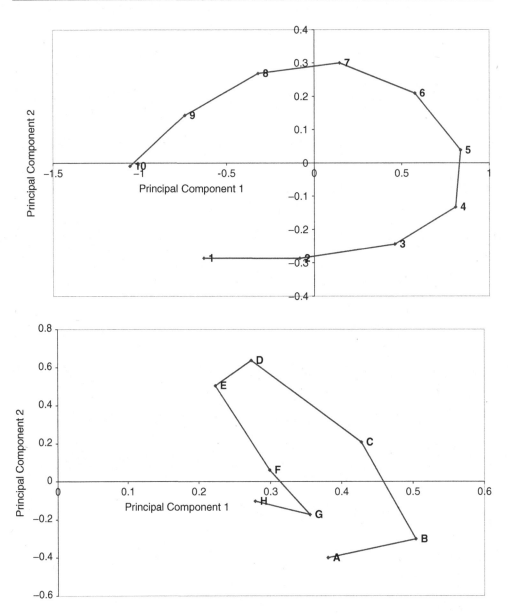

Figure 5.15 Scores and loadings plots corresponding to Figure 5.14 but for column mean centred data

10 in this example (which is the number of objects in the dataset): the population standard deviation (Section 3.3.1) is usually employed as the aim is data scaling and not parameter estimation. Figure 5.16 represents the new PC plots. Whereas the scores plot hardly changes in appearance, there is a dramatic difference in the appearance of the loadings. The reason is that standardization puts all the variables on approximately the same scale. Hence variables (such as wavelengths) of low intensity assume equal significance to those of high intensity,

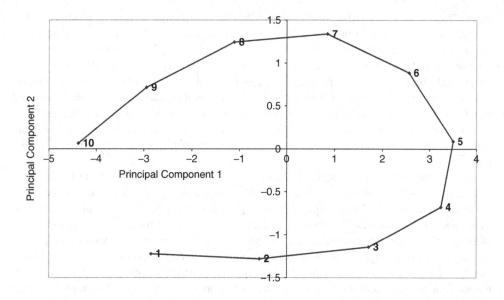

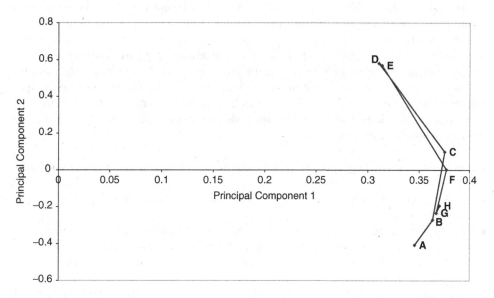

Figure 5.16 Scores and loadings plots corresponding to Figure 5.14 but for column standardized data

and, in this case all variables are roughly the same distance away from the origin, on an approximate circle (this looks distorted simply because the horizontal axis is longer than the vertical one in the graph).

Standardization can be important in many real situations. Consider, for example, a case where the concentrations of 30 metabolites are monitored in a series of organisms. Some metabolites might be abundant in all samples, but their variation is not very significant. The change in concentration of the minor compounds might have a significant relationship to the underlying biology. If standardization is not performed, PCA will be dominated by the most intense compounds. In some cases standardization (or closely similar types of scaling) is essential. In the case of using physical properties to look at relationships between elements as discussed in Section 5.3, each raw variable is measured on radically different scales and standardization is required so that each variable has an equal influence. Standardization is useful in areas such as quantitative structure – property relationships, where many different pieces of information are measured on very different scales, such as bond lengths and dipoles.

Row scaling involves scaling the rows to a constant total, usually 1 or 100 (this is sometimes called normalization but there is a lot of confusion and conflicting terminology in the literature: usually normalization involves the sum of squares rather than the sum equalling 1, as we use for the loadings – see Section 5.2.2). This is useful if the absolute concentrations of samples cannot easily be controlled. An example might be biological extracts: the precise amount of material might vary unpredictably, but the *relative* proportions of each chemical can be measured. Row scaling introduces a constraint which is often called *closure*. The numbers in the multivariate data matrix are proportions and some of the properties have analogy to those of mixtures (Sections 2.12 and 9.5).

The result of row scaling is presented in Table 5.3(d) and the PC plots are given in Figure 5.17. The scores plot appears very different from those of previous figures. The data points now lie on a straight line (this is a consequence of there being exactly two components in this particular dataset and does not always happen). The 'mixed' points are in the centre of the straight line, with the pure regions at the extreme ends. Note that sometimes if extreme points are primarily influenced by noise, the PC plot can be quite distorted, and it can be important to select carefully an appropriate region of the data.

There are a very large battery of other methods for data preprocessing, although the ones described above are the most common.

- It is possible to combine approaches, for example, first to row scale and then standardize a dataset.
- Weighting of each variable according to any external criterion of importance is sometimes employed.
- Logarithmic scaling of measurements might be useful if there are large variations in intensities, although there can be problems if there are missing or zero intensity measurements.
- Selective row scaling over part of the variables can sometimes be used. It is even possible to divide the measurements into blocks and perform row scaling separately on each block. This could be useful if there were several types of measurement, for example, a couple of spectra and one chromatogram, each constituting a single block, and each of equal importance, but recorded on different physical scales.

Undoubtedly, however, the appearance and interpretation not only of PC plots but the result of almost all chemometric techniques, depends on data preprocessing. The influence of preprocessing can be quite dramatic, so it is essential for the user of chemometric software to understand and question how and why the data has been scaled prior to interpreting the result from a package.

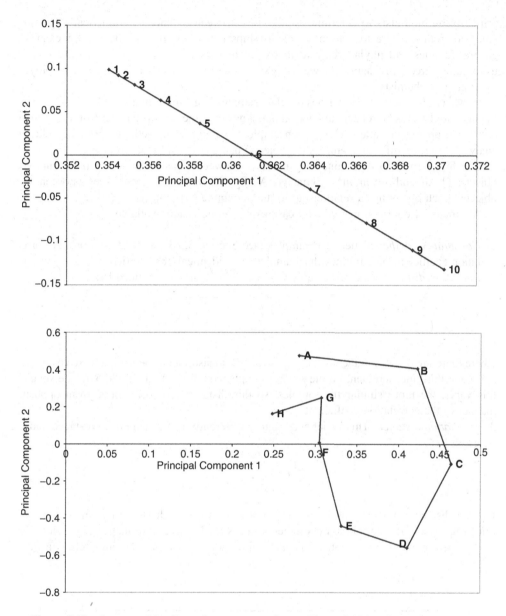

Figure 5.17   Scores and loadings plots corresponding to Figure 5.14 but for row scaled data

5.6 UNSUPERVISED PATTERN RECOGNITION: CLUSTER ANALYSIS

Exploratory data analysis such as PCA is used primarily to determine general relationships between data. Sometimes more complex questions need to be answered such as, do the samples fall into groups? Cluster analysis is a well established approach that was developed primarily by biologists to determine similarities between organisms. Numerical taxonomy emerged from a desire to determine relationships between different species, for example genera, families and phyla. Many textbooks in biology show how organisms are related using family trees. In Chapter 11 we will expand on how cluster analysis can be employed by biological chemists.

However, the chemist also wishes to relate samples in a similar manner. Can the chemical fingerprint of wines be related and does this tell us about the origins and taste of a particular wine? Unsupervised pattern recognition employs a number of methods, primarily cluster analysis, to group different samples (or objects) using chemical measurements.

The first step is to determine the similarity between objects. Table 5.4 represents six objects, (1–6) and seven measurements (A–G). What are the similarities between the objects? Each object has a relationship to the remaining five objects.

A number of common numerical measures of similarity are available.

1. *Correlation coefficient* between samples (see Section 3.3.3 for the definition). A correlation coefficient of 1 implies that samples have identical characteristics.
2. *Euclidean distance*. The distance between samples k and l is defined by:

$$d_{kl} = \sqrt{\sum_{j=1}^{J} (x_{kj} - x_{lj})^2}$$

where there are j measurements, and x_{ij} is the jth measurement on sample i, for example, x_{23} is the third measurement on the second sample, equalling 0.6 in Table 5.4. The smaller this value, the more similar the samples, so this distance measure works in an opposite manner to the correlation coefficient.
3. *Manhattan distance*. This is defined slightly differently to the Euclidean distance and is given by:

$$d_{kl} = \sum_{j=1}^{J} |x_{kj} - x_{lj}|$$

Once a distance measure has been chosen, a similarity (or dissimilarity) matrix can be drawn up. Using the correlation coefficients (measure 1 above), the matrix is presented in Table 5.5, for our dataset. Notice that the correlation of any object with itself is always

Table 5.4 Simple example for cluster analysis; six objects (1–6) and seven variables (A–G)

Objects	Variables						
	A	B	C	D	E	F	G
1	0.9	0.5	0.2	1.6	1.5	0.4	1.5
2	0.3	0.2	0.6	0.7	0.1	0.9	0.3
3	0.7	0.2	0.1	0.9	0.1	0.7	0.3
4	0.5	0.4	1.1	1.3	0.2	1.8	0.6
5	1.0	0.7	2.0	2.2	0.4	3.7	1.1
6	0.3	0.1	0.3	0.5	0.1	0.4	0.2

Table 5.5 Correlation matrix for the six objects in Table 5.4

	1	2	3	4	5	6
1	1					
2	−0.338	1				
3	0.206	0.587	1			
4	−0.340	0.996	0.564	1		
5	−0.387	0.979	0.542	0.990	1	
6	−0.003	0.867	0.829	0.832	0.779	1

1, and that only half the matrix is required, because the correlation of any two objects is always identical no matter which way round the coefficient is calculated. This matrix gives an indication of relationships: for example, object 5 appears very similar to both objects 2 and 4, as indicated by the high correlation coefficient. Object 1 does not appear to have a particularly high correlation with any of the others.

The next step is to link the objects. The most common approach is called *agglomerative* clustering whereby single objects are gradually connected to each other in groups.

- From the raw data, find the two most similar objects, in our case the objects with the highest correlation coefficient (or smallest distance). According to Table 5.5, these are objects 2 and 4, as their correlation coefficient is 0.996.
- Next form a 'group' consisting of the two most similar objects. The original six objects are now reduced to five groups, namely objects 1, 3, 5 and 6 on their own and a group consisting of objects 2 and 4 together.
- The tricky bit is to decide how to represent this new grouping. There are quite a few approaches, but it is common to change the data matrix from one consisting of six rows to a new one of five rows, four corresponding to original objects and one to the new group. The numerical similarity values between this new group and the remaining objects have to be recalculated. There are three principal ways of doing this:
  - *Nearest neighbour.* The similarity of the new group from all other groups is given by the *highest* similarity of either of the original objects to each other object. For example, object 6 has a correlation coefficient of 0.867 with object 2, and 0.837 with object 4. Hence the correlation coefficient with the new combined group consisting of objects 2 and 4 is 0.867.
  - *Farthest neighbour.* This is the opposite to nearest neighbour, and the *lowest* similarity is used, 0.837 in our case.
  - *Average linkage.* The average similarity is used, 0.852 in our case. There are, in fact, two different ways of doing this, according to the size of each group being joined together. Where they are of equal size (e.g. each group consists of one object), both methods are equivalent. The two different ways are as follows. *Unweighted* linkage involves taking the each group size into account when calculating the new similarity coefficient, the more the objects the more significant the similarity measure is whereas *weighted* linkage ignores the group size. The terminology indicates that for the unweighted method, the new similarity measure takes into consideration the number of objects in a group, the conventional terminology possibly being the opposite to what is expected. For the first link, each method provides identical results.

Table 5.6 First step of clustering of data from Table 5.5, with the new correlation coefficients indicated as shaded cells, using nearest neighbour linkage

	1	2 and 4	3	5	6
1	1				
2 and 4	−0.338	1			
3	0.206	0.587	1		
5	−0.387	0.990	0.542	1	
6	−0.003	0.867	0.829	0.779	1

As an illustration, the new data matrix using nearest neighbour clustering is presented in Table 5.6, with the new values shaded. Remember that there are many similarity measures and methods for linking, so this table is only one possible way for handling the information.

The next steps consist of continuing to group the data just as above, until only one group, consisting of all the original objects, remains. Since there are six original objects, there will be five steps before the data are reduced to a single group.

It is normal to then determine at what similarity measure each object joined a larger group, and so which objects resemble each other most.

Often the result of hierarchical clustering is presented in a graphical form called a dendrogram: note that many biologists call this a phylogram and it differs from a cladogram where the size of the branches are the same (see Section 11.4). The objects are organized in a row, according to their similarities: the vertical axis represents the similarity measure at which each successive object joins a group. Using nearest neighbour linkage and correlation coefficients for similarities, the dendrogram is presented in Figure 5.18. It can be seen that object 1 is very different from the others. In this case all the other objects appear to form a single group, but other clustering methods may give slightly different results. A good approach is

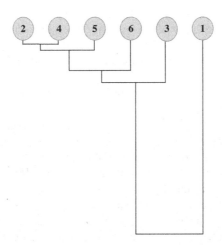

Figure 5.18 Dendrogram for data in Table 5.4, using correlation coefficients as similarity measures and nearest neighbour clustering

to perform several different methods of cluster analysis and compare the results. If similar groupings remain, no matter which method is employed, we can rely on the results.

There are a large number of books on clustering but a well recognized text, for chemists, is written by Massart and Kaufman [7]. Although quite an early vintage chemometrics text, it has survived the passage of time, and there are many clear explanations in this book. More aspects of cluster analysis are discussed in Chapter 11.

5.7 SUPERVISED PATTERN RECOGNITION

Classification (often called supervised pattern recognition) is at the heart of chemistry. Mendeleev's periodic table, the grouping of organic compounds by functionality and listing different reaction types all involve classification. Much of traditional chemistry involves grouping chemical behaviour. Most traditional texts in organic and inorganic chemistry are systematically divided into chapters according to the behaviour or structure of the underlying compounds or elements.

So the modern chemist also has a significant need for classification. Can a spectrum be used to determine whether a compound is a ketone or an ester? Can the chromatogram of a tissue sample be used to determine whether a patient is cancerous or not? Can we record the spectrum of an orange juice and decide its origin? Is it possible to monitor a manufacturing process and decide whether the product is acceptable or not? Supervised pattern recognition is used to assign samples to a number of groups (or classes). It differs from unsupervised pattern recognition (Section 5.6) where, although the relationship between samples is important, there are no predefined groups.

Although there are numerous algorithms in the literature, chemists have developed a common strategy for classification.

5.7.1 Modelling the Training Set

The first step is normally to produce a mathematical model between some measurements (e.g. spectra) on a series of objects and their known groups. These objects are called a *training set*. For example, a training set might consist of the near infrared (NIR) spectra of 30 orange juices, 10 known to be from Spain, 10 known to be from Brazil and 10 known to be adulterated. Can we produce a mathematical equation that predicts which class an orange juice belongs to from its spectrum?

Once this is done it is usual to determine how well the model predicts the groups. Table 5.7 illustrates a possible scenario. Of the 30 spectra, 24 are correctly classified. Some classes are modelled better than others, for example, nine out of 10 of the Spanish orange juices are correctly classified, but only seven of the Brazilian orange juices. A parameter %CC (percentage correctly classified) can be calculated and is 80 % overall. There appears some risk of making a mistake, but the aim of a spectroscopic technique might be to perform screening, and there is a high chance that suspect orange juices (e.g. those adulterated) would be detected, which could then be subject to further detailed analysis. Chemometrics combined with spectroscopy acts like a 'sniffer dog' in a customs checkpoint trying to detect drugs: the dog may miss some cases, and may even get excited when there are no drugs, but there will be a good chance the dog is correct. Proof, however, only comes when the suitcase is opened.

Sometimes the number of false positives or false negatives can be computed as an alternative measure of the quality of a classification technique. This, however, can only be

Table 5.7 Classification ability using a training set

Known	Predicted			Correct	%CC
	Spain	Brazil	Adulterated		
Spain	9	0	1	9	90
Brazil	1	7	2	7	70
Adulterated	0	2	8	8	80
Overall				24	80

done for what is called a 'one class classifier', i.e. one class against the rest. We may however be interested in whether an orange juice is adulterated or not. The data of Table 5.7 suggest that there are three false positives (situations where the orange juice is not adulterated but the test suggests it is adulterated) and two false negatives. This number can often be changed by making the classification technique more or less 'liberal'. A liberal technique lets more samples into the class, so would have the effect of increasing the number of false positives but decreasing the number of false negatives. Ultimately we would hope that a method can be found for which there are no false negatives at the cost of several more false positives. Whether this is useful or not depends a little on the application. If we are, for example, screening people for cancer it is better that we reduce the number of false negatives, so all suspect cases are then examined further. If, however, we are deciding whether to cut a person's leg off due to possible disease it is preferable to err on the side of false negatives so we are very sure when we cut the leg off that it is really necessary. More discussion of this approach is provided in Section 10.5.

5.7.2 Test Sets, Cross-validation and the Bootstrap

It is normal that the training set results in good predictions, but this does not necessarily mean that the method can safely be used to predict unknowns. A recommended second step is to test the quality of predictions often using a *test* set. This is a series of samples that has been left out of the original calculations, and is a bit like a 'blind test'. These samples are assumed to be unknowns at first. Table 5.8 is of the predictions from a test set (which does not necessarily need to be the same size as the training set), and we see that now only 50 % are correctly classified so the model is not particularly good.

Using a test set to determine the quality of predictions is a form of *validation*. The test set could be obtained, experimentally, in a variety of ways, for example, 60 orange juices might be analysed in the first place, and then randomly divided into 30 for the training set and 30 for the test set. Alternatively, the test set could have been produced in an independent laboratory.

An alternative approach is *cross-validation*. Only a single training set is required, but what happens is that one (or a group) of objects is removed at a time, and a model determined on the remaining samples. Then the prediction on the object (or set of objects) left out is tested. The most common approach is Leave One Out (LOO) cross-validation where one sample is left out at a time. This procedure is repeated until all objects have been left out in turn.

Table 5.8 Classification ability using a test set

| | Predicted | | | | |
Known	Spain	Brazil	Adulterated	Correct	%CC
Spain	5	3	2	5	50
Brazil	1	6	3	6	60
Adulterated	4	2	4	4	40
Overall				15	50

For example, it would be possible to produce a class model using 29 out of 30 orange juices. Is the 30th orange juice correctly classified? If so this counts towards the percentage correctly classified. Then, instead of removing the 30th orange juice, we decide to remove the 29th and see what happens. This is repeated 30 times, which leads to a value of %CC for cross-validation. Normally the cross-validated %CC is lower (worse) than the %CC for the training set.

Finally, mention should be made of a third alternative called the *bootstrap* [8]. This is a half way house between cross-validation and having a single independent test set, and involves iteratively producing several internal test sets, not just removing samples once as in cross-validation, but not just having a single test set. A set of samples may be removed for example 50 times, each time including a different combination of the original samples (although the same samples will usually be part of several of these test sets). The prediction ability each time is calculated, and the overall predictive ability is the average of each iteration.

However, if the %CC obtained when samples are left out is similar to the %CC on the training set (sometimes called the autopredictive model), the model is quite probably a good one. Where investigation is necessary is if the %CC is high for the training set but significantly lower when using one of the methods for validation. It is recommended that all classification methods are validated.

Naturally it is also possible to calculate the false positive or false negative rate as well using these, and which criterion is employed to judge whether a method is suitable or not depends very much on the perspective of the scientist.

If the model is not very satisfactory there are a number of ways to improve it. The first is to use a different computational algorithm. The second is to modify the existing method – a common approach might involve wavelength selection in spectroscopy, for example, instead of using an entire spectrum, many wavelengths which are not very meaningful, can we select the most diagnostic parts of the spectrum? Finally, if all else fails, change the analytical technique.

One important final consideration to remember that some people do not always watch out for is that there are two separate reasons for using the techniques described in this section. The first is to optimize a computational model. This means that different models can be checked and the one that gives the best prediction rate is retained. In this way the samples left out are actually used to improve the model. The second is as an independent test of how well the model performs on unknowns. This is a subtly different reason and sometimes both motivations are mixed up, which can lead to over-optimistic predictions of the quality of a model on unknowns, unless care is taken. This can be overcome by dividing the data into a training and test set, but then performing cross-validation or the bootstrap on the training set, to find the best model for the training set and testing its quality on the test set. Using iterative methods, this can be done several times, each time producing a different test set, and the predictive ability averaged.

5.7.3 Applying the Model

Once a satisfactory model is available, it can then be applied to unknown samples, using analytical data such as spectra or chromatograms, to make predictions. Usually by this stage, special software is required that is tailor made for a specific application, and measurement technique. The software will also have to determine whether a new sample really fits into the training set or not. One major difficulty is the detection of outliers that belong to none of the previously studied groups, for example if a Cypriot orange juice sample was measured when the training set consists just of Spanish and Brazilian orange juices. In areas such as clinical or forensic science outlier detection can be quite important, indeed an incorrect conviction or inaccurate medical diagnosis could be obtained otherwise. Multivariate outlier detection is discussed in Section 3.12.3.

Another problem is to ensure stability of the method over time, for example, instruments tend to perform slightly differently every day. Sometimes this can have a serious influence on the classification ability of chemometrics algorithms. One way around this is to perform a small test of the instrument on a regular basis and only accept data if the performance of this test falls within certain limits. However, in some cases such as chromatography this can be quite difficult because columns and instruments do change their performance with time, and this can be an irreversible process that means that there will never be absolutely identical results over a period of several months. In the case of spectroscopy such changes are often not so severe and methods called calibration transfer can be employed to overcome these problems often with great success.

There have been some significant real world successes of using classification techniques, a major area being in industrial process control using NIR spectroscopy. A manufacturing plant may produce samples on a continuous basis, but there are a large number of factors that could result in an unacceptable product. The implications of producing substandard batches may be economical, legal and environmental, so continuous testing using a quick and easy method such as on-line spectroscopy is valuable for rapid detection whether a process is going wrong. Chemometrics can be used to classify the spectra into acceptable or otherwise, and so allow the operator to close down a manufacturing plant in real time if it looks as if a batch can no longer be assigned to the group of acceptable samples.

5.8 STATISTICAL CLASSIFICATION TECHNIQUES

The majority of statistically based software packages contain substantial numbers of procedures, called by various names such as discriminant analysis and canonical variates analysis. It is important to emphasize that good practice requires methods for validation and optimization of the model as described in Section 5.7, together with various classification algorithms as discussed below.

5.8.1 Univariate Classification

The simplest form of classification is *univariate* where one measurement or variable is used to divide objects into groups. An example may be a blood alcohol reading. If a reading on a meter in a police station is above a certain level, then the suspect will be prosecuted for drink driving, otherwise not. Even in such a simple situation, there can be ambiguities, for example measurement errors and metabolic differences between people.

5.8.2 Bivariate and Multivariate Discriminant Models

More often, several measurements are required to determine the group a sample belongs to. Consider performing two measurements, and producing a graph of the values of these measurements for two groups, as in Figure 5.19. The objects denoted by squares are clearly distinct from the objects denoted by circles, but neither of the two measurements, alone, can discriminate between these groups, therefore both are essential for classification. It is, however, possible to draw a line between the two groups. If above the line, an object belongs to the group denoted by circles (class A), otherwise to the group denoted by squares (class B).

Graphically this can be represented by *projecting* the objects onto a line at right angles to the discriminating line as demonstrated in Figure 5.20. The projection can now be converted to a position along a single line (line 2). Often these numbers are converted to a *class distance* which is the distance of each object to the centre of the classes. If the distance to the centre of class A is greater than that to class B, the object is placed in class A and vice versa.

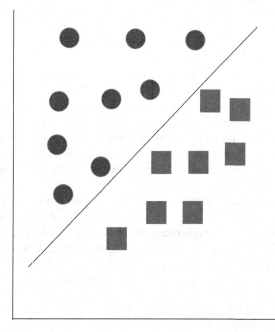

Figure 5.19   Bivariate classification where no measurement alone can distinguish groups

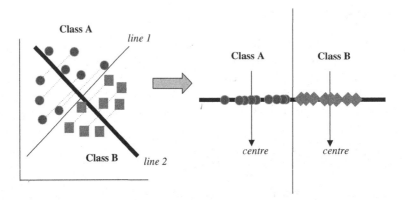

Figure 5.20 Projections

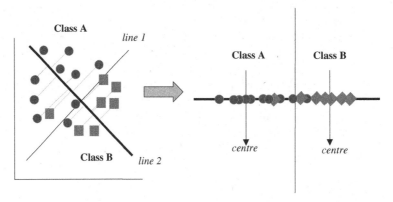

Figure 5.21 Projections where it is not possible to unambiguously classify objects

Sometimes the projected line is called a canonical variate, although in statistics this can have quite a formal meaning.

It is not always possible to exactly divide the classes into two groups by this method (see Figure 5.21) but the mis-classified samples should be far from the centre of both classes, with two class distances that are approximately equal. The data can be presented in the form of a class distance plot where the distance of each sample to the two class centres are visualized, which can be divided into regions as shown in Figure 5.22. The top right-hand region is one in which classification is ambiguous.

Figure 5.22 is rather simple, and probably does not tell us much that cannot be shown from Figure 5.21. However, the raw data actually consist of more than one measurement, so it is possible to calculate the class distance using the raw two-dimensional information, as shown in Figure 5.23. The points no longer fall onto straight lines, but the graph can still be divided into four regions.

- Top left: almost certainly class A.
- Bottom left: unambiguous membership.
- Bottom right: almost certainly class B.
- Top right: unlikely to be a member of either class, sometimes called an outlier.

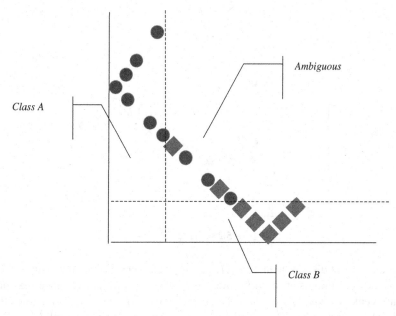

Figure 5.22 A simple class distance plot corresponding to the projection in Figure 5.21: horizontal axis = distance from class A, vertical axis = distance from class B

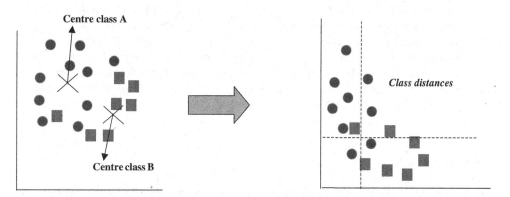

Figure 5.23 Class distance plot using two-dimensional information about centroids

In chemistry, these four divisions are perfectly reasonable. For example, if we try to use spectra to classify compounds into ketones and esters, there may be some compounds that are both or neither. If, on the other hand, there are only two possible classifications, for example whether a manufacturing sample is acceptable or not, a conclusion about objects in the bottom left or top right is that the analytical data are insufficiently good to allow us to conclusively assign a sample to a group. This is a valuable conclusion, for example it is helpful to tell a laboratory that their clinical diagnosis or forensic test is inconclusive and that if they want better evidence they should perform more experiments or analyses.

It is easy to extend the methods above to multivariate situations where instead of two variables many (which can run to several hundreds in chromatography and spectroscopy) are used to form the raw data.

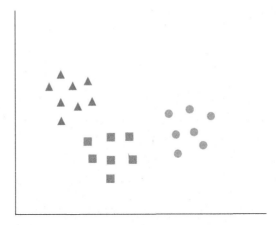

Figure 5.24 Three classes

Most methods for discriminant analysis can contain a number of extensions. The most common is to scale the distances from the centre of a class by the variance (or spread) in the measurements for a particular class. The greater this variance, the less significant a large distance is. Hence in Figure 5.23, Class A is more dispersed compared with Class B, and so a large distance from the centre is indicative of a poor fit to the model. The class distance plot can be adjusted to take this into account. The most common distance measure that takes this into account is the *Mahalanobis distance* which is contrasted to the *Euclidean distance* above; the principles are described in greater detail in the context of biological pattern recognition in Section 10.4 but are generally applicable to all classification procedures.

In most practical cases, more than two variables are recorded, indeed in spectroscopy there may be several hundred measurements, and the aim of discriminant analysis is to obtain projections of the data starting with much more complex information. The number of 'canonical variates' equals the number of classes minus one, so, in Figure 5.24 there are three classes and two canonical variates.

5.8.3 SIMCA

The SIMCA method, first advocated by the Swedish organic chemist Svante Wold in the early 1970s, is regarded by many as a form of soft modelling used in chemical pattern recognition. Although there are some differences with discriminant analysis as employed in traditional statistics, the distinction is not as radical as many would believe. However, SIMCA has an important role in the history of chemometrics so it is important to understand the main steps of the method.

The acronym stands for *Soft Independent Modelling of Class Analogy* (as well as the name of a French car). The idea of soft modelling is illustrated in Figure 5.25. Two classes can overlap (hence are 'soft'), and there is no problem with an object belonging to both (or neither) class simultaneously: hence there is a region where both classes overlap. When we perform hard modelling we insist that an object belongs to a discrete class. For example, a biologist trying to sex an animal from circumstantial evidence (e.g. urine samples), knows that the animal cannot simultaneously belong to two sexes at the same time, and a forensic scientist trying to determine whether a banknote is forged or not, knows that there can

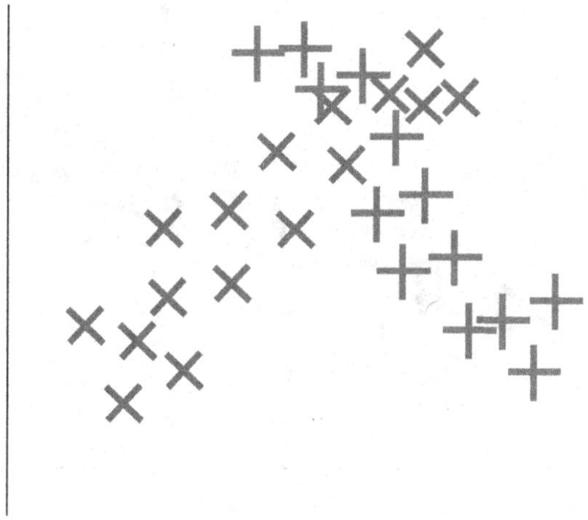

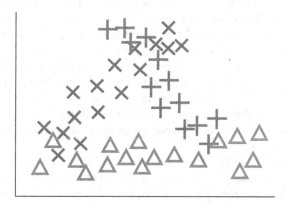

Figure 5.25 Overlapping classes

be only one answer: if this appears not to be so, the problem lies with the quality of the evidence. The original philosophy of soft modelling was that, in many situations in chemistry, it is entirely legitimate for an object to fit into more than one class simultaneously, for example a compound may have an ester and an alkene group, so will exhibit spectroscopic characteristics of both functionalities, hence a method that assumes the answer must be either a ketone or an alkene is unrealistic. In practice, there is not such a large distinction between hard (traditional discriminant analysis) and soft models and it is possible to have a class distance derived from hard models that is close to two or more groups.

Independent modelling of classes, however, is a more useful feature. After making a number of measurements on ketones and alkenes, we may decide to include amides in the model. Figure 5.26 represents a third class (triangles). This new class can be added independently to the existing model without any changes. This contrasts to some other methods of classification in which the entire modelling procedure must be repeated if different numbers of groups are employed.

The main steps of SIMCA are as follows.

Figure 5.26 Three classes

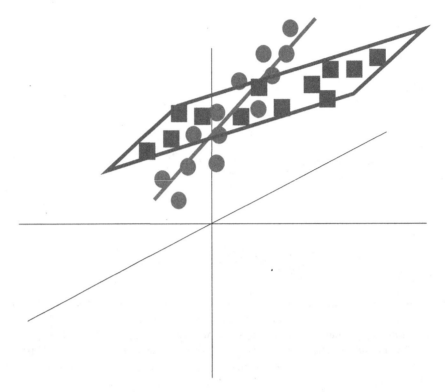

Figure 5.27   Two groups, one modelled by one PC and one by two PCs

Each group is independently modelled using PCA. Note that each group could be described by a different number of PCs. Figure 5.27 represents two groups each characterized by three raw measurements, which may, for example, be chromatographic peak heights or physical properties. However, one group falls mainly on a straight line, which is defined as the first PC of the group. The second group falls roughly on a plane: the axes of this plane are the first two PCs of this group. This way of looking at PCs (axes that best fit the data) are sometimes used by chemists, and are complementary to the definitions introduced previously (Section 5.2.2). It is important to note that there are a number of proposed methods for determining how many PCs are most suited to describe a class, of which the original advocates of SIMCA preferred cross-validation (Section 5.10).

The class distance can be calculated as the geometric distance from the PC models (see Figure 5.28). The unknown is much closer to the plane formed from the group represented by squares than the line formed by the group represented by circles, and so is tentatively assigned to this class. A rather more elaborate approach is in fact usually employed in which each group is bounded by a region of space, which represents 95 % confidence that a particular object belongs to a class. Hence geometric class distances can be converted to statistical probabilities.

Sometimes it is interesting to see which variables are useful for discrimination. There are often good reasons, for example in gas chromatography-mass spectrometry we may have hundreds of peaks in a chromatogram and be primarily interested in a very small number

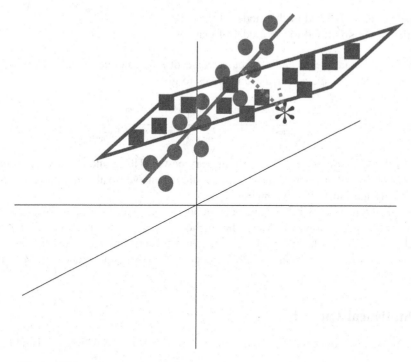

Figure 5.28 Distance of an unknown sample (asterisk) to two known classes

of, for example, biomarkers that are used to distinguish two groups, so this interpretation can have a chemical basis.

The *modelling power* of each variable in each class is defined by:

$$M_j = 1 - s_{jresid}/s_{jsraw}$$

where s_{jraw} is the standard deviation of the variable in the raw data, and s_{jresid} the standard deviation of the variable in the residuals given by:

$$E = X - T.P$$

which is the difference between the observed data and the PC model as described earlier. The modelling power varies between 1 (excellent) and 0 (no discrimination). Variables with M below 0.5 are of little use.

Another second measure is how well a variable discriminates between two classes. This is distinct from modelling power – being able to model one class well does not necessarily imply being able to discriminate two groups effectively. In order to determine this, it is necessary to fit each sample to both class models. For example, fit sample 1 to the PC model of both class A and class B. The residual matrices are then calculated, just as for discriminatory power, but there are now four such matrices:

1. Samples in class A fitted to the model of class A.
2. Samples in class A fitted to the model of class B.

3. Samples in class B fitted to the model of class B.
4. Samples in class B fitted to the model of class A.

We would expect matrices 2 and 4 to be a worse fit than matrices 1 and 3. The standard deviations are then calculated for these matrices to give:

$$D_j = \sqrt{\frac{\text{class A model B} \, s_{jresid}^2 + \text{class B model A} \, s_{jresid}^2}{\text{class A model A} \, s_{jresid}^2 + \text{class B model B} \, s_{jresid}^2}}$$

The bigger the value the higher the discriminatory power. This could be useful information, for example if clinical or forensic measurements are expensive, so allowing the experimenter to choose only the most effective measurements.

The original papers of SIMCA have been published by Wold and coworkers [9,10]. It is important not to get confused between the method for supervised pattern recognition and the SIMCA software package which, in fact, is much more broadly based. An alternative method proposed in the literature for soft modelling is UNEQ developed by Massart and coworkers [11].

5.8.4 Statistical Output

Software packages produce output in a variety of forms, some of which are listed below:

- The distances for each object from each class, suitably scaled as above.
- The most appropriate classification, and so per cent correctly classified (see Section 5.7).
- Probability of class membership, which relates to class distance. This probability can be high for more than one class simultaneously, for example if a compound exhibits properties both of a ketone or ester.
- Which variables are most useful for classification (e.g. which wavelengths or physical measurements), important information for future analyses.
- Variance within a class: how spread out a group is. For example, in the case of forgeries, the class of nonforged materials is likely to be much more homogeneous than the forged materials.

Information is not restricted to the training set, but can also be used in an independent test set or via cross-validation, as discussed above.

5.9 K NEAREST NEIGHBOUR METHOD

The methods of SIMCA (Section 5.8.3) and discriminant analysis (Section 5.8.2) discussed above involve producing statistical models, such as PCs and canonical variates. Nearest neighbour methods are conceptually much simpler, and do not require elaborate statistical computations.

The K Nearest Neighbour (KNN) method has been with chemists for over 30 years. The algorithm starts with a number of objects assigned to each class. Figure 5.29 represents five objects belonging to two classes, class A (diamonds) and class B (squares), recorded using two measurements which may, for example, be chromatographic peak areas or absorption intensities at two wavelengths.

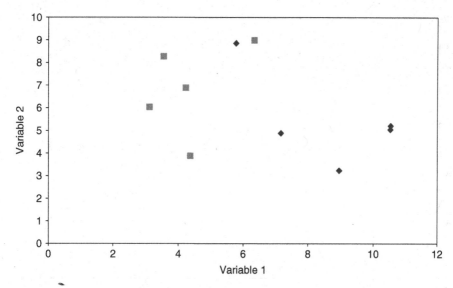

Figure 5.29   Objects in two classes

Table 5.9   Example for KNN calculations

Class	Measurement 1	Measurement 2	Distance to unknown	Rank
A	5.77	8.86	3.86	6
A	10.54	5.21	5.76	10
A	7.16	4.89	2.39	4
A	10.53	5.05	5.75	9
A	8.96	3.23	4.60	8
B	3.11	6.04	1.91	3
B	4.22	6.89	1.84	2
B	6.33	8.99	4.16	7
B	4.36	3.88	1.32	1
B	3.54	8.28	3.39	5
unknown	4.78	5.13		

The method is implemented as follows:

1. Assign a training set to known classes.
2. Calculate the distance of an unknown to all members of the training set (see Table 5.9). Usually the simple geometric or Euclidean distance is computed.
3. Rank these in order (1 = smallest distance and so on).
4. Pick the K smallest distances and see what classes the unknown in closest to. The case where $K = 3$ is illustrated in Figure 5.30. All objects belong to class B.
5. Take the 'majority vote' and use this for classification. Note that if $K = 5$, one of the five closest objects belongs to class A.
6. Sometimes it is useful to perform KNN analysis for a number of different values of K, e.g. 3, 5 and 7, and see if the classification changes. This can be used to spot anomalies.

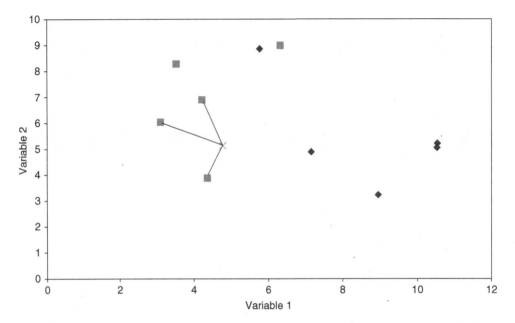

Figure 5.30   Classifying an unknown using KNN (with $K = 3$)

If, as is usual in chemistry, there are many more than two measurements, it is simply necessary to extend the concept of distance to one in multidimensional space, each axis representing a variable. Although we cannot visualize more than three dimensions, computers can handle geometry in an indefinite number of dimensions, and the idea of distance is easy to generalize. In the case of Figure 5.30 it is not really necessary to perform an elaborate computation to classify the unknown, but when a large number of measurements have been made, it is often hard to determine the class of an unknown by simple graphical approaches.

This conceptually simple approach works well in many situations, but it is important to understand the limitations.

The first is that the numbers in each class of the training set should be approximately equal, otherwise the 'votes' will be biased towards the class with most representatives. The second is that for the simplest implementations, each variable is of equal significance. In spectroscopy, we may record hundreds of wavelengths, and some will either not be diagnostic or else be correlated. A way of getting round this is either to select the variables or else to use another distance measure. The third problem is that ambiguous or outlying samples in the training set can cause major problems in the resultant classification. Fourth, the methods take no account of the spread or variance in a class. For example, if we were trying to determine whether a forensic sample is a forgery, it is likely that the class of forgeries has a much higher variance to the class of nonforged samples.

It is, of course, possible to follow procedures of validation (Section 5.7) just as in all other methods for supervised pattern recognition. There are quite a number of diagnostics that can be obtained using these methods.

However, KNN is a very simple approach that can be easily understood and programmed. Many chemists like these approaches, whilst statisticians often prefer the more elaborate methods involving modelling the data. KNN makes very few assumptions, whereas methods based on modelling often inherently make assumptions such as normality of noise distributions that are not always experimentally justified, especially when statistical tests are employed to provide probabilities of class membership. In practice, a good strategy is to use several different methods for classification and see if similar results are obtained. Often the differences in performance of different approaches are not due to the algorithm itself but in data scaling, distance measures, variable selection, validation method and so on. In this chemometrics probably differs from many other areas of data analysis where there is much less emphasis on data preparation and much more on algorithm development.

5.10 HOW MANY COMPONENTS CHARACTERIZE A DATASET?

One of the most controversial and active areas in chemometrics, and indeed multivariate statistics, is the determination of how many PCs are needed to adequately model a dataset. These components may correspond to compounds, for example, if we measure a series of spectra of extracts of seawater, how many significant compounds are there? In other cases these components are simply abstract entities and do not have physical meaning.

Ideally when PCA is performed, the dataset is decomposed into two parts, namely, meaningful information and error (or noise). The transformation is often mathematically described as follows:

$$X = T.P + E = \hat{X} + E$$

where $\hat{X}$ is the 'estimate' of X using the PC model. Further details have been described previously (Section 5.2.2).

There are certain important features of the PC model. The first is that the number of columns in the scores matrix and the number of rows in the loadings matrix should equal the number of significant components in a dataset. Second the error matrix E, ideally, should approximate to measurement errors. Some chemists interpret these matrices physically, for example, one of the dimensions of T and P equals the number of compounds in a series of mixtures, and the error matrix provides information on instrumental noise distribution, however, these matrices are not really physical entities. Even if there are 20 compounds in a series of spectra, there may be only four or five significant components, because there are similarities and correlations between the signals from the individual compounds.

One aim of PCA is to determine a sensible number of columns in the scores and loadings matrices. Too few and some significant information will be missed out, too many and noise will be modelled or as many people say, the data are over-fitted. The number of significant components will never be more than the *smaller* of the number of variables (columns) or objects (rows) in the raw data. So if 20 spectra are recorded at 200 wavelengths, there will never be more than 20 nonzero components. Preprocessing (Section 5.5) may reduce the number of possible components still further.

In matrix terms the number of significant components is often denoted the 'rank' of a matrix. If a 15×300 X matrix (which may correspond to 15 UV/visible spectra recorded at 1 nm intervals between 201 nm and 500 nm) has a rank of 6, the scores matrix T has six columns, and the loadings matrix P has six rows.

Many approaches for determining the number of significant components relate to the size of successive eigenvalues. The larger an eigenvalue, the more significant the component. If each eigenvalue is defined as the sum of squares of the scores of the corresponding PC, then the sum of all the nonzero eigenvalues equals the overall sum of squares of the original data (after any preprocessing). Table 5.1 illustrates this. The eigenvalues can be converted to percentages of the overall sum of squares of the data, and as more components are calculated, the total approaches 100 %. Statisticians often preprocess their data by centring the columns, and usually define an eigenvalue by a variance, so many softwares quote a percentage variance which is a similar concept, although it is important not to get confused by different notation.

A simple rule might be to retain PCs until the cumulative eigenvalues account for a certain percentage (e.g. 95 %) of the data, in the case of Table 5.1, this means that the first three components are significant.

More elaborate information can be obtained by looking at the size of the error matrix. The sum of squares of the matrix E is simply the difference between the sum of squares of the matrices X and $\hat{X}$. In Table 5.1, after three components are calculated the sum of squares of $\hat{X}$ equals 639 (or the sum of the first 3 eigenvalues). However, the sum of square of the original data X equals 670. Therefore, the sum of squares of the error matrix E equals $670 - 639$ or 31.

This is sometimes interpreted physically. For example:

- if the dataset of Table 5.1 arose from six spectra recorded at 20 wavelengths:
- the error matrix is of size 6×20, consisting of 120 elements;
- so the root mean square error is equal to $(31/120)^{1/2} = 0.508$.

Is this a physically sensible number? This depends on the original units of measurement and what the instrumental noise characteristics are. If it is known that the root mean square noise is about 0.5 units, then it seems sensible. If the noise level, however, is around 5 units, far too many PCs have been calculated, as the error is way below the noise level and so the data have been over-fitted.

These considerations can be extended, and in spectroscopy, a large number of so-called 'indicator' functions have been proposed, many by Malinowski, whose text on factor analysis [12] is a classic in this area. Most functions involve producing graphs of functions of eigenvalues, and predicting the number of significant components using various criteria. Over the past decade several new functions have been proposed, some based on distributions such as the F-test. For more statistical applications, such as quantitative structure – activity relationships, these indicator functions are not so applicable, but in spectroscopy and certain forms of chromatography where there are normally a physically defined number of factors and well understood error (or noise) distributions, such approaches are valuable.

A complementary series of methods are based on cross-validation which has been introduced previously (Section 5.7.2) in a different context of classification. When performing PCA, as an increasing number of components is calculated, for prediction of the training set (often called 'autoprediction') the error reduces continuously, that is the difference between the X matrix predicted by PCA and the observed matrix reduces the more the components employed. However, if the later components correspond to error, they will not predict effectively an 'unknown' that is left out of the original training set. Cross-validation involves predicting a portion of the dataset using information from the remainder of the samples. The residual error using cross-validation should be a minimum as the correct number of components are employed, and unlike autoprediction will increase again afterwards, because later PCs correspond to noise and will not predict the data that is left out well.

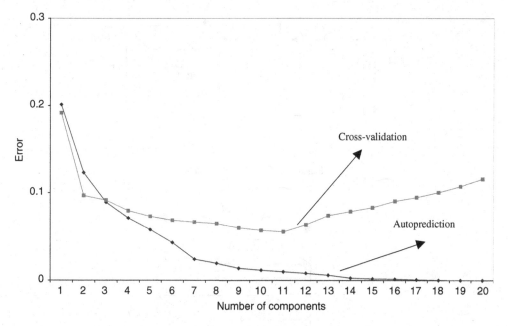

Figure 5.31 Cross-validation and autoprediction errors

Figure 5.31 shows the autopredictive (on the full training set) and cross-validated errors for a typical dataset as an increasing number of components is calculated. Whereas the autopredictive error reduces all the time, the cross-validated error is a minimum at 11 components, suggesting that later PCs model mainly noise. Cross-validation is a good indicator of the quality of modelling whereas autoprediction often forces an unrealistically optimistic answer on a system. The cross-validated graphs are not always as straightforward to interpret. Of course are many different methods of cross-validation but the simplest (LOO) 'leave one out' at a time approach is normally adequate in most chemical situations. The bootstrap as discussed in section 5.7.2 in the context of PCA is an alternative but less common approach for determining the number of significant components.

Validation is very important in chemometrics and is also discussed in the context of classification in Section 5.7 and calibration in Section 6.7. It is always important to recognize that there are different motivations for validation, one being to optimize a model and the other to determine how well a model performs on an independent set of samples, and sometimes a clear head is required not to mix up these two reasons.

5.11 MULTIWAY PATTERN RECOGNITION

Most traditional chemometrics is concerned with two-way data, often represented by matrices. Yet over the past decade there has grown a large interest in what is often called three-way chemical data. Instead of organizing the information as a two-dimensional array [Figure 5.32(a)], it falls into a three-dimensional 'tensor' or box [Figure 5.32(b)]. Such datasets are surprisingly common.

Consider, for example, an environmental chemical experiment in which the concentrations of six elements are measured at 20 sampling sites on 24 days in a year. There will be

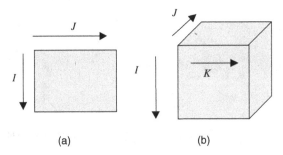

Figure 5.32 Multiway data

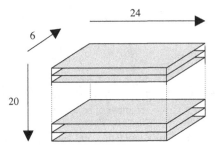

Figure 5.33 Example of three-way data from environmental chemistry. Dimensions are elements, sampling sites and sampling days

$20 \times 24 \times 6$ or 2880 measurements, however, these can be organized as a 'box' with 20 planes each corresponding to a sampling site, and of dimensions 24×6 (Figure 5.33). Such datasets have been available for many years to psychologists and in sensory research. A typical example might involve a taste panel assessing 20 food products. Each food could involve the use of 10 judges who score eight attributes, resulting in a $20 \times 10 \times 8$ box. In psychology, we might be following the reactions of 15 individuals to five different tests on 10 different days, possibly each day under slightly different conditions, so have a $15 \times 5 \times 10$ box. These problems involve finding the main factors that influence the taste of a food or the source of pollutant or the reactions of an individual, and are a form of pattern recognition.

Three-dimensional analogies to PCs are required. The analogies to scores and loadings in PCA are not completely straightforward, so the components in each of the three dimensions are often called 'weights'.

There are a number of methods available to tackle this problem.

5.11.1 Tucker3 Models

These models involve calculating weight matrices corresponding to each of the three dimensions (e.g. sampling site, date and element), together with a 'core' box or array, which provides a measure of magnitude. The three weight matrices do not necessarily have the same dimensions, so the number of components for sampling sites may be different to those for date, unlike normal PCA where one of the dimensions of both the scores and

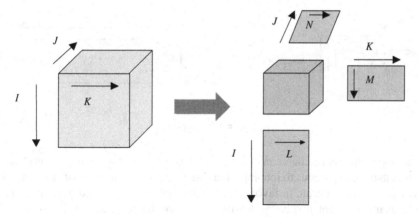

Figure 5.34 Tucker3 models

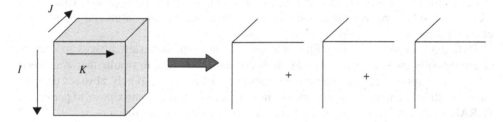

Figure 5.35 PARAFAC models

loadings matrices must be identical. This model is represented in Figure 5.34. The easiest mathematical approach is by expressing the model as a summation:

$$x_{ijk} \approx \sum_{l=1}^{L} \sum_{m=1}^{M} \sum_{n=1}^{N} a_{il} b_{jm} c_{kn} z_{lmn}$$

where z represents the core array. Some authors use the concept of 'tensor multiplication' being a three-dimensional analogy to 'matrix multiplication' in two dimensions, however, the details are confusing and it is conceptually probably best to stick to summations, which is what computer programs do well.

5.11.2 PARAFAC

Parallel Factor Analysis (PARAFAC) differs from Tucker3 models in that each of the three dimensions contains the same number of components. Hence, the model can be represented as the sum of contributions due to g components, just as in normal PCA, as illustrated in Figure 5.35 and represented algebraically by:

$$x_{ijk} \approx \sum_{g=1}^{G} a_{ig} b_{jg} c_{kg}$$

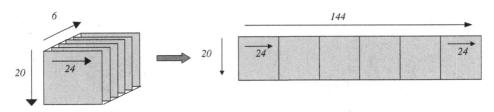

Figure 5.36 Unfolding

Each component can be characterized by a vector that is analogous to a scores vector and two vectors that are analogous to loadings, but some keep to the notation of 'weights' in three dimensions. Components can, in favourable circumstances, be assigned a physical meaning. A simple example might involve following a reaction by recording a diode array HPLC chromatogram at different reaction times. A box whose dimensions are *reactiontime × elutiontime × wavelength* is obtained. If there are three factors in the data, this would imply three significant compounds in a cluster in the HPLC (or three significant reactants), and the weights should correspond to the reaction profile, the chromatogram and the spectrum of each compound.

PARAFAC, however, is quite difficult to use and, although the results are easy to interpret, is conceptually more complex than PCA. It can, however, lead to results that are directly relevant to physical factors, whereas the factors in PCA have a purely abstract meaning. Note that there are many complex approaches to scaling the data matrix prior to performing PARAFAC, which must be taken into account when using this approach.

5.11.3 Unfolding

Another approach is simply to 'unfold' the 'box' to give a long matrix. In the environmental chemistry example, instead of each sample being represented by a 24 × 6 matrix, it could be represented by a vector of length 144, each element consisting of the measurement of one element on one date, e.g. the measurement of Cd concentration on 15 July. Then a matrix of dimensions 20 (sampling sites) × 144 (variables) is produced (see Figure 5.36) and subjected to normal PCA. Note that a box can be divided into planes in three different ways (compare Figure 5.33 and Figure 5.36).

This comparatively simple approach is sometimes sufficient but the PCA calculation neglects to take into account the relationships between the variables. For example, the relationship between concentration of Cd on 15 July and that on 16 July, would be considered to be no stronger than the relationship between Cd concentration on 15 July and Hg on November 1 during the calculation of the components. However, after the calculations are performed it is still possible to regroup the loadings and sometimes an easily understood method such as unfolded PCA can be of value.

For more details, a tutorial review by Smilde is an excellent starting point in the literature [13].

REFERENCES

1. K. Pearson, On lines and planes of closest fit to systems of points in space. *Philosophical Magazine*, 2 (6) (1901), 559–572.

2. A.L. Cauchy, *Oeuvres, IX* (2) (1829), 172–175

3. R.J. Adcock, A problem in least squares, *Analyst*, 5 (1878), 53–54

4. H. Hotelling, Analysis of a complex of statistical variables into principal components. *Journal of Educational Psychology*, 24 (1933), 417–441 and 498–520

5. P. Horst, Sixty years with latent variables and still more to come, *Chemometrics and Intelligent Laboratory Systems*, 14 (1992), 5–21

6. S. Dunkerley, J. Crosby, R.G. Brereton, K.D. Zissis and R.E.A. Escott, Chemometric analysis of high performance liquid chromatography – diode array detector - electrospray mass spectrometry of 2- and 3-hydroxypyridine, *Chemometrics and Intelligent Laboratory Systems*, 43 (1998), 89–105

7. D.L. Massart and L. Kaufman, *The Interpretation of Analytical Chemical Data by the use of Cluster Analysis*, John Wiley & Sons, Inc., New York, 1983.

8. B. Efron and R.J. Tibshirani, *An Introduction to the Bootstrap*, Chapman and Hall, New York, 1993

9. S. Wold, Pattern Recognition by means of disjoint Principal Components models, *Pattern Recognition*, 8 (1976), 127–139

10. C. Albano, W.J. Dunn III, U. Edland, E. Johansson, B. Norden, M. Sjöström and S. Wold, 4 levels of Pattern Recognition, *Analytica Chimica Acta*, 103 (1978), 429–433

11. M.P. Derde and D.L. Massart, UNEQ – a disjoint modeling technique for Pattern Recognition based on normal-distribution, *Analytica Chimica Acta*, 184 (1986), 33–51

12. E.R. Malinowski, *Factor Analysis in Chemistry*, 3rd Edn, John Wiley & Sons, Inc., New York, 2002

13. A.K. Smilde, 3-way analyses – problems and prospects, *Chemometrics and Intelligent Laboratory Systems*, 15 (1992), 143–157

6

Calibration

6.1 INTRODUCTION

Calibration is a major application of chemometrics, and normally involves using one type of measurement to predict the value of an underlying parameter or property. A simple example might be to use the peak area in chromatography to predict a concentration, the area being related to concentration. More complex problems involve using structural parameters (such as bond lengths, angles and dipoles) to predict biological activity of drugs, or spectroscopic properties to predict the taste of a food.

Historically, analytical chemists have been performing calibration for over a century, using tests to determine concentrations of analytes in samples such as water, blood or urine; the performance of a test is related or 'calibrated' to concentration. However a quite distinct school of thought emerged in the 1970s based on multivariate methods, especially the Partial Least Squares (PLS) algorithm proposed first by the statistician Hermann Wold a decade before. The origins of PLS were in economics, but the main modern day applications are in the field of chemistry. Instead of calibrating a single variable (e.g. a peak height) to another single variable (e.g. a concentration), several variables (e.g. spectroscopic absorbances at 50 wavelengths or structural parameters) are calibrated to one or more variables.

There have been a number of historically important driving forces for the development of multivariate calibration, but a major economic influence has been near infrared spectroscopy (NIR) especially of food and as applied to industrial process control. For example, consider the use of wheat in baking. It may be desirable to determine the amount of protein in wheat. With each batch, the quality of raw materials will vary, e.g. due to time of year, the initial manufacturing process, the supplier and growth conditions. How can this be assayed? Conventional methods are slow and time consuming. A cheaper method involves using spectroscopy so that measurements can be made almost continuously, but the chemical signals are 'mixed up' together, and some form of computational procedure is required to convert these spectroscopic measurements to concentrations, called calibration. Applications to food chemistry are discussed further in Section 13.3.

In its simplest form, calibration could be regarded as a form of curve fitting, but has different aims. A physical chemist may wish to form a model of a reaction as a function of temperature, e.g. to determine how the rate of a reaction varies with temperature. From this, he or she may then interpret the model in terms of fundamental molecular parameters and gain an insight into the reaction mechanism. It is not the prime aim to be able to predict the

Applied Chemometrics for Scientists   R. G. Brereton
© 2007 John Wiley & Sons, Ltd

reaction rate at any conceivable temperature, but, rather, to obtain an understanding of the molecular basis of reactivity. However, many analytical chemists are interested in prediction, rather than detailed mechanism. A process chemist might wish to monitor the quality of products on a continuous basis by spectroscopy. He or she will be interested in how the composition changes with time, and so to take corrective action if there are significant deviations from the accepted limits. The aim is to calibrate the spectra to concentrations, but there is no interest in a physical model between the two types of information. The spectra are used purely to predict physical parameters.

Multivariate calibration is distinct from univariate calibration in that the experimental data consist of several variables (Figure 6.1). The ability to use multivariate methods on a common basis goes hand in hand with the increasing availability of desktop laboratory computers able to acquire and handle large quantities of data rapidly.

There are a whole series of problems in chemistry for which calibration is appropriate, but each is very different in nature. Many of the most successful applications have been in the application of multivariate calibration to the spectroscopy or chromatography of mixtures.

- The simplest is calibration of the concentration of a single compound using a spectroscopic or chromatographic method, an example being determining the concentration of a plant pigment by ultraviolet (UV)/visible spectroscopy. Univariate calibration might involve determining the concentration using a single wavelength (or extinction coefficient) whereas multivariate methods involve using several wavelengths in the model, and often involve an improvement in the estimate due to the effect of averaging.
- A more complex situation involves a multicomponent mixture where all pure standards are available. It is possible to control the concentration of the reference compounds, so that a number of carefully designed mixtures can be produced in the laboratory. Sometimes the aim is to see whether a spectrum of a mixture can be employed to determine individual concentrations, and, if so, how reliably. The aim may be to replace a slow and expensive chromatographic method by a rapid spectroscopic approach. Another rather different aim might be impurity monitoring, how well the concentration of a small impurity be determined, for example, buried within a large chromatographic peak.

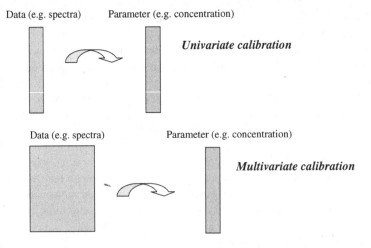

Figure 6.1   Univariate and multivariate calibration

- A quite different approach is required if only the concentrations of some components in a mixture are known, for example, an example being to detect polyaromatic hydrocarbons in samples of water as potential pollutants. In natural samples there may be hundreds of unknowns, but only a few can be quantified and calibrated. The unknowns cannot necessarily be determined and it is not possible to design a set of samples in the laboratory containing all the potential components in real samples. Multivariate calibration is effective providing the range of samples used to develop the model is sufficiently representative of all future samples that may be analysed by this model.
- A final case is where the aim of calibration is not so much to determine the concentration of a particular compound but determine an overall prediction of a property or parameters. There are no longer pure standards available, but the training set must consist of a sufficiently representative group of samples. An example is to determine the concentration of a class of compounds in food, such as protein in wheat. This situation also occurs, for example, in quantitative structure–property relationships (QSPR) or quantitative structure–activity relationships (QSAR), the properties being statistical functions of molecular parameters; multivariate calibration is a good alternative to more conventional molecular modelling. Calibration has also been used for classification, for example, by regressing spectra onto a parameter that relates to class membership, and also has an important role in sensory analysis where composition can be calibrated to taste or other features such texture.

A classic, highly cited, and much reprinted, text is by Martens and Næs [1]. Brereton's article [2] expands on topics within this chapter.

6.2 UNIVARIATE CALIBRATION

The most traditional, but still very widely used, method of calibration in chemistry is univariate, involving calibrating a single variable (e.g. a spectroscopic intensity) to another variable (e.g. a concentration). The literature goes back a century, however, with the ready availability of matrix based computer packages ranging from spreadsheets to more sophisticated programming environments such as Matlab, we can approach this area in new ways.

There is much confusion in the chemical literature as to how to denote the variables. Classically an 'independent' variable such as concentration is called 'x', whereas a 'dependent' variable such as spectroscopic intensity is called 'y'. However, many chemometricians reverse this. The spectra are often referred to as 'x' with the concentrations as 'y'.

In order to reach a compromise we will reference the spectra as the 'x' block, and the concentrations (or parameters of interest such as physical or biological properties) by the 'c' block. The aim of calibration is to use information from the 'x' (or measurement) block (e.g. spectra, chromatograms, electrochemical measurements, molecular descriptors) to predict parameters in the 'c' block. In univariate calibration, both blocks consist of a single variable.

6.2.1 Classical Calibration

Most chemists are familiar with this approach. An equation is obtained of the form:

$$x = b \, c$$

relating, for example, a chromatographic peak height or spectroscopic intensity (usually represented by the vertical axis of a graph) to a concentration (the horizontal axis). Sometimes an intercept term is introduced:

$$x = b_0 + b_1\ c$$

and it is also possible to include squared and higher order terms.

There are several difficulties with classical methods. The first is that the aim of calibration is to predict c from x and not vice versa: the classical method involves obtaining a model of x from c, it may be what a physical chemist would want but not necessarily what an analytical or process or biological chemist wants to do. The second, and more serious, difficulty is that the greatest source of error is generally in sample preparation such as dilution, weighing and extraction, rather than instrumental reproducibility, so the measurement of a concentration (even when determining a calibration model) is likely to be less certain than the measurement of spectral intensity. Twenty years ago many instruments were less reproducible than now, but the dramatic increase in instrumental performance has not been accompanied by an equally dramatic improvement in the quality of measurement cylinders, pipettes and balances. Classical calibration assumes that all errors are in the x block [Figure 6.2(a)].

6.2.2 Inverse Calibration

An alternative approach is to assume that the errors are in the c block, or the horizontal axis [Figure 6.2(b)]. This means fitting an equation of the form:

$$c = a\ x$$

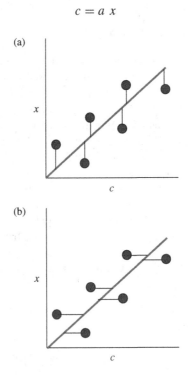

Figure 6.2   (a) Classical and (b) inverse calibration: errors are indicated

or

$$c = a_0 + a_1 \, x$$

Both in terms of error analysis and scientific motivation, this is more rational than classical calibration, but most chemists still use classical methods even though they are often less appropriate.

If, however, a calibration dataset is well behaved and does not have any significant outliers, straight lines obtained using both forms of calibration should be roughly similar, as illustrated in Figure 6.3. Sometimes there are outlying points, which may come from atypical or erroneous samples, and these can have considerable influence on the calibration lines (see Figure 6.4): there are many technical ways of handling these outliers, which are often said to have high leverage, as discussed in Sections 3.11 and 3.12. Generally the two

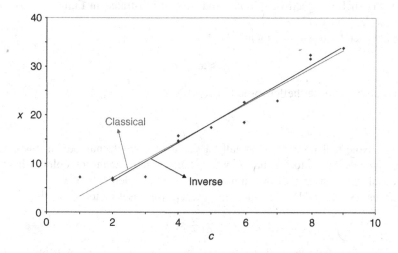

Figure 6.3 Classical and inverse calibration lines that agree well

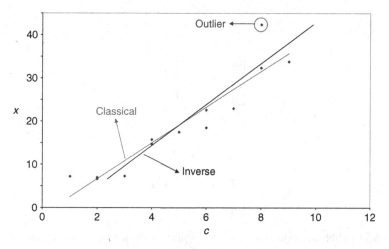

Figure 6.4 Classical and inverse calibration lines that do not agree well due to the presence of an outlier

types of best fit straight line are influenced in different ways, and this discrepancy can be eliminated by removing any outliers.

6.2.3 Calibration Equations

Calibration is best understood in terms of matrices and vectors which are introduced in Section 2.4.

We can express the inverse calibration equation in the form of vectors as follows:

$$c = x \ a$$

where a is a constant or scalar, c a column vector of concentrations and x a column vector of measurements such as spectroscopic absorbances, as illustrated in Figure 6.5(a). Knowing x and c can we deduce a?

Using the pseudoinverse x^{+}, then:

$$a \approx x^{+} . c$$

In the case above we use the left-pseudoinverse, so:

$$a \approx (x' . x)^{-1} . x' . c$$

in this case. Note that $(x'.x)^{-1}$ is actually a scalar or a single number in this case (when extending the calculation to a situation where there is more than one column in 'x' this is not so), equal to the inverse of the sum of squares of the elements of x.

Sometimes it is desirable to include an intercept (or baseline term) of the form:

$$c = a_0 + a_1 \ x$$

Now the power of matrices and vectors becomes evident. Instead of using a single column vector for x we use a matrix whose first column consists of 1s and second of the

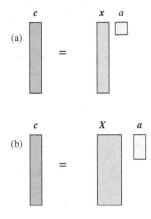

Figure 6.5 Calibration expressed in matrix form (a) without and (b) with an intercept term

experimentally determined x values, and a becomes a column vector [Figure 6.5(b)] of two elements, each element being one coefficient in the equation: these ideas have also been introduced in the context of experimental design (Chapter 2). The equation can now be expressed as:

$$c = X \cdot a$$

so that:

$$a \approx X^+ \cdot c$$

which is an easy way of calculating the coefficients computationally, and saves complex equations.

6.2.4 Including Extra Terms

It is easy to include further terms in the model. Suppose, for example we wish to extend this to squared terms so that:

$$c = a_0 + a_1 \, x + a_2 \, x^2$$

we simply extend the vector a by one element and matrix X by a column.

This is illustrated by data in Table 6.1 which includes some nonlinearity, where the c block is modelled by three terms to give:

$$c \approx 0.161 + 0.131 \, x - 0.000466 \, x^2$$

and the estimates presented in the table using the equation above.

6.2.5 Graphs

A large amount of graphical output can be obtained from regression, all providing valuable insights, and will be illustrated with the data of Table 6.1. Probably the most common is a graph of the estimated line superimposed over the experimental points (Figure 6.6), the closer these are, the more faithful the model. However, another good way is to plot the predicted against true values of c. This graph should ideally be linear, no matter what model is used provided the fit is good, and is presented in Figure 6.7. A third and quite useful graph is a residual plot of the difference between the estimated and true c block data, which is presented in Figure 6.8. A similar type of graph is introduced in Section 3.12.2 in the context of outlier detection. Sometimes there may be obvious outliers or systematic trends that can be picked up in the residuals. In certain cases errors are heteroscedastic, that is the larger the value of x, the larger the errors, so the residuals would increase from left to right in the graph.

The expert in univariate calibration will use several different types of graph, in fact there are probably a hundred or more ways of looking at the fitted data, but the three illustrated above are probably the simplest, and it is generally recommended to try to visualize the fitted data by each of these methods before making a decision as to whether a calibration curve is suitable or not.

Table 6.1 Predictions of c for a quadratic model

Concentration (c block)	Response (x block)
1	5.871
2	16.224
2	16.628
3	21.900
4	32.172
4	33.006
5	44.512
6	53.285
6	55.985
7	68.310
8	91.718
8	88.906
9	104.403

X matrix

1s	x	x^2
1	5.871	34.474
1	16.224	263.232
1	16.628	276.506
1	21.900	479.590
1	32.172	1035.015
1	33.006	1089.364
1	44.512	1981.339
1	53.285	2839.266
1	55.985	3134.302
1	68.310	4666.189
1	91.718	8412.149
1	88.906	7904.359
1	104.403	10899.931

Pseudoinverse of X

$5.85*10^{-1}$	$3.29*10^{-1}$	$3.20*10^{-1}$	$2.11*10^{-1}$	$3.45*10^{-2}$	$2.24*10^{-2}$	$-1.11*10^{-1}$	$-1.71*10^{-1}$	$-1.82*10^{-1}$	$-1.90*10^{-1}$	$-6.27*10^{-3}$	$-4.19*10^{-2}$	$2.01*10^{-1}$
$-2.17*10^{-2}$	$-9.27*10^{-3}$	$-8.84*10^{-3}$	$-3.64*10^{-3}$	$4.48*10^{-3}$	$5.02*10^{-3}$	$1.07*10^{-2}$	$1.28*10^{-2}$	$1.30*10^{-2}$	$1.17*10^{-2}$	$-1.24*10^{-3}$	$1.06*10^{-3}$	$-1.41*10^{-2}$
$1.65*10^{-4}$	$6.02*10^{-5}$	$5.66*10^{-5}$	$1.31*10^{-5}$	$-5.31*10^{-5}$	$-5.74*10^{-5}$	$-1.00*10^{-4}$	$-1.13*10^{-4}$	$-1.13*10^{-4}$	$-9.23*10^{-5}$	$4.34*10^{-5}$	$2.04*10^{-5}$	$1.70*10^{-4}$

Estimates of c

True	Estimated
1	0.914
2	2.165
2	2.212
3	2.808
4	3.895
4	3.979
5	5.072
6	5.822
6	6.038
7	6.939
8	8.261
8	8.130
9	8.764

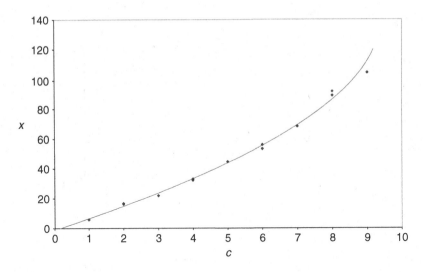

Figure 6.6 Best fit line for calibration model, using quadratic terms for the data in Table 6.1

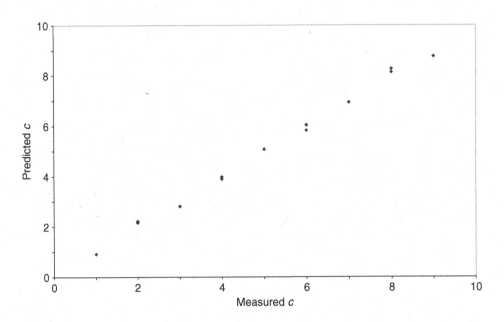

Figure 6.7 Graph of predicted versus true observations from calibration model of Table 6.1

6.3 MULTIVARIATE CALIBRATION AND THE SPECTROSCOPY OF MIXTURES

In scientific data analysis, new approaches can become quite widespread with hardly anybody noticing. Multivariate calibration is one such technique that has made a significant

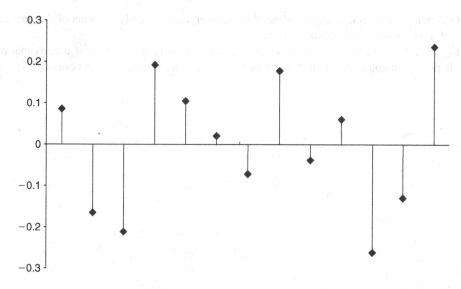

Figure 6.8 Residuals after calibration

impact in all areas of chemistry, but especially analytical spectroscopy of mixtures. An indication of the interest comes from citations. At the time of writing this book, Geladi and Kowalski's 1986 paper on PLS [3] has been cited over 1300 times, whereas Martens and Næs' 1989 text [1] over 2600 times. Why is there such an interest?

The chemistry of mixtures plays an important role in modern science. Indeed most manufactured products ranging from drugs to paints are mixtures. Environmental samples such as diesel exhausts or factory fumes contain a mixture of compounds. Clinical and pharmaceutical tests on urine essentially aim to determine the composition of a mixture of chemicals dissolved in water.

It is important to know the composition of such samples. A good example is in synthetic chemistry for the pharmaceutical industry. It is difficult, often impossible, to obtain a compound at 100 % purity levels, and small quantities of by-products, generally isomers, usually remain. How can we check this and measure the level of these impurities? Although they may be present in less than 1 %, the consequences could be devastating. Thalidomide is a well known example, which led to a tightening of controls. Can we detect these isomers, and can we quantify them? In particular can we do this on a continuous basis as new batches are produced? Another example is in environmental studies. We may be concerned about the composition of the air in a factory. It is possible to take spectra on a regular basis of particles caught on a filter, but what is their composition? There could be 20 or more significant potential pollutants. What are they? Are they carcinogenic and are they below occupational exposure limits? Finally, how about monitoring a reaction with time? There may be four or five products. Can we take spectra in 'real time' so follow the build up (and degradation) of products and hence study kinetics on-line?

Conventional slow (and expensive) methods usually require off-line chromatography, or very specific tests for individual compounds. Analysis has moved on from the days of flame tests, and modern quantitative analytical techniques, in particular, spectroscopy, are easy to automate and link to a computer, so it is feasible to perform measurements rapidly and in real time, however, these spectra are of mixtures, often consisting of between two or three and a hundred or so compounds, and need to be interpreted, usually in terms of the amount of each component in the mixture.

Figure 6.9 is of three UV/visible spectra, and Figure 6.10 a spectrum of a combination of all three compounds. We may want to know how much of each component is in the

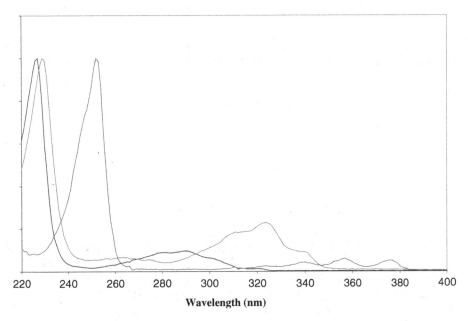

Figure 6.9 Spectra of three compounds

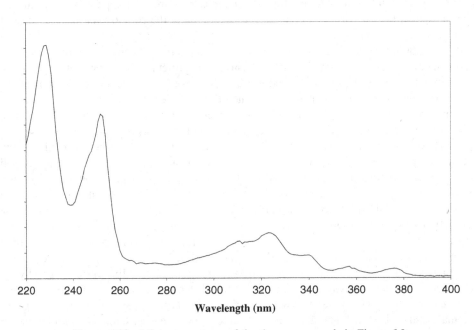

Figure 6.10 Mixture spectrum of the three compounds in Figure 6.9

mixture spectrum. We can formulate this problem in terms of vectors, so that:

$$x \approx c_1 s_1 + c_2 s_2 + c_3 s_3$$

where x represents a mixture spectrum, being a row vector each of whose elements corresponds to a wavelength, c_1 the concentration (e.g. in mM) of compound 1, s_1 the spectrum of 1 mM of this compound and so on. This equation can also be expressed in terms of matrices so that:

$$x \approx S \cdot c$$

where c is now a vector consisting of the three concentrations, and S a matrix consisting of the three spectra. Our interest is to determine c from x.

If there is a single component in the spectrum this equation can be simplified still further to:

$$x \approx s \cdot c$$

where c is now a single number (or a scalar), and s the spectrum of one unit of this compound. This equation is analogous to that of univariate calibration (Section 6.2), except s is now measured at several wavelengths (e.g. 100) rather than one. There is an advantage in extending the number of wavelengths because the noise (or errors) across the spectrum are averaged out, so a better estimate can often be obtained, even of a single component, by using multivariate approaches.

If the aim is to determine the concentration of more than one compound in a mixture, it is essential to measure at more than one wavelength, for example, at least three wavelengths are required for three compounds, but it is best to use many more wavelengths, and there is no reason why 100 wavelengths cannot be employed for this purpose. Multivariate calibration can be used to estimate the concentration of three components in a spectrum consisting of 100 wavelengths.

Sometimes not all the spectra of the individual components in a mixture are known. For example, we may wish to use multivariate calibration to determine the amount of vitamin C in orange juice using UV/visible spectroscopy. There will be many other spectroscopically active compounds, and it would be a large task to isolate, purify and characterize each of these. Multivariate calibration combined with spectroscopy becomes a very helpful tool in such situations. The unknown compounds will probably obey certain statistical properties, for example, their concentrations are likely to be restricted to a given range, the presence of many of the compounds will be correlated, e.g. a high concentration of a particular protein might imply a high concentration of another protein, and also they may absorb at characteristic wavelengths. These mean that there are certain characteristics of the background which can be used to advantage to produce sound statistical models.

A typical calibration experiment involves recording a series of spectra (X) containing a mixture of compounds of known concentrations (C). This is called a 'training' or 'calibration' set (as introduced in Section 5.7 in the context of pattern recognition). The concentrations of the compounds can be determined experimentally in various ways, either by preparation of reference samples (adding compounds of known weight or concentration to the mixture) or by using an independent technique which separates the compounds such as chromatography which then provides independent estimates of concentrations. The aim is to determine a *model* between the spectra and concentrations. There are numerous methods, which will be described below, the choice depending partly on how much information is known about the system.

Once the model has been established on the training set, then its quality can be assessed on an independent 'test' set, which tells us how well the model can predict unknowns. Chemometricians become very excited about how the quality of the model is predicted, which is an important step to prevent what is sometimes called over-fitting, which is indicated by a model that is too good on the training set but then very bad on an independent test set, and will be discussed further in Section 6.7.

6.4 MULTIPLE LINEAR REGRESSION

Multiple Linear Regression (MLR) is one of the commonest techniques for calibration and regression both in chemistry and statistics. Some chemometricians distinguish between Classical Least Squares (CLS) and Inverse Least Squares (ILS) models, in the context of MLR as have already been introduced for univariate regression in Section 6.2, but we will simplify the notation in this text. There are many types of terminology in chemometrics and no single group has a monopoly.

In the context of the spectroscopy of mixtures, the principle is that a series of mixture spectra can be characterized by:

(a) the spectra of each individual component;
(b) the concentrations of these components;
(c) noise (or experimental error).

Mathematically this information can be expressed in matrix format:

$$X = C \cdot S + E$$

We have already come across this equation before, in the context of Principal Components Analysis (PCA) (Section 5.2). The X matrix might, for example, consist of a series of 30 UV spectra of a reaction of the form:

$$A \longrightarrow B + C$$

collected at 15 min intervals over a period of 7.5 h. If the spectra are recorded over 200 nm, at 2 nm intervals then: X is a 30×100 matrix; C is a 30×3 matrix; and S is a 3×100 matrix.

A kineticist wishing to determine the rate constants, or an organic chemist wishing to determine the yield with time, is primarily interested in the contents of the matrix C, i.e. the concentrations of each compound during the course of the reaction. Sections 8.3 and 8.4 extend the applications and techniques for reaction monitoring.

There are many ways of using MLR to obtain such information. The simplest is not really calibration, but involves knowing the spectra of a unit concentration of each pure component, S. Using the pseudoinverse (Section 2.4) we can write:

$$C \approx X \cdot S^+$$

or

$$C \approx X \cdot S' \cdot (S \cdot S')^{-1}$$

and so find the concentrations knowing the spectra.

Life is not always so simple. For example, UV spectra are not always transferable between machines, so each compound must be obtained pure and their spectra measured on the specific system. It is not sufficient to download a reference spectrum from a database, indeed quite serious errors can creep in. Furthermore, instrumental performance can vary on a day by day (or hour by hour) basis, so a pure spectrum recorded on a Monday may be different to the same pure spectrum recorded on a Tuesday. A common problem is that pure reference standards are not always available. For example, compound C might be a by-product of no real interest, except that it absorbs in the mixture spectra, so a chemist may not wish to spend days or even weeks purifying this compound just to be able to take its spectrum.

Another way of doing this is to measure spectra of the mixtures in known concentrations and use these as a calibration or training set. If pure A, B and C are indeed available, then calibration samples containing a mixture of A, B and C, often using a systematic experimental design (e.g. for multivariate calibration see Section 2.16) of known concentrations, can be prepared. The spectra of these standard mixtures are measured. Under such circumstances, the information available consists of the spectra of the training set together with the known concentrations of the constituents. The spectra can now be 'estimated' by:

$$\hat{S} = C^+ \cdot X_{calibration}$$
$$= (C'.C)^{-1}.C'.X_{calibration}$$

note the 'hat' over the S which indicates that the spectra are estimated and not directly recorded.

These estimated spectra may not always be very good, for example there may be overlapping peaks, solvent interferents or noisy regions of the spectrum, however it is possible to check at this stage that the concentrations of the calibration set are estimated well: we discuss validation of calibration models in Section 6.7. Quite often concentrations can be estimated quite well even if spectral features are not, this is because there can always be a few 'bad' wavelengths providing good quality information is available at some of the more intense wavelengths in the spectra. At this stage some people select only a few wavelengths and carry on with the calibration using these best ones, but we will assume below that all wavelengths are retained. The next phase is to use these to estimate the concentrations in the training set:

$$\hat{C} = X \cdot \hat{S}^+ = X.\hat{S}'.(\hat{S}.\hat{S}')^{-1}$$

The predicted and actual values of concentrations can be compared, and the error between these computed, often using a criterion of root mean square error. If this looks small enough, it is usual to then test this out on a test set, whose spectra were kept out of the original model.

If we are satisfied with the calibration model we can then predict the concentrations in the unknown samples:

$$\hat{C} = X_{reaction} \cdot \hat{S}^+$$

and so obtain the profiles of each compound with time.

If pure standards are not available, it is possible to perform an independent measurement on certain 'calibration' samples, for example by high performance liquid chromatography (HPLC). The peak areas can be used to provide concentration estimates and for a selected number of samples can be used to determine concentrations, as in Figure 6.11, then this is the basis of a calibration model which can be used for the remainder of the samples. The advantage is that it is possible to use spectroscopy in real-time, but chromatography is employed only for developing the model (or at regular intervals if there are significant

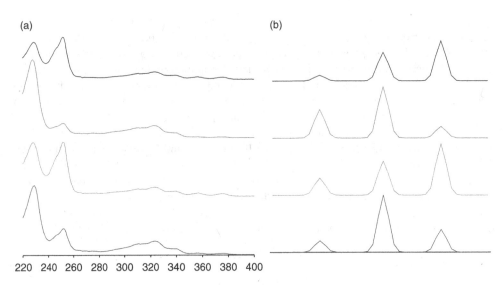

Figure 6.11   Calibrating (a) spectra to (b) HPLC peak areas

changes in instrumental performance). For routine analysis chromatography can be expensive and time consuming, so for example, in the case of continuous process monitoring, it is possible to obtain regular spectroscopic information without the need for off-line analysis, after the calibration model has been established.

MLR can be extended to include nonlinear terms such as squared terms, which can be useful under some circumstances, for example if there are nonlinearities due to overloaded spectra (which may occur if concentrations in a reaction mixture, for example, are outside the region where the Beer–Lambert law is obeyed). Extra columns are simply added to the X matrix equal to squares of the intensities. Another common modification is weighted regression, where certain wavelengths have more significance than others.

However, MLR has the disadvantage that concentrations of all significant compounds in the training set should be known to obtain the regression model. If, for example, we have information on the concentrations of only two out of five compounds in a training set then the two predicted spectra will contain features of the spectra of the remaining three compounds, distributed between the two known components (dependent on the design) and the concentration estimates will contain large and fairly unpredictable errors. It is sometimes possible to overcome this problem if there are selective wavelengths where only two compounds absorb, so that the X matrix consists of only certain selective regions of the spectrum. However, there are alternative approaches, discussed below, where good models can be obtained using information when only a proportion of the compounds in a mixture can be characterized.

6.5 PRINCIPAL COMPONENTS REGRESSION

Principal Components Regression (PCR) is considered by some as a form of factor analysis. Whereas the two concepts are interchangeable in the vocabulary of some chemometricians, especially spectroscopists, it is important to recognize that there are widespread differences in useage of the term 'factor analysis' by different authors, groups, and software vendors.

Malinowski pioneered the concept of factor analysis in chemical spectroscopy in the 1970s and there is a strong group of followers of his work who use terminology developed by this pioneering scientist.

PCR has a major role in the spectroscopy of mixtures. It is most useful where only some compounds can be identified in the mixture. For example, we may be interested in the concentration of a carcinogenic chemical in a sample of sea-water, but have no direct knowledge or interest in the identities and concentrations all the other chemicals. Can we determine this by spectroscopy of mixtures? MLR works best when all the significant components are identified.

The first step, as always, is to obtain a series of spectra (the calibration or training set), for example, the UV spectra of 20 samples recorded at 200 wavelengths, represented by a 20×200 matrix X.

The next step is to perform PCA (Section 5.2) on these spectra, to obtain scores and loadings matrices, often denoted by T and P. A crucial decision is to determine how many PCs should be retained (Section 5.10): most people use Leave One Out (LOO) cross-validation. Ideally this should equal the number of compounds in a mixture, for example, if there are 10 compounds we should retain the first 10 PCs, so that the matrix T is of dimensions 20×10 and P of dimensions 10×200, in this context. However, in complex real world situations this information is rarely known exactly. In many cases the number of Principal Components (PCs) needed to model the data is very different from the number of significant (real) components in a series of mixtures. This is because of correlations between compound concentrations, spectral similarities, and experimental design which often reduce the number of components, and noise and instrumental features such as baselines that may increase the number of components.

The next step is called 'regression', 'transformation' or 'rotation' according to author. In calibration, a common approach is to find a relationship between the scores and the true concentrations of each compound in the mixture, of the form:

$$c = T \cdot r$$

where c corresponds to the known concentrations in the mixture spectra (obtained by an independent technique for the training set), and r is a column vector whose length equals the number of PCs (10 in this example). This is represented diagrammatically in Figure 6.12. The equation above can also be expressed as:

$$c = t_1 r_1 + t_2 r_2 + \cdots + t_{10} r_{10}$$

if there are 10 PCs, where t_1 is the score of the first PC, and r_1 its corresponding coefficient obtained using regression.

Knowing c and T it is not too difficult to estimate r using the pseudoinverse (Section 2.4) by:

$$r = T^+ \cdot c$$

This permits the concentration of an individual compound to be estimated by knowing the scores. Further applications of PCR are discussed in the context of chromatography in Section 7.9.

If there are several compounds of interest, it is a simple procedure to expand 'c' from a vector to a matrix, each column corresponding to a compound, so that:

$$C = T \cdot R$$

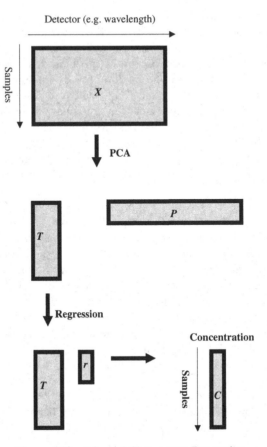

Figure 6.12 Principal Components Regression

Now R is a matrix; if there are 20 samples, 10 PCs and four compounds, the dimensions of R will be 10×4. As usual we can determine R by:

$$R = T^+ . C$$

The extreme case is when the number of PCs exactly equals the number of known compounds in the mixture, in which case R becomes a square matrix.

As usual the quality of prediction is then demonstrated on a test set and finally the method is used on real data of unknowns. It is also useful to be able to preprocess the data under certain circumstances, either by mean centring or standardization, dependent on the type of problem as discussed in Section 5.5.

The mathematics above may seem complicated, but the principles are fairly straightforward and can be summarized as follows:

1. Conventional approaches are inadequate if only some components in a mixture are known.
2. PCR requires only an estimate of the number of significant components in a series of mixture spectra not the identities and concentrations of all the compounds.

3. The method involves first performing PCA and then regressing the components onto the concentrations of one or more known calibrants.

Such an approach represents a significant improvement over MLR when only partial information is available about the number of components in a mixture. Its application is not, of course restricted to spectroscopy, PCR can be employed, for example, in quantitative structure–property relationships, where PCA is performed on structural parameters and regression is onto one or more physical properties.

6.6 PARTIAL LEAST SQUARES

PLS has a long and venerated history in the annals of chemometrics. This technique was originally proposed by the Swedish statistician, Herman Wold, whose interests were primarily in economic forecasting. In the 1960s his son Svante Wold, together with a number of Scandinavian scientists, most notably Harald Martens, advocated its use in chemistry. Possibly no technique in chemometrics is so mired in controversy, with some groups advocating PLS for almost everything (including classification) with numerous extensions over the last two decades, with other groups agreeing that PLS is useful, but one of very many techniques in the toolbox of the chemometrician. Probably one confusion is between the availability of good software, there is currently some excellent public domain software for PLS, and the desirability (independently of software) for using PLS. In many areas of calibration PCR serves almost the same purpose, however, despite this PLS has been the subject of more extensions and articles than probably any other chemometrics method, and there is a large hard-core of believers who solve most problems (usually quite successfully) by PLS.

One principle guiding PLS is that modelling the concentration (or c) information, is as important as modelling the experimental (or x) information. A drawback of PCR, as discussed in Section 6.5, is that the PCs are calculated exclusively on the x block, and do not take account of the c block. In PLS, components are obtained using both x and c data simultaneously. Statisticians like to think about maximizing the covariance (between the two blocks) as opposed to variance (for the x block as happens in PCA). Most conventional least squares approaches involve finding a variable that maximizes the value of x^2, or the square of the modelled experimental data or variance. PLS finds a variable that maximizes xc, or the product of the modelled experimental data with the modelled concentrations, often called the covariance. In physical terms PLS assumes that there are errors in both blocks which are of equal importance. This makes some sense: the concentrations used in calibration are subject to error (e.g. dilution and weighing) just as much as the spectra or chromatograms. MLR and PCR as commonly applied in chemistry assume all the errors are in the measured data and that the concentrations in the calibration set are known exactly.

Often PLS is presented in the form of two equations. There are a number of ways of expressing these, a convenient one being:

$$X = T.P + E$$

$$c = T.q + f$$

as illustrated in Figure 6.13. X represents the experimental measurements (e.g. spectra) and c the concentrations. The first equation above (page 211) appears similar to that of PCA, but the scores matrix also models the concentrations, and the vector q has some analogy to

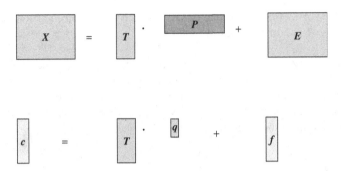

Figure 6.13 PLS principles

a loadings vector. The matrix T is common to both equations. E is an error matrix for the x block and f an error vector for the c block. The scores are orthogonal, but the loadings (P) are not orthogonal, unlike in PCA, and usually they are not normalized. There are various algorithms for PLS, and in many the loadings are not normalized either. PLS scores and loadings differ numerically from PCA scores and loadings, and are dependent both on experimental measurements (e.g. spectra) and the concentrations. In PCA, the scores and loadings depend only on the spectra. The product $T.P$ does not provide the best fit estimate of the X block. The size of each successive component (as assessed by the sum of squares of each vector in T) does not always decline, later vectors may be larger than earlier ones.

Some people in chemometrics call PLS components 'principal components' and call the sum of squares of the scores vectors 'eigenvalues', but this causes confusion when compared with PCA, as these matrices do not have similar properties, although there are some analogies. In PLS there often is another matrix called a 'weights' matrix W which has analogies to the P matrix and is a consequence of the algorithm. Some people like to use W instead of or in addition to P. In the PLS literature it is important to read each paper closely and determine precisely which algorithm is being employed.

The quality of the models can be determined by the size of the errors, normally the sum of squares of E and f is calculated. The number of significant PLS components can be estimated according to the size of these errors, often using cross-validation, the bootstrap or test sets although it can be done on the training set (often called autoprediction) (Section 6.7). It is important to recognize that there are a huge number of criteria for determining the number of significant components which can cause major arguments. If a particular piece of work relies completely on one approach, this may cause problems, especially if the approach results in 'flat' surfaces, for example finding an error minimum when there is only a small amount of variation over a wide range of components. Under such circumstances it is probably best to choose the criterion that gives the sharpest minimum.

An important feature of PLS is that it is possible to determine how well the data have been modelled either by using c or x blocks. Figure 6.14 illustrates the change in autopredictive (training set) error as different numbers of components are calculated for both x and c in a typical dataset. This means that two different answers for the optimal number of components can be obtained. Most chemometricians prefer to base their conclusions using the errors in the estimate of c, but it is important always to check the method employed.

There are several extensions to PLS.

The standard algorithm described above is sometimes called PLS1. One extension, first proposed in the 1980s is PLS2. In this case, the concentration (or c) block consists of more

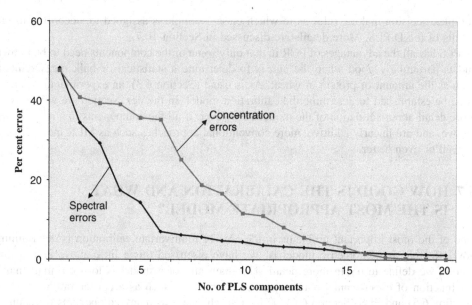

Figure 6.14   PLS autopredictive errors using both the concentration and spectral estimates

than one variable, an example being a case in which there are three components in a mixture, hence C becomes a matrix with three columns, and the calibration equation for the c block (page 211) changes to:

$$C = T.Q + F$$

In analytical spectroscopy PLS2 is not very popular and often gives worse results to PLS1. However, if there are interactions between variables that correspond to the columns of C this method can have advantages. In QSAR, for example, it may be that the properties of a compound are not independent functions of each structural feature separately, in which case PLS2 will have advantages. In most cases in analytical chemistry, however, mixture spectra or chromatograms are linearly additive and independent. An original advantage of PLS2 also was that it was faster than performing several PLS1 calculations, but with modern computing power this is no longer serious. A second advantage is that only one PLS2 model is required however many c variables are measured, whereas separate PLS1 models (often with different numbers of significant components) are required for each c variable.

Another common extension is to include nonlinear (in most case squared) terms in the PLS equation. This is very similar in effect to extending MLR models by including extra terms, and has been discussed in terms of experimental design previously (Section 2.11). Biologists frequently encounter such situations. Often, however, if a problem is highly nonlinear, completely different approaches such as Support Vector Machines (Section 10.6) may be more appropriate for calibration.

PLS can also be used in supervised pattern recognition (see Section 10.7 in the context of biological pattern recognition). The x block or experimental data is regressed onto a number indicating class membership, for example, class A might be indicated by -1 and class B by $+1$. This is called discriminant PLS or D-PLS. A cut-off (e.g. a value of 0) is determined as a decision boundary according to which class a sample is assigned to. If there are several classes, a number of D-PLS models must be computed, and there are often complex decision making rules as to which class a sample is assigned to, according to the results of the D-PLS. More details are discussed in Section 10.7.

PLS has all the advantages of PCR in that only some of the components need to be known and is particularly good when the aim is to determine a statistical or bulk measurement, such as the amount of protein in wheat. As is usual (Section 6.3), an experimental strategy must be established to determine the calibration model. In the next section we will look in more detail about validation of the models. However, if all the components in a mixture are known, and are linearly additive, more conventional approaches such as MLR may perform as well or even better.

6.7 HOW GOOD IS THE CALIBRATION AND WHAT IS THE MOST APPROPRIATE MODEL?

One of the most important problems in the area of multivariate calibration is determining how effective a chemometric model is. We have discussed this a little above and in this section we define terms in more detail. The main aim of a model is to use a mathematical function of experimental data to predict a parameter such as a concentration. In PCR (Section 6.5) and PLS (Section 6.6) we have to choose how many components to retain in the model, which represent one of the dimensions of the scores and loadings matrices. The size of these matrices influences the quality of predictions.

6.7.1 Autoprediction

As the number of PLS and PCR components is increased, the prediction error of the calibration or the 'training' set reduces. For example, if only one component is used in the model there may be a 50 % error, if five are used the error might reduce to 10 %, and for 10 components the error may be 1 %. Usually the *root mean square* error between predicted and known data is calculated.

In many cases there is no advance knowledge of how many components should be retained. In complex, natural or industrial mixtures, it is often impossible to determine the number of significant compounds in a series of spectra, so the normal approach is to look at how the error reduces with increasing components. There are two ways of checking this. The first and most common is to look at the change in prediction of concentration (c) as more components are calculated and the second to look at how well the spectral or experimental (x) block is estimated.

Sometimes it is sufficient to select the optimum number of components according to a percentage error criterion, for example, 1 %. If eight components give an error of 1.7 % and nine components 0.6 %, choose nine components. Often the error graphs level off, and it is possible to choose the components according to this criterion. However, it is important to recognize that there may be two different answers according to which block of data is employed. It is most common to use errors in the c block for definitive decisions.

One apparent dilemma is that the more the components are employed the better the prediction. Why not select 15 or even 20 components? Ultimately the data will apparently be modelled perfectly. The weakness is that later components model primarily noise, and so can produce artifacts when applied to unknowns. How can we determine which component models noise?

6.7.2 Cross-validation

Many chemometricians employ cross-validation (see also Section 5.10) for determining the optimum number of components. Unlike applications in PCA for pattern recognition, in calibration, most prefer to use the c block performance as a criterion. Typically cross-validation errors level off and then increase after the true number of components has been calculated. In contrast, autoprediction errors always decrease with increasing number of components.

The reason for this is that the later components model noise, and so samples left out of the original calibration set are predicted worse when using more components. Hence the shapes of graphs for autoprediction and cross-validation will diverge at a certain point. This information is important because, to be effective, calibration models will be used to determine the concentration of unknowns, so the user must understand at which point he or she is 'over-fitting' the data.

Over-fitting can be very dangerous. Consider, for example, using spectroscopy to determine the concentration of an additive to a drug in process control, ensuring that the quality is within legally defined limits. If the PLS or PCR model contains too many components, there is no guarantee that the concentration estimate on data that has not been part of the original calibration (or training set) will be accurate, and in the long run this can have serious implications.

6.7.3 Test Sets

One serious problem with both autoprediction and cross-validation is that the calibration model is tested internally. For example, an aim might be to measure the quantity of vitamin C in orange juice by spectroscopy. If the calibration (or training) set consists only of orange juices from Spain, even if it has been cross-validated, will it be effective if applied to Brazilian orange juices? It is generally a good idea to test out a model against a totally independent 'test' set of samples, in addition to cross-validation. Figure 6.15 compares errors from autoprediction, cross-validation and an independent test set for a typical model. The autoprediction errors reduce as expected, the cross-validation errors reach a minimum after 10 components, and the test set errors are high throughout. This suggests that the test set has quite different characteristics to the training set. Whether this matters depends on whether we really are likely to encounter samples of the type represented by the test set or not. If the answer is no (for example we are not interested ever in Brazilian orange juices) then this does not matter, although is valuable information to us.

Sometimes a poor test set error is due to a bad design in the training set. There is a great deal of debate as to how big or representative a dataset must be. Are 10 spectra adequate to test out and determine a model or should we use 20? In many situations it is difficult to control this information, for example, if the aim is to develop a method for orange juices it is not possible to control the correlation between concentrations of different compounds in the samples. A solution advocated by some is to use a large number of samples, but this may not always be feasible, asking an analyst to check a thousand samples will take too much time, it also may risk the problems of instrumental instability over a period of time. There are ways of overcoming this by, for example, using calibration transfer, however this always poses dilemmas and requires careful planning. In the early days of chemometrics, in the late 1980s and early 1990s, these methods often got a bad name, probably due to unscrupulous salesmen selling NIR instruments with multivariate calibration as the answer

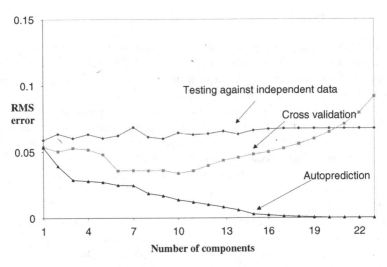

Figure 6.15 Autopredictive, cross-validated and test set error

to all problems but without at the same time insisting on understanding of the approaches and how they should be validated. This of course does represent a dilemma, does one insist that a customer who purchases a NIR spectrometer with PLS software go on and pass a course on chemometrics? Of course not, that would break the market, but then there are many people around who use chemometrics methods that do not have the correct appreciation of how to safely use these approaches.

There can be a real danger that a future unknown sample will not be representative of the calibration set and the estimated concentrations could then be grossly in error. Why is this so? In any series of mixtures of reasonable complexity there are inevitable correlations between concentrations and occurrences of different compounds. Take, for example, an experiment to monitor the level of polycyclic aromatic hydrocarbons (PAHs) in air samples. There may be 50 or more PAHs arising from industrial emissions. However, it is unlikely that the concentration of each PAH will be independent of each other, there will be correlations. Sometimes there may be groups of PAHs that are detected arising from specific processes. Hence for most conceivable real samples it is entirely appropriate to form a calibration model on a dataset which contains correlations. Where problems occur is if a sample is produced that does not have this structure. This could be particularly important in process control. The aim might be to monitor for 'strange' samples that occur if a process is malfunctioning, but the particular deviation from the norm may never have occurred before and would not, therefore, be part of the original calibration or 'training' set. Neither autoprediction nor cross-validation will be able to guide the experimenter as to the quality of the model. For this reason, in certain areas of calibration, there is much interest in 'outlier' detection, involving detecting samples whose spectra are incompatible with the original calibration set, a brief discussion of some methods is provided in Section 3.12.3.

In some cases it is possible to control the design of calibration or 'training' set (Section 2.16) when mixtures are created artificially in the laboratory. Under such circumstances, the results of cross-validation should be reliable, but the experiment depends on knowing the identities of all significant components in a mixture and cannot be used if only a proportion are available.

6.7.4 Bootstrap

An alternative half way house to cross-validation and using a single independent test set, is the bootstrap which involve repeatedly taking out sets of samples, using many different iterations [4]. At each iteration a different selection of samples is removed: the number of significant components (for PLS or PCR) can be assessed from those samples remaining (some of which are included more than once in the training set), and then the quality of prediction on the samples left out. Each iteration leaves out different samples, and since several (typically 100) iterations are performed, each sample is normally left out many times. The prediction ability is the average of all the iterations. The bootstrap is also described in the context of pattern recognition in Section 5.7.2.

6.8 MULTIWAY CALIBRATION

Above we discussed univariate calibration where a concentration is calibrated to a single parameter such as a chromatographic peak area or multivariate calibration where one or more parameters such as concentrations or molecular properties are calibrated to a series of measurements such as a spectrum. Chemists also have the capacity to record multiway data. Instead of a single spectrum (or vector) per sample, it is possible to record a two-dimensional set of data such as a coupled chromatogram (e.g. diode array HPLC or gas chromatography – mass spectrometry) or florescence excitation emission spectra. This type of data is also described in Section 5.11 in the context of pattern recognition.

A calibration dataset may then consist, for example, of a series of two way chromatograms, calibrated to a concentration, such as in Figure 6.16 which represents 20 samples recorded at 30 elution times and 100 wavelengths. The aim of calibration would be to take a two way chromatogram of an unknown and then determine its concentration from this data.

6.8.1 Unfolding

In analogy to multiway pattern recognition a simple approach is to unfold the data matrix, in our example to a matrix of 20 rows (corresponding to the samples) and 3000 ($= 30 \times 100$) columns (each corresponding to a single wavelength/elution time combination), and perform normal calibration such as PLS or PCR on the data.

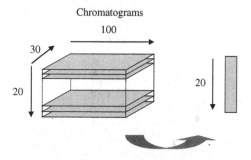

Figure 6.16   Multiway calibration

This simple approach can work well but has certain disadvantages. It neglects, for example, that wavelengths and points in time are connected. For example, in unfolded calibration the absorbance at time 15 and wavelength 80 has no specific link to the other absorbances at time 15, they are treated as completely independent so some information is lost. In addition many combinations of wavelength and time are uninformative and represent noise, so the majority of the 3000 variables in our case are redundant. Of the original 130 variables (100 wavelengths and 30 points in time) it is much more likely that the majority are informative, as they apply to the entire dataset. Finally there are often quite elaborate methods for data scaling that are necessary to make a success of this method.

However, in certain circumstances this quite straightforward approach works well, especially if sensibly combined with suitable variable selection and data preprocessing.

6.8.2 Trilinear PLS1

Another approach is to treat the experimental data as a three-way block (called by some a tensor). A modification to PLS1 can be used, called trilinear PLS1 (or 3-PLS1). The method is often represented diagrammatically as in Figure 6.17, replacing 'squares' or matrices by 'boxes' or tensors, and replacing, where necessary, the dot product ('.') used in normal PLS by something called a tensor product ('⊗').

However, the algorithms are quite straightforward and, as in bilinear PLS (Section 6.6), with only small extensions. A scores matrix is found that is common both to the experimental data and the concentration 'tensor', and if there are sufficient components should provide good predictions of c. There are two 'weights' matrices corresponding to loadings that can

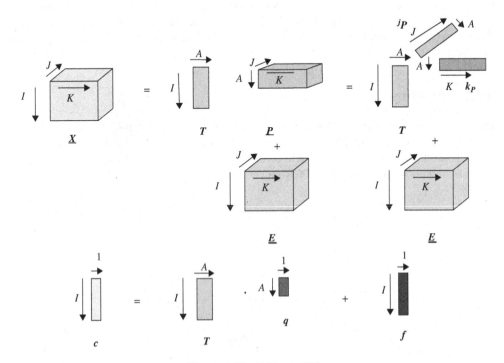

Figure 6.17 Trilinear PLS

also be determined. All the usual protocols of validation (Section 6.7) can be employed. Scaling (and mean-centring) can be quite complicated in three-way methods and at first it is best to run the algorithms without any preprocessing.

One of the conceptual problems of trilinear PLS is that the scores and weights (there is some difference in terminology according to authors) are no longer orthogonal unlike in normal PLS. This slightly technical property means that the contribution of each component to the model is not additive. In practical terms, the numbers in the vector q differ according to the number of components used in the model. In bilinear PLS, if one component is used in the model and the value of q_1 is 0.5, then this value will remain the same for a two or more component model, so we can say that the concentration is modelled by independent terms due to each successive component. In trilinear PLS this is not so. If the value of q_1 using one component is 0.5, it may be something completely different e.g. 0.3, when using two components. This property also means that there are difficulties modelling the experimental (x) data. Nevertheless if the main aim is to use the extra measurement to determine concentrations well, multiway methods have a good potential in chemometrics.

As in many chemometrics methods, often certain specific types of data are suitable for analysis using trilinear PLS. One potential example is coupled chromatography, but there are limitations in that the elution times of a series of chromatograms must be quite stable, or at least, must be able to be aligned carefully (a way of overcoming this is to divide a chromatogram into windows over several data points for which alignment errors are not so serious however this loses resolution in time). Fluorescence excitation emission spectra are another important source of data in which one dimension is the excitation wavelength and the other the emission wavelength. Reaction kinetics is an application of increasing interest. For example, it is possible to record the spectrum of a reaction mixture with time using different starting conditions. Each 'plane' in the box may represent a series of spectra (of mixtures of reactants and products) recorded in time. As conditions are changed, such as temperature and pH, another series of spectra are obtain. These could be calibration to a c block consisting of a maximum yield, or an enzymic activity parameter.

6.8.3 N-PLSM

The 'N' in N-PLSM determines how many dimensions are in the x block, whereas the 'M' how many in the c block. Conceptually, a natural extension is to increase the dimensionality of each block. For example, there is no reason why the concentration block is only one-dimensional, it could contain the concentrations of three or four compounds, in which case $M = 2$, to reflect that there are two dimensions in the c block. Chemical data with $M > 2$ are rare.

In addition there is no reason to restrict the x block to $N = 3$ dimensions. An example might be of a series of reactions in which two way fluorescence excitation emission spectra are recorded with time. The dimensions may be 20 (number of reactions) $\times$ 30 (sampling points in time per reaction) $\times$ 50 (excitation wavelengths) $\times$ 50 (emission wavelengths). If more than one parameter is to be calibrated (e.g. the yield of three products), 4-PLS2 is required.

The extension of calibration to several dimensions is one of active interest in the chemometrics community. Although the algorithms are not too difficult to understand, the methods for preprocessing of the data are crucial and have to be understood quite carefully, and there are as yet a shortage of datasets of complexity above 3-PLS2. However, with rapid on-line spectroscopy such as UV/visible and NIR available these methods are making a cautious entrée to the menu of the industrial chemometrician. For further literature, see Bro [5]. Although the methods for multiway factor analysis and calibration are somewhat different, many of the advocates of these approaches are the same.

REFERENCES

1. H. Martens and T. Næs, *Multivariate Calibration*, John Wiley & Sons, Ltd, Chichester, 1989
2. R.G. Brereton, Introduction to multivariate calibration in analytical chemistry, *Analyst*, 125 (2000), 2125–2154
3. P. Geladi and B.R. Kowalski, Partial Least Squares regression – a tutorial, *Analytica Chimica Acta*, 185 (1986), 1–17
4. B. Efron and R. Tibshirani, *An Introduction to the Bootstrap*, Chapman & Hall, New York, 1993
5. R. Bro, Multiway calibration. Multilinear PLS, *Journal of Chemometrics*, 10 (1993), 47–61

7

Coupled Chromatography

7.1 INTRODUCTION

Many of the classical applications in chemometrics were in the area of hyphenated chromatography which poses specific challenges in data analysis, and so has catalysed the development of methods of the past 20 years with specific chemical applications. Such type of information is increasingly common: in the modern laboratory some of the most widespread applications are in the area of coupled chromatography, such as diode array high performance liquid chromatography (diode array HPLC), liquid chromatography-mass spectrometry (LC-MS) and liquid chromatography-nuclear magnetic resonance (LC-NMR). A chromatogram is recorded whilst a ultraviolet (UV)/visible, mass spectrometry (MS) or nuclear magnetic resonance (NMR) spectrum is recorded. The information can be presented in matrix form, with time in the rows and wavelength, mass number or frequency in the columns, as indicated in Figure 7.1, often called two-way data: in Sections 5.2, 5.3 and 5.5 we have already been introduced to such datasets. This chapter will deal with more specialized approaches that are closely targeted towards coupled chromatography. With modern laboratory computers it is possible to obtain huge quantities of information very rapidly. For example, spectra sampled at 1 nm intervals over a 200 nm region can be obtained every second using modern chromatography, hence in an hour 3600 spectra in time × 200 spectral frequencies or 720 000 pieces of information can be produced from a single instrument. A typical medium to large industrial site may contain a hundred or more coupled chromatographs, meaning the potential of acquiring 72 million data points per hour.

In Chapters 5 and 6 we have discussed a number of methods for multivariate data analysis, but these methods do not normally take into account the sequential nature of the information. When performing Principal Components Analysis (PCA) on a data matrix, the order of the rows and columns is irrelevant and since PCA and most other classical methods for pattern recognition would not distinguish these sequences, there is a need for new approaches.

In most cases there are underlying factors which correspond to individual compounds in a chromatographic mixture which have certain characteristics, for example, the elution profiles are unimodal in time, they have one maximum, and are positive. An aim is to deconvolute the experimentally observed sum of chromatographic profiles into individual components and determine the features of each component, such as their spectra. The change in spectral characteristics across the chromatographic peak can be used to provide this information.

Applied Chemometrics for Scientists   R. G. Brereton
© 2007 John Wiley & Sons, Ltd

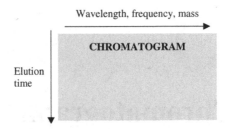

Figure 7.1 Evolutionary two-way data as occurs in coupled chromatography

To the practicing analytical chemist, there are three main questions that can be answered by applying multivariate techniques to coupled chromatography, of increasing difficulty:

- *How many peaks in the chromatogram*? Can we detect small impurities? Can we detect metabolites against a background? Can we determine whether there are embedded peaks and analyse overlapping peaks in cluster?
- *What are the characteristics of each pure compound*? What are their spectra? Can we obtain mass spectra and NMR spectra of embedded chromatographic peaks at low levels of sufficient quality that we can be confident of their identities?
- *What are the quantities of each component*? Can we quantitate small impurities? Could we use chromatography of mixtures for reaction monitoring and kinetics? Can we say with certainty the level of dope or a potential environmental pollutant when it is detected in low quantities buried within a major peak?

There are a large number of 'named' methods in the literature but they are based mainly around certain main principles, whereby underlying factors corresponding to individual compounds whose profiles evolve in time (elution profiles). Such methods are not only applicable to coupled chromatography but in areas such as pH dependence of equilibria, whereby the spectra of a mixture of chemical species can be followed with change of pH, and kinetic analysis, and further illustrations and extensions of some of the methods described below will be described in Chapter 8.

7.2 PREPARING THE DATA

7.2.1 Preprocessing

The first step in analysing two-way data such as from coupled chromatography is preprocessing, involving scaling variables. We have been introduced to a number of methods previously (Section 5.5) but there are several additional tricks of the trade, which will be discussed briefly below.

7.2.1.1 Baseline correction

A preliminary first step before applying most methods in this chapter is often baseline correction. The reason for this is that many chemometric approaches look at variation above a baseline, and if this is not done artifacts can be introduced.

Baseline correction is best performed on each variable (such as a mass or wavelength) independently. There are many ways of doing this, but it is first important to identify regions of baseline and of peaks between chromatographic peak clusters. Then normally a function is fitted to the baseline regions. This can simply involve the average intensity in this region or else, more commonly, a linear or polynomial best fit line. Sometimes estimating the baseline both before and after a peak cluster is useful, but if the cluster is quite sharp, this is not essential. Sometimes the baseline is calculated over the entire chromatogram, in other cases separately for each region of the chromatogram. After that, a simple mathematical model for the baseline which is usually a low order polynomial is subtracted from the entire region of interest, separately for each variable. In the examples in this chapter it is assumed that either there are no baseline problems or correction has already been performed.

7.2.1.2 Scaling the rows

Each subsequent row in a data matrix formed from a coupled chromatogram corresponds to a spectrum taken at a given elution time. One of the simplest methods of scaling involves summing the rows to a constant total. The influence of this procedure on Principal Component (PC) scores plots has already been discussed (Section 5.5). As shown for a cluster of two peaks, the elution times corresponding to each compound fall at the ends of a straight line after this procedure (Figure 5.17). In the case of two-way data, a problem with row scaling is that the correct region of a chromatogram must be selected. The optimum region depends on the size and nature of the noise.

Summing each row to a constant total is not the only method of dealing with individual rows or spectra. Two variations below can be employed:

1. *Selective normalization.* This allows each portion of a row to be scaled to a constant total, for example, it might be interesting to scale the wavelength regions 200–300 nm, 400–500 nm and 500–600 nm each to 1. Or perhaps the wavelengths 200–300 nm are more diagnostic than the others, so why not scale these to a total of 5, and the others to a total of 1? Sometimes more than one type of measurement can be performed on a chromatogram, such as UV/visible and mass spectrometry, each data block could be scaled to a constant total. When doing selective normalization it is important to consider very carefully the consequences of the preprocessing.
2. *Scaling to a base peak.* In some forms of measurement, such as mass spectrometry [e.g. LC-MS or gas chromatography-mass spectrometry (GC-MS)] it is possible to select a base peak and scale to this, for example, if the aim is to analyse the LC-MS of two isomers, rationing to the molecular ion can be performed. In certain cases the molecular ion can then be discarded. This method of preprocessing will study how the ratio of fragment ions varies across a cluster. Alternatively an internal standard may have been added to the mixture of known concentration and scaling to this allows true concentrations to be compared across different samples.

7.2.1.3 Scaling the columns

In many cases it is useful to scale down the columns, e.g. each wavelength or mass number or spectral frequency. This will put all the variables on a similar scale. Mean centring, involving subtracting the mean of each column is the simplest approach. Many packages do this automatically, but in the case of signal analysis may be inappropriate, because the interest is about variability above the baseline rather that around an average. Conventionally statisticians are mainly concerned with deviations from an average whereas much of chemometric analysis of spectra and chromatograms is about deviation above zero.

Standardization is a common technique that has already been discussed (Section 5.5) and is often also called *autoscaling*. Notice that it is conventional to divide by the number of variables rather than one minus this number in this application. If doing the calculations using many statistical packages use the 'population' rather than 'sample' standard deviation (Section 3.3). This procedure can be quite useful, for example, in mass spectrometry where the variation of an intense peak (such as a molecular ion of isomers) is no more significant than that of a much less intense peak, such as a significant fragment ion or an ion deriving from a significant impurity or biomarker. However, standardization will also emphasize variables that are pure noise, and if there are, for example, 200 mass numbers of which 180 correspond to noise, this could substantially degrade the analysis.

Sometimes weighting by the standard deviation can be performed without centring and it is also possible to use any weighting criterion for the columns. For example, the weights may relate to noise content or significance of a variable. Sometimes quite complex criteria are employed. In the extreme if the weight equals 0, in practice this becomes a form of variable selection, which will be discussed in Section 7.2.2.

It is possible to scale both the rows and columns simultaneously, first by scaling the rows and then the columns. Note that the reverse (scaling the columns first) is rarely meaningful and standardization followed by normalization has no physical meaning. Double centring is also sometimes used, to do this, for each element, create a new element which is

the value of the old element minus the value of the mean of its corresponding row minus the value of the mean of its corresponding column plus the overall mean

or, in mathematical terms, where there are I rows and J columns in matrix X:

$$^{new}x_{ij} = x_{ij} - \sum_{i=1}^{I} x_{ij}/I - \sum_{j=1}^{J} x_{ij}/J + \sum_{i=1}^{I}\sum_{j=1}^{J} x_{ij}/IJ$$

For graphical visualization of two way data such as by PC plots, it is very useful to employ a variety of methods of scaling and see what happens. More on the influence of data scaling on the visualization of this type of data is described in Section 5.5.

7.2.2 Variable Selection

Variable selection has an important role throughout chemometrics, but will be described in this chapter in the context of coupled chromatography. This involves keeping only a portion of the original variables, selecting only those such as wavelengths or masses that are most relevant to the underlying problem. Although there are some generic approaches, on the whole specific methods have been developed primarily according to the nature of the underlying spectroscopy. MS, UV/visible and NMR tend to be the most common and have specific features which have a diversity of approaches.

Coupled chromatography and MS such as LC-MS and GC-MS represent a rich source of data. However, raw data from LC-MS (or GC-MS) form what is sometimes called a *sparse* data matrix, in which the majority of data points are zero or represent noise. In fact often only a small percentage (perhaps 5 % or less) of the measurements are of any interest. The trouble with this is that if multivariate methods are applied to the raw data, often the results are nonsense, dominated by noise, since the majority of data points consist of noise. Consider the case of recording the LC-MS of two closely eluting isomers, whose fragment ions are of principal interest. The most intense peak might be the molecular ion, but in order to study the fragmentation ions, a method such as standardization is required to place equal significance on all the ions. Unfortunately, not only are perhaps 20 or so significant fragment ions increased in importance, but so are 200 or so ions that represent pure noise, so the data gets worse not better. Typically, out of 200 or 300 masses, there may be around 20 significant ones, and the aim of variable selection is to find these key measurements. However, too much reduction has the disadvantage that the dimensions of the multivariate matrices are reduced and key information may be lost. It is important to find an optimum size as illustrated in Figure 7.2. Some of the tricks used for reducing the number of masses are as follows.

1. A simple approach is to remove masses outside a certain range, for example, below m/z 50. It is also possible to find interesting regions of a spectrum and keep only these windows. However, this involves having a level of knowledge about the system, and probably if everything is known in advance, there is no point in the chemometrics. Nevertheless removing masses at extreme values (high or low m/z) is a useful first step.
2. Sometimes there can be strong background peaks such as due to a solvent, or due to known interferents, e.g. siloxanes, in sample preparation, and removing these together with noise is a useful exercise.
3. Variability of the intensity at each mass is another criterion. The more it varies, the more likely it is to represent an interesting ion. However, it is best before this to first select a region of interest containing the peak cluster and look at variability over this region only. There are a number of approaches for looking at this variability. The simplest is to take the raw standard deviation or variance, the greater the more useful the variable, the second is to take the ratio of the standard deviation to the mean (normally after baseline correction),

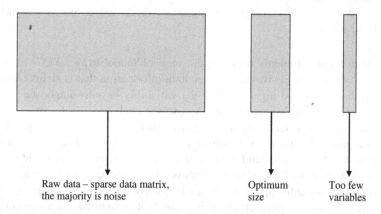

Raw data – sparse data matrix, the majority is noise

Optimum size

Too few variables

Figure 7.2 Finding an optimum number of variables

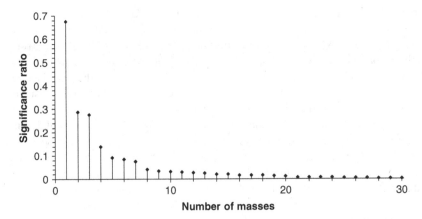

Figure 7.3 Ordering masses according to significance; the lower the significance the less important the masses

which allows for mass channels corresponding to small peaks that may be significant secondary metabolites or impurities to be retained. The masses can be ordered according to the values obtained (Figure 7.3) and the top ones retained. Sometimes an additional offset is added to the mean so that the ratio used is *standard deviation*/(*mean* + *offset*): this reduces the importance of small peaks whose may have a high relative standard deviation simply because their mean is small, and the value of the offset has an important influence on the nature of the variables selected.

4. It is possible to measure the noise content of the chromatograms at each mass, simply by looking at the standard deviation of the noise region. The higher the noise the less significant the mass statistically.

5. There are a variety of approaches for measuring information content at each mass, that take noise and smoothness of the mass channel into account of which the Component Detection Algorithm (CODA) method is probably the best known. The principles are illustrated in Figure 7.4. Each m/z value is given an information content (MCQ) varying between 0 and 1, in the example m/z 955 has a value of 0.231 (it mainly consists of spikes), m/z 252 of 0.437 and m/z 207 of 0.844. The higher this value the more useful the mass and a cut-off is employed to determine which masses are most significant and so should be retained.

Somewhat different problems occur in the case of diode array HPLC. In this case a much higher proportion of the wavelengths contain information that is significant compared with MS, but there many be regions where the variability is most diagnostic of a specific compound, so choosing these regions can improve the analysis.

One approach is to calculate a variety of purity curves for each wavelength. If a cluster consists of two peaks, the spectrum at each elution time will change across the chromatogram as the proportion of each compound alters. For each wavelength it is possible to determine the relative absorbance as a contribution to the overall spectrum, for example as a proportion of the absorbance of the entire spectrum. Diagnostic wavelengths will exhibit a smooth curve such as Figure 7.5 over the peak cluster (for example its magnitude as a percentage of the normalized spectral intensity at each elution time). Wavelengths that are primarily noise or

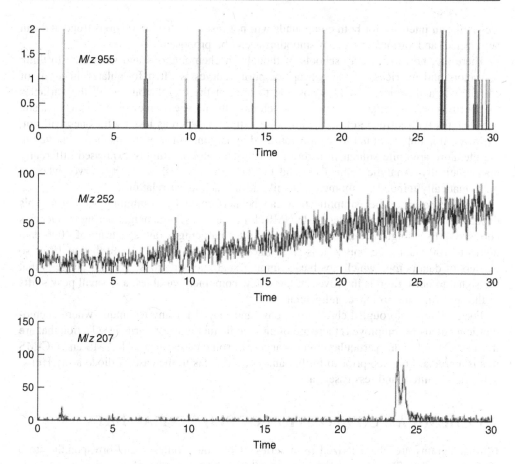

Figure 7.4 Information content of three m/z values in a chromatogram

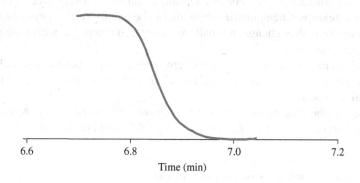

Figure 7.5 Purity curve for a specific wavelength over 30 s chromatographic elution time in a diode array HPLC

are similar in intensity for both compounds will not result in such an obvious trend and can be rejected, and various criteria of smoothness can be proposed.

There are, however, many schools of thought in chemometrics and some like to think in vectors and matrices, so the concept of purity curves is often formulated in terms of correlations and projections. It is possible to look at the correlations of all the variables (or spectroscopic wavelengths) and see which have the most extreme behaviour. Diagnostic wavelengths will correlate closely with these extremes, whereas less useful ones will not. However, it is important not to get too confused by jargon. Many of the methods reported in the literature are quite similar in nature to each other and are simply expressed differently mathematically. Analytical chemists tend to like concepts such as purity curves whereas mathematically minded chemometricians like ideas such as correlations.

In some cases further simplification can be performed by combining variables. This depends on the type of spectroscopy. In NMR, for example, several neighbouring frequencies correspond to a single peak, so, for example, after reducing a raw spectrum of 4096 data points to 100 diagnostic points, it is sometimes sensible to reduce this further to 10 or so clusters of data points, which involves summing contiguous regions. The advantage is that the signal to noise ratio is improved of these new combined variables, and small peak shifts in the spectrum are no longer influential.

For each type of coupled chromatography (and indeed for any technique where chemometric methods are employed) there are often specific methods for variable selection that are most successful under particular circumstances. In some cases such as LC-MS and GC-MS this is a crucial first step prior to further analysis, whereas in the case of diode array HPLC it is often omitted, and less essential.

7.3 CHEMICAL COMPOSITION OF SEQUENTIAL DATA

Chromatograms are characterized by a series of elution profiles each corresponding to a single compound. The observed chromatographic response at each elution time is obtained from the spectra of different numbers of compounds. Many chromatographers try to identify selective regions where only one compound elutes, and so to obtain spectra of each pure component in a mixture. Chemometrics methods can be employed to analyse these complex spectra of mixtures as they change, normally with time, to determine where the profiles for each species start and end.

As is usual, there is no commonly agreed terminology. Some people approach the problem from the perspective of practising analytical chemists and others from a statistical or computational viewpoint.

One concept is the idea of 'composition' which we will use below. A chromatogram of overlapping components can be split into regions of different composition:

- A composition 0 region is where no compounds elute.
- Composition 1 is where only one compound elutes.
- Composition 2 is where two compounds coelute.

In some cases regions of composition 3 or higher can be found. This is illustrated in Figure 7.6.

Mathematically minded chemometricians like to think more in terms of matrices, and often refer to the composition as the rank. The mathematical definition of rank of a

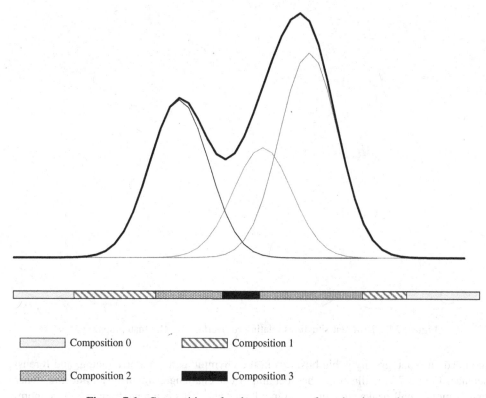

Composition 0   Composition 1

Composition 2   Composition 3

Figure 7.6   Composition of a chromatogram of overlapping peaks

matrix equals the number of nonzero PCs, so a matrix of rank two has two nonzero PCs
(Section 5.2.2). Hence in an ideal situation a composition 2 region is also a region of rank
2. Because of noise, this rank is approximate, but specific methods have been developed for
determining the number of significant components in the context of coupled chromatography
as will be discussed in Sections 7.4–7.7.

Chromatographers often prefer to call composition 1 regions selective regions, those of
composition 2 or above regions of coelution. It is important to read each paper carefully to
find out what terminology the authors use.

There are many different situations in which overlapping peaks occur. The simplest is
illustrated in Figure 7.7(a), where there are two partially overlapping peaks. Chemometrics
is used to complete the jigsaw by providing the elution profile in the composition 2 region,
the composition 1 information being easily obtained by conventional means. In this case
knowing the profiles in the region of overlap can help, for example, in quantitation of partially overlapping peaks. A more challenging situation is that of Figure 7.7(b) in which one
or more compounds do not have any composition 1 regions. However, the spectral features
of these so-called embedded peaks may be important for identification as well as quantitative purposes. In addition multivariate methods can be used to answer how many peaks
characterize a cluster. Are there several compounds buried within a region of the chromatogram? This can have real practical implications, a good example being the monitoring
of potential drugs in urine or blood samples, where there may be 50 or more peaks a chromatogram. Can we uncover hidden peaks? Such information can be important in athletics

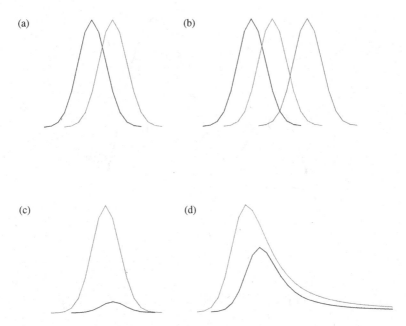

Figure 7.7   Different situations relating to overlapping chromatographic peaks

(coupled chromatography is big business in the Olympic games), horse racing, and forensic science. Figure 7.7(c) illustrates another important challenge, impurity monitoring, where one or more small peaks are embedded in a larger one. This can have tremendous economic and legal consequences in the pharmaceutical industry. When a new drug is produced it is often a requirement to ensure that impurities are below a certain limit (0.5 % or 0.1 %) and this is commonly done using coupled chromatography, so it is important to determine whether observed peaks are pure or not. Finally the problem of tailing peaks, illustrated in Figure 7.7(d), occurs in many practical situations. Unlike in Figure 7.7(a), where each compound has a clear selective region, only the faster eluting component exhibits selectivity in this example: these situations commonly occur using many methods for chromatography, especially rapid separations which can speed up a series of analyses several fold and many conventional approaches fail under such circumstances.

Below we will look at some of the main methods for determining where peaks are in a chromatogram and how many can be detected; most can be readily extended to other forms of two-way evolutionary multivariate data such as in equilibria studies (Section 8.1).

7.4 UNIVARIATE PURITY CURVES

Many laboratories depend on using analytical measurements that have significant economic and legal implications. A common example is the detection of dope, for example, in athletes and even in horse or greyhound racing. As drugs become more potent, smaller quantities have the desired effect, and in fact, in many cases it is not the parent drug but a metabolite that is potent. There are often huge legal arguments about the quality of evidence presented.

A common approach is to employ coupled chromatography of urine and blood samples. These chromatograms can be quite complex consisting of 50 or more peaks. The active compound is often buried underneath a major peak.

We need to be confident that we have actually found this additional compound in the chromatogram. False positives can involve major loss of revenue due to large law suits and unfavourable publicity. False negatives are equally unhelpful as they allow the criminal to get away with their act. So determining how many components characterize a cluster of chromatographic peaks and so whether any unexpected minor component is detectable is an important application.

We will illustrate the methods in this chapter using a small simulation of two closely eluting peaks in diode array HPLC whose three-dimensional and summed elution profiles are given in Figure 7.8. The slower eluting compound is represented by a shoulder on the right-hand of the main peak in the profile. Note that there is also some information in the second spectroscopic dimension. Can we be certain that there are two peaks? More

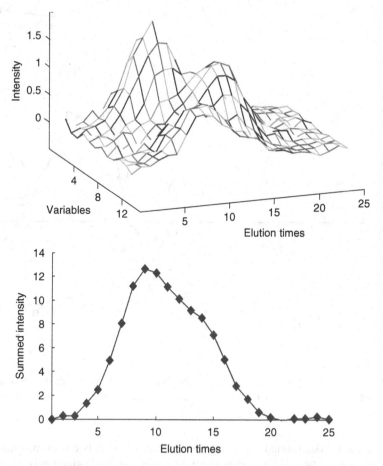

Figure 7.8 Two closely eluting peaks recorded over 12 spectroscopic variables and 25 elution times: three-dimensional plot and chromatographic profile

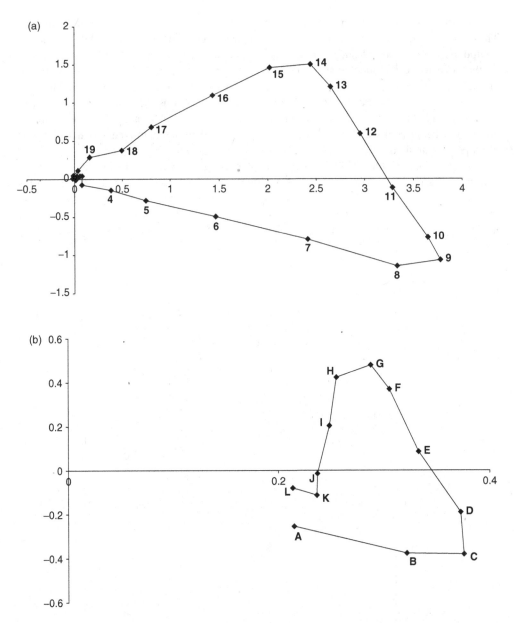

Figure 7.9 Scores (a) and loadings (b) plot of the raw unscaled data from the chromatogram in Figure 7.8, with elution times numbered and wavelengths indicated by letters; the horizontal axis corresponding to PC1 and the vertical axis to PC2

interestingly can we determine the points in time where each analyte is of maximum purity? If so, it is possible to find the spectra at these two points in time, which may give us useful diagnostic information. Note that in diode array HPLC the UV/visible spectra may not be particularly informative in terms of structure elucidation, but in LC-MS and LC-NMR the corresponding spectra can be very helpful.

A first step is to find wavelengths or variables that are most diagnostic of each compound. There are various ways of doing this, but a common approach is to employ PC plots (see also Sections 5.3 and 5.5). Figure 7.9 is the scores and loadings plots of the raw unscaled data. Comparing these, we will take wavelengths D and F as diagnostic of the fastest and slowest eluting compound, respectively (the alternative of wavelengths C and G are not always good choices as the intensity at one wavelength may be close to 0 for one of the compounds – choosing extreme wavelengths can pose problems if the aim is to calculate ratios between them). Many instrument manufacturers' software uses rather less sophisticated approaches or allow the user to select his or her favourite wavelengths. In fact, if looking for specific impurities whose spectra are known, using known wavelengths is probably best.

Once these wavelengths have been chosen, a simple next step is to calculate the ratio of intensities at each of the wavelengths with time, see Figure 7.10(a). At first this appears somewhat discouraging, this is because at the edges of the data the spectra are dominated by noise so almost any ratio could be obtained, distorting the picture, for example the negative

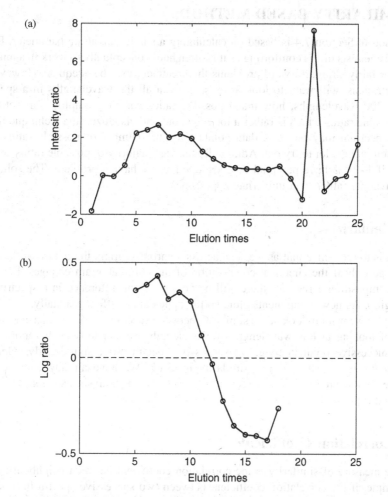

Figure 7.10 Ratios of intensities of characteristic wavelengths (a) over all elution times and (b) the logarithm of the ratio at the central part of the peak

ratios are of no physical meaning and arise from noise. Figure 7.10(b), however, is of a smaller region in the middle of the chromatogram, presenting the ratios on a logarithmic scale for clarity, and the pattern is now reasonably clear, the flat regions at the ends correspond to areas of purity (or selectivity) for each compound, the changing ratio in the middle reflects a region of overlap, where the composition of the chromatogram changes and so the ratio between intensities at the two wavelengths changes rapidly. This curve is sometimes described as a purity curve. If a chromatographic peak is pure, peak ratios do not vary much, but if there is an impurity or an unexpected coeluent, then there will be a systematic variation.

The method described above is intuitively easy for the chromatographer, however as usual, there are limitations. One problem is when there is very noisy data, where the response is not very reproducible. Using UV detectors the spectra are normally quite stable, but other, often more diagnostic, approaches such as MS can result in much more bumpy peak shapes and so intensities even in a region of purity are often not clear.

7.5 SIMILARITY BASED METHODS

The method in Section 7.4 is based on calculating a single univariate parameter. For many analytical chemists this is comforting as it is often hard conceptually to work in a multivariate world. For fairly straightforward problems this treatment may be adequate. However, many chemometricians, will want to look at most or even all the wavelengths in a spectrum. If there are 200 wavelengths, how many possible ratios can they employ? In fact there are 19900 possible ratios which is rather a lot to test out and visualize. In a technique like NMR where a spectrum may be 16 K data points long, the number of possible ratios becomes 134 million, quite a lot to try out. Admittedly the vast majority of possible ratios are useless, but even if $1-10\%$ have potential diagnostic value, we have a problem. The solution is to use multivariate rather than univariate approaches.

7.5.1 Similarity

The trick is to recognize that as we scan across a chromatogram, the nature of the spectrum will change only if the chemical composition of the chromatogram changes. For selective or pure composition 1 regions, there will be no significant difference in a spectrum over a given region. As new components elute so the appearance differs gradually.

Therefore we want to look at the similarity between successive spectra in a chromatogram. Instead of looking at this wavelength by wavelength, we can look at the entire spectrum at each successive point in time. Do successive spectra differ significantly? One way of checking this is by calculating a similarity measure. We have already come across this technique in the area of pattern recognition and cluster analysis (Section 5.6), and can apply similar principles in chromatographic purity assessment.

7.5.2 Correlation Coefficients

A simple measure of similarity is the correlation coefficient between neighbouring spectra. For example, if the correlation coefficient between two successive spectra in time is close to 1, they are similar. If there is a dip in correlation, there are likely to be two peaks overlapping with the composition of the chromatogram differing at successive elution times.

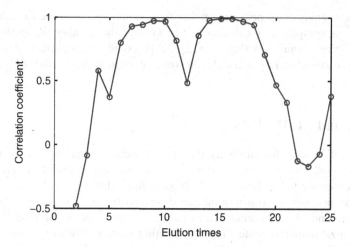

Figure 7.11 Graph of correlation coefficient between successive spectra in the chromatogram in Figure 7.8 against elution time

One problem is that the correlation coefficient over regions of noise can vary quite wildly, but almost certainly will not be close to 1. Figure 7.11 represents the graph of the correlation coefficient of between successive points in the chromatographic peak cluster of Figure 7.8. This would suggest that:

- points 7–9 are composition 1 or selective;
- points 10–13 are composition 2, or a region of overlap;
- points 14–18 are composition 1.

For more complex peak clusters, the flat plateaux help us find where there are regions of purity, and so guide us to the spectra of the pure components. This can be very valuable where four or five peaks closely elute.

Correlation coefficients can also be used in a slightly different way. Some people try to identify the single point in a region of a chromatogram where we are most confident that one pure component elutes. In the example of Figure 7.11, take the point at time 15 s. Then calculate the correlation coefficient of every other point in the chromatogram with this one. The spectrum that is least similar will be the one that is most likely to originate from another compound. Noise regions often have correlations close to 0. This procedure could be repeated for complex clusters, by first identifying one component, then the next by the most negative correlation with this first component. A third component could be determined by looking at which elution times are most dissimilar to the two already found and so on until the dissimilarity is not very great (as judged by the correlation coefficient) suggesting we have found the right number of components.

7.5.3 Distance Measures

Correlation coefficients are not the only similarity measures. As in cluster analysis, there are several distance measures. The normalized Euclidean distance is a common alternative. Each spectrum is first normalized so that its sum of squares totals 1, this is because we are not interested in intensity of peaks but rather their similarity. Then the Euclidean distance between successive spectra is calculated. The smaller the number the more similar the spectra, so a graph similar to that of Figure 7.11 could be calculated. In addition this approach can be extended, just as correlation coefficients, to looking at the similarity between pure spectra.

7.5.4 OPA and SIMPLISMA

The concepts of looking for similarity throughout a chromatogram have been developed further in two well known methods, namely OPA (Orthogonal Projection Approach) and SIMPLISMA (Simple to use Interactive Self Modelling Multivariate Analysis).

In OPA the first step is to determine the dissimilarity between the normalized spectra at each elution time and the normalized average spectrum over the entire chromatographic region (normalization in this context involves dividing each spectrum by its root mean square intensity rather than row scaling to a constant total and allows for negative intensities). This dissimilarity is calculated by setting up a matrix Y of dimensions $J \times 2$ for each elution time whose rows consist of the J spectral wavelengths recorded and whose two columns consist of the average spectrum and the spectrum at each specific elution time to be tested. The determinant of $Y' \cdot Y$ is higher the more different the spectra and can be plotted against elution time. The point that has a maximum is taken as the spectrum for the first pure component, which is the elution time whose characteristics are most different from the average. Then the average spectrum is replaced by the spectrum of this first component, and the calculation repeated to reveal a second component, being defined as the spectrum at the elution time most different from the first component. In the third step, the matrix Y has three columns, two consisting of the best spectra for the first two components and the third consisting of the test spectrum at each elution time. The number of rows of this matrix are extended each time a new pure spectrum is revealed. The algorithm stops either when a fixed number of pure components is identified or when the plot of the dissimilarity value appears to consist of noise, which is usually determined visually.

SIMPLISMA is based on fairly similar principles except that a purity value is computed rather than spectral dissimilarity.

7.6 EVOLVING AND WINDOW FACTOR ANALYSIS

Evolving Factor Analysis (EFA) is often loosely used as a name for a series of methods of which the first step is to use eigenvalues of principal components calculated over a region of the data matrix to determine which analytes have signals in which region. Many chemometricians enjoy using matrices and PCs, so EFA fits naturally into their conceptual framework. The idea is that the approximate rank of a matrix corresponding to a portion of a chromatogram equals the number of compounds in this region. We have already introduced the idea of the number of significant components or rank of a matrix (Sections 5.2 and 5.10), and a simple indicator is given by the size of successive eigenvalues. So it is possible to perform PCA on a region of the chromatogram and determine the size of the largest eigenvalues. If the first two are very large compared with the rest, this is an indication that there are only two components in this region. The region is then changed ('the window') by moving along the chromatogram and the calculation repeated. There are, however, many variations on the theme. Note that there is a lot of different terminology in the literature, and the original description of EFA involved a second step of transformation (or rotation or resolution) which is regarded by many as a completely separate phase and so is not described in this Section: methods for resolution are described in Sections 7.8–7.10.

7.6.1 Expanding Windows

Many of the earliest methods reported in the literature involved expanding windows. A simple implementation is as follows:

1. Perform PCA (usually uncentred) on the first few data points of the series, e.g. points 1–4. This would involve starting with a $4 \times J$ matrix where J wavelengths are recorded.
2. Keep the first few eigenvalues of this matrix, which should be more than the number of components expected in the mixture and cannot be more than the smallest dimension of the starting matrix, e.g. 3.
3. Extend the matrix by an extra point in time to a matrix of points 1–5 in this example, and repeat PCA, keeping the same number of eigenvalues as in step 2.
4. Continue until the entire data matrix is covered, so the final step involves performing PCA on a data matrix of dimensions $I \times J$ if there are I points in time.
5. Produce a list of the eigenvalues against matrix size. This procedure is often called *forward expanding window factor analysis.*
6. Next take a matrix starting at the opposite end of the dataset, from points $I - 3$ to I and perform steps 1 and 2 on this matrix.
7. Expand the matrix backwards, so the second calculation is from points $I - 4$ to I, and so on.
8. Produce a list similar to that in step 5. This procedure may be called *backward expanding window factor analysis.*

The eigenvalues are normally plotted on a logarithmic scale, against the end of the window. From the chromatographic profile of Figure 7.8, we obtain the eigenvalue plot of Figure 7.12. From these graphs we can tell how many components are in the mixture and where their elution windows are as follows:

1. Although the third eigenvalue increases slightly this is largely due to the data matrix increasing in size and does not indicate a third component, so there are two components in the mixture.
2. In the forward plot, it is clear that the fastest eluting component has started to become significant in the matrix by time 4, so the elution window starts at time 4.
3. In the forward plot, the slowest component starts to become significant by time 10.
4. In the backward plot, the slowest eluting component starts to become significant at time 19.
5. In the backward plot, the fastest eluting component starts to become significant at time 14.

This allows us to determine approximate elution windows of each component, namely the fastest elutes between times 4 and 14, and the slowest between times 10 and 19.

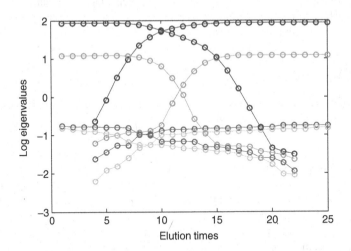

Figure 7.12   Expanding window eigenvalue plot for data in Figure 7.8

7.6.2 Fixed Sized Windows

Fixed sized window methods involve moving a window of a few data points along the chromatogram rather than increasing the size of a window and are implemented as follows:

1. Choose a window size, usually a small odd number such as 3 or 5. This window should not be more than the maximum composition (or number of coeluents at any point in time) expected at any region in the chromatogram, ideally somewhat larger. It does not need to be as large as the number of components expected in the overall system, only the maximum overlap anticipated.
2. Perform PCA (usually uncentred) on the first points of the chromatogram corresponding to the window size.
3. Record the first few eigenvalues of this matrix, which should be no more than the highest composition expected in the mixture and cannot be more than the smallest dimension of the starting matrix.
4. Move the window successively along the chromatogram. In most implementations, the window is not changed in size.
5. Produce a list of the eigenvalues against the centre of each successive matrix.

Similar to expanding window methods, an eigenvalue plot can be obtained often presented on a logarithmic scale. Figure 7.13 involves using a three point window for the calculations, using logarithmic scaling for presentation, and can be interpreted as follows:

1. It appears fairly clear that there are no regions where more than two components elute.
2. The second eigenvalue appears to become significant between points 10 and 14. However, since a three point window has been employed this would suggest that the chromatogram

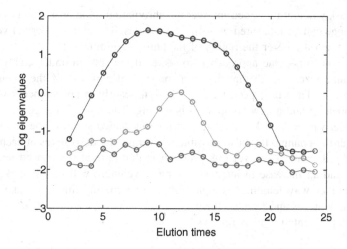

Figure 7.13 Fixed sized window (size = 3 data points) eigenvalue plot for data in Figure 7.8

is composition 2 between points 9 and 15, compatible with the interpretation of the expanding window graphs discussed above.
3. The first eigenvalue mainly reflects the overall intensity of the chromatogram.

Elution limits for each compound can be obtained just as for expanding window methods.

7.6.3 Variations

There are several variations on this theme. Examples involve both moving and changing the window size at the same time. Variable selection in the spectroscopic direction prior to performing local PCA is quite common, either choosing the best wavelengths (e.g. UV/visible spectroscopy) or in some cases such as NMR performing factor analysis on a spectroscopic peak cluster. Some people use statistical approaches such as the F-test (Section 3.8) to give an indication of the significance of the eigenvalues, comparing their size with eigenvalues in noise regions, but this depends on the assumption that the noise is normally distributed.

7.7 DERIVATIVE BASED METHODS

Many spectroscopists employ derivatives, common examples being in electron spin resonance and UV/visible spectroscopy. The aim is to resolve out signals, taking advantage of the signal narrowing in higher derivatives. Combined with a smoothing function (e.g. Savitzky–Golay) noise is also reduced as already described (Section 4.8). However, in the case of two-way signals, derivatives can be used for quite different purposes, to follow the changes in spectral characteristics over a chromatographic peak cluster. If the spectrum changes in nature this is a sign that a new analyte is appearing, if the spectrum does not change much, the peak is probably pure. To illustrate the use of derivatives in this way, return to the example of peak purity. Is a chromatographic peak pure? Are there significant impurities buried within the main peak and can we detect them? One way of answering this question is to ask whether the spectrum over the chromatographic peak changes significantly in nature. Change can be measured by using derivatives. The closer a derivative is to 0, the less the change and the purer the region of the chromatogram.

How can we do this? One approach is to scale all the spectra under a chromatographic peak to the same size, e.g. so that the sum (or sum of squares) of the intensity at each sampling time is 1. Then if a spectrum changes significantly in appearance, something suspicious is happening and a new component is eluting. The rate of change can be measured by taking the derivative at each wavelength in the row scaled spectra. The larger the magnitude, the bigger the change. It is also possible to average the derivatives at each wavelength to give an overall value. However one problem is that derivatives can be negative rather than positive, and an increase in intensity at one wavelength will be counterbalanced by a decrease at another wavelength. A simple way of overcoming this is to take the absolute value of the derivatives and average them.

A simple implementation is as follows:

1. Row scale the spectra at each point in time. For pure regions, the normalized spectra should not change in appearance.
2. Calculate the first derivative at each wavelength and each point in time. Normally the Savitzky–Golay method, described in Section 4.8, can be employed. The simplest case is a 5-point quadratic first derivative. Note that it is not always appropriate to choose a 5-point window, this depends very much on the nature of the raw data.
3. Convert to the absolute value of the derivative calculated in step 2.
4. To get an overall consensus, average the absolute value of the derivatives at each wavelength or measurement. There are a number of ways of doing this, but if each wavelength is to be assumed to have approximate equal significance, and providing absolute values are computed, this can be done by scaling the sum of values in step 3 above for each wavelength over all points in time to 1 and subsequently averaging these values over all wavelengths. This step of scaling all values in time for each wavelength to a constant sum is not particularly crucial if all variables are of approximately the same magnitude.

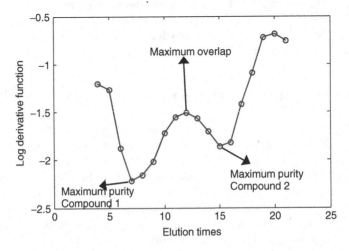

Figure 7.14 Use of derivatives to determine areas of maximum purity and overlap for the data in Figure 7.8

It is often easiest to plot the derivative function on a logarithmic scale. The regions of maximum purity then are fairly obvious. For the chromatogram of Figure 7.8, we see obvious features in the derivative plot of Figure 7.14. The 'best regions' can then be used to pull out the spectra of the pure compounds if possible. Regions of overlap might indicate impurities or in this case simply a composition 2 region between the two pure peaks.

There are many other uses of derivatives, often buried within complex algorithms for determining the numbers of components in a mixture. Derivative based approaches to multivariate data can be employed in many other applications, for example, if we are monitoring a process with time, does the composition of the product vary significantly? These approaches are complementary to multivariate methods and are often very effective, for example, when there are severely tailing peak shapes, where eigenvalue based methods tend to break down.

7.8 DECONVOLUTION OF EVOLUTIONARY SIGNALS

Most real world materials contain a huge mixture of compounds, for example, blood, urine, environmental and forensic samples. It is common to use either chromatography or spectroscopy of these rather complex mixtures to determine their composition.

Above we have discussed methods for determining, for example, how many compounds characterize a cluster of peaks, and where they elute. Many people want more detailed information, usually a spectrum to allow them to identify each individual compound, and sometimes a quantitative measurement. To do this it is often important to deconvolute (or resolve or decompose) a chromatographic mixture into its individual chemical constituents using computational methods. Many of the classical applications have been first applied to coupled chromatography, but there are also some well established examples in spectroscopy, an interesting one being pyrolysis MS.

There are several pieces of information required.

1. *Profiles* for each resolved compound. These are the elution profiles (in chromatography).
2. *Spectra* of each pure compound. This may allow identification or library searching. In some cases, this procedure merely uses the multivariate signals to improve on the quality of the individual spectra which may be noisy, but in other cases, such as an embedded chromatographic peak, genuinely difficult information can be gleaned.
3. *Quantitative information.* This involves using the resolved data to provide concentrations (or relative concentrations when pure standards are not available) often from peak areas of the resolved components. This can be quite important. For example, there is a legal requirement that impurities in pharmaceuticals are below a certain concentration, can we determine this by chromatography? In forensic science, what is the concentration of a potential performance enhancer? Note that calibration (Chapter 6) is a common alternative, but requires some independent information about pure standards or samples.
4. *Automation.* Complex chromatograms may consist of 50 or more peaks, some of which will be noisy and overlapping. Speeding up procedures, for example, using rapid chromatography in a matter of minutes resulting in considerable overlap, rather than taking half an hour per chromatogram, often results in embedded or overlapping peaks. Chemometrics can ideally pull out the constituents' spectra and profiles. A good example involves the GC-MS of environmental extracts, can we automatically pull out the spectra of all identified components? This could save a technician many hours going through a chromatogram manually.

To achieve this we perform deconvolution, involving pulling out the individual profiles and spectra of each compound. This step normally involves some form of regression either Multiple Linear Regression (MLR) (Section 6.4) or Principal Components Regression (PCR) (Section 6.5). Often the algorithms or packages are called by quite sophisticated names but can be broken down into a series of simple steps, and with an understanding of basic chemometric principles it is possible to appreciate the basis of the main steps.

Most methods first try to determine some information about each compound in a mixture. Typical examples are as follows:

1. *Spectra* of some or all pure compounds, as obtained from the selective or composition 1 regions.
2. *Pure or key variables* e.g. diagnostic masses in MS or resonances in NMR for each compound in a mixture.
3. *Selective regions* of the chromatogram or spectra for each pure compound.
4. *Elution profiles* of one or more pure compounds, in coupled chromatography, this is sometimes possible after first determining pure variables for example one may have diagnostic m/z values in LC-MS for each component even though their chromatography coelutes.

Where there is a comparable bit of information for each significant compound in the mixture, MLR is often be used to fill in the picture, rather like completing a jigsaw. If there is complete information for only some compounds (e.g. an embedded peak in diode array HPLC does not have any regions of purity) then PCR is an approach of choice. Over the last few years new methods have been developed based on constraints, that is knowledge of properties of the spectra and elution profiles, e.g. that they should be positive and that chromatographic peaks corresponding to pure compounds should be unimodal. There is however no universally accepted approach. Some will be described briefly below.

7.9 NONITERATIVE METHODS FOR RESOLUTION

Many methods for resolution of mixtures involve finding selective variables for each compound: these could be wavelengths, m/z values or elution times. There are two fundamental situations. The first is where every component in a mixture contains some selective or unique information, that is used as a handle for factor analysis, and the second where there are some components without selective information.

7.9.1 Selectivity: Finding Pure Variables

Where all components have some selective information, the first step is to find selective or key variables for each component in the mixture.

On approach is to look at the chromatogram. In Section 7.3 we discussed the idea of composition and then subsequently outlined a number of approaches as to how to determine the composition of each region of a chromatogram. In coupled chromatography such as diode array HPLC, in favourable cases, each component in a mixture has a selective region. This means that some information such as the spectrum of each pure component can be obtained. Be warned it is not always so simple, because the selective regions may be quite small, they may be on the tails of peaks or they may be noisy. However, by identifying such a region using one or more of the methods described above, it is possible to get a good estimate of the spectrum of each component for the next step. Sometimes the average spectrum over the selective region is calculated, but sometimes PCA on each selective region can be employed as a form of smoothing, the loadings of the first PC being an estimate of the spectrum. Alternatively the regions of the chromatogram which are selective can be employed in their own rights as starting estimates for the next stage.

An alternative approach, in certain types of coupled chromatography, such as GC-MS, is to find diagnostic spectroscopic variables, such as m/z values for each component, even if some chromatographic peaks are embedded and have no composition 1 region. The profile at each m/z value can be used as an estimate of the elution profile for each component, the trick is to find which masses correspond to which compound. Sometimes this can be done manually, for example, by knowledge of the chemical structures. However, a tricky problem, often encountered in pharmaceutical chemistry, is the resolution of isomers: in this case two (or more) closely eluting compounds may have very similar fragmentation patterns and it can be difficult to find what is often not a very intense ion diagnostic of each isomer. A completely different application is in automated resolution of complex chromatograms, chemometrics helps speed up the process. One interesting way of finding the m/z values is by a PCA plot where the masses characteristic of each component cluster at the extremes. However, use of such graphs depends on judicious use of scaling often requiring standardization of the masses.

7.9.2 Multiple Linear Regression

Once we have found our diagnostic variables, what is the next stage? MLR (Section 6.4) has been introduced several times before and is a common way to complete the picture.

If the spectra (S) of all pure compounds are known, then it is a simple procedure to determine elution (or concentration) profiles using MLR since:

$$X = C \cdot S$$

so

$$C \approx X \cdot S^+$$

where S^+ is the pseudoinverse (Section 2.4). This approach works well, providing the spectra have been estimated well. This allows full elution profiles to be built up, and chemometrics is used to fill in the unknown portions of the data.

A different application is where elution or concentration profiles are known (or can be approximated) using key variables such as m/z values, or wavenumbers. Then we can estimate the spectra by:

$$S \approx C^+ \cdot X$$

This can be very useful, for example in obtaining the MS or NMR of an unknown, and is especially helpful in the spectroscopy of mixtures. One problem is that the estimate of the elution or concentration profile may not necessarily be very good if some of the variables are noisy: however, there are ways round it. A common one is to take several selective variables for each component and average the profile (or perform PCA on the selected variables for each component and take the first PC) and then use these as the estimates. Another involves having several variables and increasing the number of variables in each step, starting with one diagnostic variable per component, obtaining a first guess of the profiles, increasing the number of variables to obtain a better estimate and so on. Iterative approaches will be described in Section 7.10 and are complementary to the approaches in this section.

7.9.3 Principal Components Regression

Some more mathematically oriented chemometricians prefer PCR (Section 6.5). Readers might wish to refresh themselves by reading Section 5.2.2 to understand the relationship between PCs (or abstract factors) and true factors (often related to specific chemical species).

There are several sophisticated algorithms in the literature, but a common basis of most of the approaches is to try to find a rotation or transformation or regression matrix between the scores of a portion of the data and the known information. The first step involves performing PCA over the entire chromatographic region of interest, and keeping a scores matrix T which has as many columns as compounds are suspected to be in the mixture. For example, if we know the composition 1 regions of all the compounds in a portion of chromatogram, then we can estimate their elution profiles in their composition 1 region. Calling c_{1a}, the concentration profile in the composition 1 region of compound a, try to find

$$c_{1a} = T_{1a} . r_{1a}$$

where T_{1a} is the portion of a scores matrix on the overall matrix T that is restricted to the composition 1 region of compound a using the pseudoinverse of T_{1a}. The vector r_{1a} provides the transformation from scores (often called abstract factors) to concentration profiles. Find a similar vector for each of the compounds in the mixture, and combine each vector to obtain a matrix R each column representing a separate compound. Then take the overall scores and loadings matrices to determine

$$C \approx T . R$$

and

$$S \approx R^{-1} . P$$

as estimates of the elution profiles and spectra, respectively. Notice that the result of transforming the loadings by this mechanism is likely to produce smoother spectra than simply taking an average of each composition 1 region, as PCA acts as a method for noise reduction.

Some PCA based approaches are also called Evolving (or Evolutionary) Factor Analysis but there is a significant terminology problem in the literature, as EFA (Section 7.6) also denotes the first step of using eigenvalue plots to determine how many components are in a mixture and when they start and end. There is a lot of history in the names chemometricians use for methods in this area, but the main thing is to keep a very cool head and understand the steps in each approach rather than become confused by the huge variety of names.

7.9.4 Partial Selectivity

Somewhat harder situations occur when only some components exhibit selectivity. A common example is a completely embedded peak in diode array HPLC. In the case of LC-MS or

LC-NMR this problem is often solved by finding pure variables, but because spectral information in UV/visible spectra of two different compounds can be completely overlapping it is not always possible to treat data in this manner.

Fortunately PCA comes to the rescue. In Section 6.5 we discussed PCR and showed that it is not necessary to have information about the concentration of every component in the mixture, simply a good idea of how many significant components there are. So in the case of resolution of two-way data these approaches can easily be extended. There are several different ways of exploiting this.

One approach uses the idea of a zero concentration, or composition 0, window. The first step is to identify compounds that we know have selective regions, and determine where they do *not* elute. This information may be obtained from a variety of approaches as discussed above. In this region we expect the intensity of the data to be zero, so it is possible to find a vector r for each component so that:

$$0 = T_0 . r$$

where 0 is a vector of zeros, and T_0 is the portion of the scores in this region; normally one excludes the region where no peaks elute, and then finds the zero component region for each component. For example, if we record a cluster over 50 data points and we suspect that there are three peaks eluting between points 10 and 25, 20 and 35 and between points 30 and 45, then there are three T_0 matrices, for the fastest eluting component this is between points 26 and 45. Notice that the number of PCs should be made equal to the number of compounds suspected to be in the overall mixture. There is one problem in that one of the elements in the vector r must be set to an arbitrary number, usually the first coefficient is set to 1, which this does not have a serious effect on the algorithm. It is possible to perform this operation on any embedded peaks because these also exhibit zero composition regions. If one is uncertain about the true number of compounds (or PCs) it is possible simply to change this and see what happens with different guesses and see whether sensible results are obtained. A matrix R can be set up as above, and the estimated profiles of all the compounds obtained over the entire region by $\hat{C} = T.R$ and the spectra by $\hat{S} = R^{-1}.P$.

We will illustrate this using the data of Figure 7.15 which corresponds to three peaks, the middle one being completely embedded in the others. From inspecting the data we might conclude that compound A elutes between times 4 and 13, B between times 9 and 17 and C between times 13 and 22. This information could be obtained using methods described above which we will assume has been done already. Hence the zero composition regions are as follows:

- compound A: points 14–22;
- compound B: points 4–8 and 18–22;
- compound C: points 4–12.

Obviously different approaches may identify slightly different regions. The result using the method described above is given in Figure 7.16. There is no pure region for the middle peak.

Many a paper and thesis has been written about this problem, and there are a large number of modifications to this approach, but here we will use one of the best established approaches.

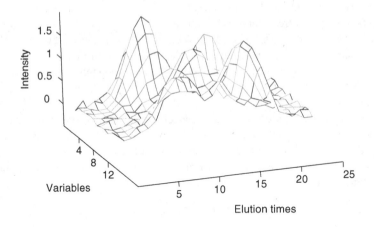

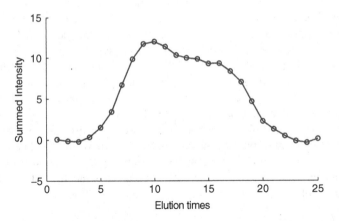

Figure 7.15   Three closely eluting peaks recorded over 12 spectroscopic variables and 25 elution times: three-dimensional plot and chromatographic profile

7.10 ITERATIVE METHODS FOR RESOLUTION

The methods above are noniterative, that is they have a single step that comes to a single solution. However, over the years, chemometricians have refined these approaches to develop iterative methods, that are based on many of the principles above but do not stop at the first answer. These methods are particularly useful where there is partial selectivity, not only in chromatography, but also in areas such as reaction monitoring or spectroscopy as described further in Chapter 8.

Common to all iterative methods is to start with a guess about some characteristic of each component. The methods described in Section 7.9 are based on having good information about each component, whether it is a spectrum or composition 0 region. If the information is poor the methods will fall over, as there is only one chance to get it right, based on the starting information. In contrast, iterative methods are not so restrictive as to what the first

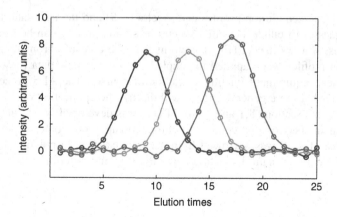

Figure 7.16 Reconstructed profiles from data in Figure 7.15 using method of Section 7.9.4

guess is. If perfect knowledge is obtained about each component at the start, then this is useful, but iterative methods then may not be necessary.

There are however many other approaches for the estimation of some properties of each component in a mixture: remember that for iterative methods these estimates do not need to be very good and are just a start. One simple method is called a 'needle' search. For each component in a mixture, the data points of maximum purity in the chromatogram are taken as the initial guesses of the elution profiles, these are simply spikes at this first stage, one spike for each component. Statistical methods such as varimax rotation, discussed in Section 13.6 in the context of the food industry, can also be used to provide first guesses, the scores of the first varimax rotated components corresponding to the first guesses of the elution profiles (in coupled chromatography). OPA and SIMPLISMA (Section 7.5.4) can also be employed, each pure variable being the starting guess of each component. A big variety of other methods have also been used. None of these approaches are likely to provide perfect estimates of elution profiles.

ITTFA (Iterative Target Transform Factor Analysis) is then commonly used to improve these guesses. PCA is performed first on the dataset, and a rotation or transformation matrix R is found to relate the scores to the first guesses of the spectra, just as above. This then allows the spectra to be predicted in turn. The principle is discussed in Section 7.9.3 for a noniterative implementation.

However, the next stage is the tricky part. The predicted spectral matrix $\hat{S}$ is examined. This matrix will, in the initial stages, usually contain elements that are not physically meaningful, for example, negative regions. These negative regions are changed, normally by the simple procedure of substituting the negative numbers by zero, to provide new and second estimates of the spectra. This in turn results in fresh estimates of the elution profiles which can then be corrected for negativity and so on. This iteration continues until the estimated concentration profiles C show very little change: several criteria have been proposed for convergence, many originally proposed by Malinowski.

The performance of ITTFA is dependent on the quality of the initial guesses, but can be an effective algorithm. Non-negativity is the most common condition but other constraints can be added.

Alternating Least Squares (ALS) rests on similar principles, but the criterion for convergence is somewhat different and looks at the error between the reconstructed data from the product of the estimated elution profiles and spectra (*C.S*) and the true data. In addition as implemented guesses of either the initial spectra or elution profiles can be used to the start of the iterations, whereas for ITTFA it is normal, in the case of chromatography to start with the elution profiles when applied to coupled chromatographic data (however, there is no real theoretical requirement for this restriction of course). The ALS procedure is often considered somewhat more general and we will discuss the application of ALS methods in other contexts later (Sections 8.1 and 12.6). ITTFA was developed very much with analytical applications in spectroscopy coupled to chromatography or another technique in mind, whereas the idea of ALS was borrowed from general statistics. In practice neither method is radically different, the main distinction being the stopping criterion.

8

Equilibria, Reactions and Process Analytics

8.1 THE STUDY OF EQUILIBRIA USING SPECTROSCOPY

An important application of chemometrics involves studying equilibrium processes by spectroscopy, especially using ultraviolet (UV)/visible techniques for compounds with chromophores, which includes many molecules of pharmaceutical and biological interest as well as inorganic complexes of organic ligands. In the case of biologically active compounds, with several ionizable groups, protonation effects can often have a significant influence on the activity, so it is useful to be able to determine how many ionizable groups are in a compound, what their pK_as are and so their charges at different pHs. A common approach is to perform pH titrations, recording the UV/visible spectra at different pH values. As each group is deprotonated with increasing pH, often there are significant changes in resonance and so chromophores. Chemometrics methods can be employed to analyse this pH profile of spectra. Figure 8.1 is of a series of UV spectra recorded with pH of a mixture of yeast tRNAs [1]. Although appearing at first rather uninteresting, Figure 8.2 is of a projection and shows that there are definite changes in appearance with pH, which could then be interpreted in terms of chemical equilibria.

A multivariate matrix is obtained involving sample number as one dimension and spectroscopic wavelength as the other.

Many methods have been devised to look at these multivariate matrices, a good summary being in Malinowski's book [2]. Many of the principles are similar to those used in diode array high performance liquid chromatography (diode array HPLC). The first step is to find out how many components are in the data. This can be done by a variety of methods such as eigenvalue plots (Section 7.6), Principal component (PC) plots (Section 5.3), derivatives (Section 7.7) and other quite sophisticated approaches such as OPA (orthogonal projection approach) and SIMPLISMA (Section 7.5). These help identify how many unique chemical species exist in the profile and between which pHs. Once this is done, it is then usual to determine the spectrum of each pure species. In equilibrium studies there are often regions, for example, of pH, where one compound predominates or uniquely is present. If this is not so, it is possible to obtain an estimate of the purest points in the profile, and use these as a first stab of the spectrum of each species.

Applied Chemometrics for Scientists   R. G. Brereton
© 2007 John Wiley & Sons, Ltd

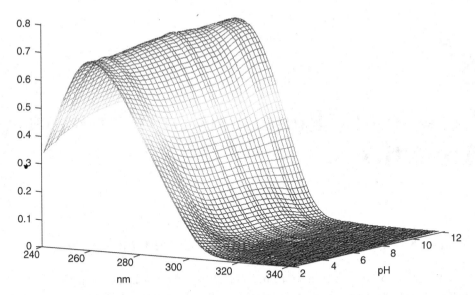

Figure 8.1   Typical pH profile of yeast tRNAs [1]

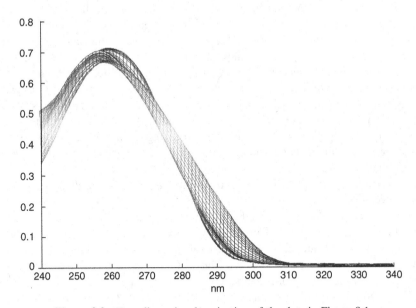

Figure 8.2   Two-dimensional projection of the data in Figure 8.1

The next step is to find the pure profiles of each compound, for example, as a function of pH. Figure 8.3 illustrates the ideal result of the chemometric analysis. If lucky and pure profiles are available, then Multiple Linear Regression (MLR) (Section 6.4) can be employed since:

$$X = C \cdot S$$

and S is known.

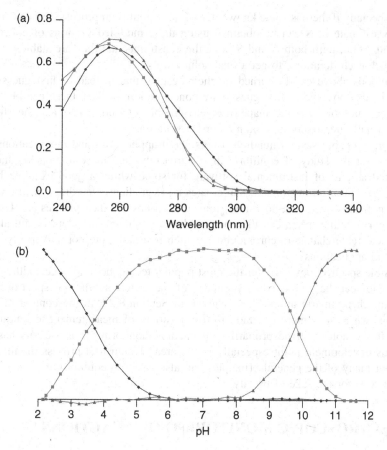

Figure 8.3 (a) Spectra and (b) titration profiles that can be obtained from an equilibrium titration

However, in many cases life is not so simple, and only an estimate of the spectra can be obtained as there may not be pure or selective regions of the spectra that can be found, in addition sometimes only a guess of the spectra can be available. This has led to a deluge of approaches for handling these data. Many are related, and go under a variety of names such as Multivariate Curve Resolution (MCR), Self Modelling Curve Resolution (SMCR), (Iterative) Target Transformation Factor Analysis [(I)TTFA], and Alternating Least Squares (ALS). All varieties are somewhat different, and have their pros and cons, but many methods differ in how they find the initial guesses or vectors prior to iteration. These methods have been introduced in the context of coupled chromatography in Section 7.10.

A feature of most of these methods is that different criteria can be employed to assess the desirability of the model. ALS is a popular choice. Consider estimating, perhaps imperfectly, the spectra above. We can now estimate in turn the concentration profiles by:

$$C = X . S^+$$

If a good model is obtained the concentration profiles should obey certain properties: the most common is non-negativity, but other constraints can be imposed such as unimodality if desired, especially if there is some knowledge of the equilibrium process. Hence C is altered and a new estimate of S can be obtained using these modified versions of concentration profiles and so on, until both C and S obey the constraints and they are stable (i.e. do not change much each iteration), to get a final solution.

The methods above can be turned on their head in that instead of first guessing pure spectra it is also possible to first guess pure concentration profiles, these can be obtained, for example, via looking at the shape of eigenvalue plots (Section 7.6), and then following the steps, usually iteratively, to obtain a correct solution.

There are probably several hundred papers about applications and modifications of factor analysis for the study of equilibria in spectroscopy, and these approaches have been incorporated as part of instrumental software for some while, a good example being an automated pK_a titration instrument, that creates a pH gradient, recording spectra as the pH is changing, then working out how many ionizable groups and their profiles are. Using conventional curve fitting, pK_as can then be established, although this depends on all groups having a detectable change in chromophore, which is usually, but not universally, the case in biological applications.

UV/visible spectroscopy is by far the most popular technique in this area, although other spectroscopies can be employed, a summary of the different types of spectroscopy and the relevant chemometric approaches is given in Section 8.2.5 in the context of reaction monitoring (where it is more common to use a variety of measurement techniques). The focus on this chapter will be primarily on reaction monitoring, because this has been a tremendous development point especially in the area of industrial process monitoring and control, but many of the general principles can also apply to equilibrium studies, but will not be repeated for the sake of brevity.

8.2 SPECTROSCOPIC MONITORING OF REACTIONS

Over the past few years there has been an increasing ability to study reactions in real time. This allows information on a process to be generated as it progresses, rather than off-line, e.g. via sampling and performing HPLC or thin layer chromatography (TLC). This means that we can now automatically generate information about the reaction every few seconds. Simple measurements such as pH, temperature, or electrochemical data are easy to obtain, but over the past few years with increasing emphasis on miniaturization, probes for several spectroscopies such as infrared (IR)(both mid and near), UV/visible and Raman have been marketed. This means that the chemist who wants to optimize a yield or monitor a process can obtain complex spectroscopic data as the reaction progresses, and look at it either on-line (in real time) or more commonly, at present, at-line in which the data are obtained during the reaction but the analysis is performed afterwards.

However, the spectra obtained contain a combination of signals from all reactants (that can be monitored spectroscopically) rather than individual compounds on chromatographic column, and so usually require data analysis to resolve the signals into their components, and to reconstruct concentration profiles. There are several common types of spectroscopy and each pose different challenges for the chemometrician. Figure 8.4 illustrates a typical experimental arrangement in which one arm of the reaction vessel contains a spectroscopic probe.

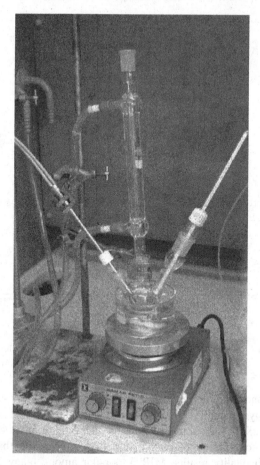

Figure 8.4 Insertion of spectroscopic probe into a reaction mixture via the left side arm of the reaction vessel

8.2.1 Mid Infrared Spectroscopy

Many organic chemists are suspicious of chemometrics approaches and like to see individual spectral peaks for each compound in a reaction mixture. Mid infrared (MIR) spectrometry has been used for some years for such purposes. The idea is that the characteristic peak for each compound that is IR active changes with time and this change then can be used to follow the reaction, as illustrated in Figure 8.5. There are often clear and diagnostic peaks that can be tracked as the reaction proceeds, which is especially useful for on-line monitoring. One advantage of MIR is that the probes can be used over a variety of temperatures, another being that the results are easy to interpret in terms of fundamental chemistry, usually visually. Peak intensities can be fit to kinetic models. But there are sometimes problems with the data, such as baselines, and lower reproducibility compared with other spectroscopies. Some compounds do not exhibit clear or easily distinguished peaks, the most suitable ones having groups such as carbonyls that have very characteristic frequencies. Sometimes compounds

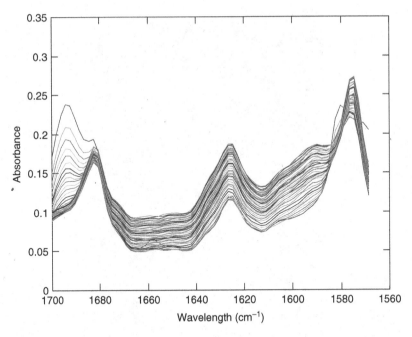

Figure 8.5 Use of MIR for reaction monitoring – the change in peak intensity can be monitored with reaction time

that occur in low concentrations are missed altogether, and MIR is not necessarily the best approach for studying small deviations from ideal behaviour. Characteristic peaks may overlap in MIR, especially for complex reactions where there are several isomers and by-products so require deconvolution to obtain quantitative information: this deconvolution does not always yield high quality results. MIR is popular among many chemists for reaction monitoring because it is easy to interpret how the appearance of peaks relates to the creation of products.

8.2.2 Near Infrared Spectroscopy

Near infrared (NIR) has been historically one of the main tools, especially of the industrial chemometrician. One reason is that cost effective NIR fibre optic probes have been available for over two decades, and have a variety of features. They are either invasive (able to be dipped in a solution), or else can simply be pointed at a reaction mixture externally. They can be used to study solids as well as liquids. One feature of NIR is that compounds often do not exhibit highly characteristic peaks, and often the changes are more subtle and in small regions of the spectrum, as in Figure 8.6. This means that it is not necessarily easy to distinguish the signal due to individual components and statistical approaches such as multivariate calibration (Chapter 6) have to be employed to obtain concentrations of individual compounds. However many of the pioneers of chemometrics cut their teeth on these approaches and most NIR instruments contain chemometrics software especially Partial Least Squares (PLS) (Section 6.6). NIR instruments tend to be much more stable than MIR instruments, although they still need regular calibrating. A method called calibration transfer

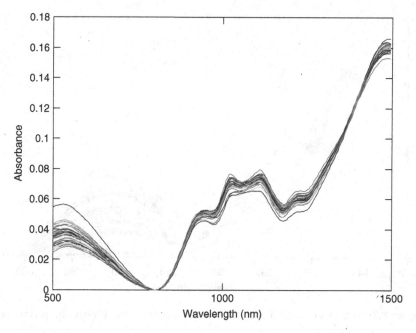

Figure 8.6 Typical set of NIR spectra as may be obtained during a reaction

can be employed to transfer a model obtained for example over 1 day or 1 week to a reaction monitored at a different time, and careful attention to these aspects are required if NIR approaches are to be employed regularly. Other common applications of NIR in combination with chemometrics are food analysis (Section 13.3) and biomedical analysis (Section 10.9).

However, in contrast to MIR, where primarily conventional physical chemical methods such as kinetics curve fitting are employed, in NIR mainly multivariate chemometrics approaches are required.

8.2.3 UV/Visible Spectroscopy

Recently probes for UV/visible or electronic absorption spectroscopy have become more commonplace, as discussed in Section 8.1 in the context of equilibria. This spectroscopic technique can be regarded as a halfway house between NIR and MIR. Like MIR each compound in a mixture can be characterized by individual peaks or fingerprints, but unlike MIR these peaks tend to be very broad relative to the overall spectrum, and so unless the reaction is quite simple, do not exhibit obvious characteristic peaks, as the signals of each chemical species overlap. Hence organic chemists tend to be a bit suspicious of this approach, but in fact UV/visible spectroscopy is quite a sensitive and very quantitative technique, although there usually is a need for deconvolution of the signals, e.g. using MLR (Section 6.4) or calibration using PLS (Section 6.6) and single wavelength monitoring is only successful for fairly simple situations, such as a first order reaction for which either the reactant or product has a clearly distinct absorbance. Quite a number of new algorithms have been developed over the past few years specifically for UV/visible spectroscopy as will be discussed below.

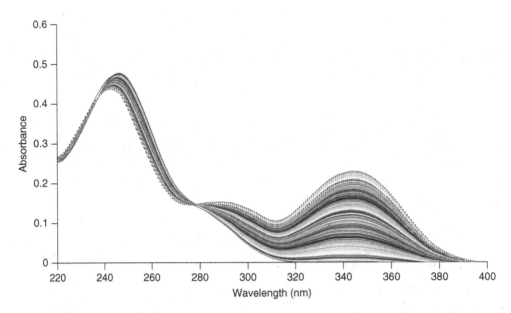

Figure 8.7 A Series of spectra obtained during a reaction using UV/visible spectroscopy

Figure 8.7 illustrates a typical series of spectra obtained using a UV/visible probe. In this example it can be seen that there are characteristic isobestic points: most of the observable wavelength range is encompassed by the spectra, unlike MIR where only certain very specific regions are interesting. There are many advantages in UV/visible spectroscopy, but one problem is that all significant compounds need chromophores, and the absorbances of each species should ideally be of approximately comparable size, plus the reaction should result in substantial differences in absorbance. In addition it is often hard to follow the reactions at low temperatures. Finally, like all spectroscopy, to use linear chemometric models, the spectroscopic intensities need to be within the Beer–Lambert law limits (a maximum absorbance of around 1.2–1.5 AUs) and if this is not so, probes of different pathlengths are required, which can be expensive.

8.2.4 Raman Spectroscopy

Raman spectroscopy probes have become fashionable recently: this technique is an alternative to MIR, but the relative intensities of bands differ in the two techniques. However many of the same methods can be employed to analyse the data.

8.2.5 Summary of Main Data Analysis Techniques

As the focus of this book is on data analysis, it is important to summarize the main data analytical techniques most common for the analysis of on-line spectroscopic data from reactions using different spectroscopies.

1. *MIR and Raman.* Generally conventional kinetic curve fitting of peak areas, but some deconvolution necessary if overlapping peaks.

2. *NIR.* Usually multivariate calibration such as PLS and a number of related classical chemometrics methods.
3. *UV/visible.* Deconvolution using MLR and pure standards, multivariate calibration, and constrained mixed multivariate models such as ALS and Iterative Target Transform Factor Analysis (ITTFA).

Below we will look in a little more detail at the methods for modelling reactions, primarily in the context of UV/visible spectroscopy.

8.3 KINETICS AND MULTIVARIATE MODELS FOR THE QUANTITATIVE STUDY OF REACTIONS

Kinetics is a subject that strikes fear into the hearts of many chemists. Deep in the bowels of most University chemistry libraries, there will be a dark corner containing shelves of large dusty hard bound books, written many decades ago with loving care by dedicated kineticists. Many of the early practitioners of kinetics were either pure physical chemists looking to understand reactions (primarily in the gas phase) in terms of fundamental phenomena, or physical organic chemists interested in reaction mechanism.

Over the past few years, kinetics has played an increasingly small role in the battery of techniques thought essential to the education of the mainstream chemist. An example is the study of organic synthesis, which students often choose mainly to avoid courses such as ones on kinetics. Synthesis is a very popular area, and one for which there is a heavy demand for graduates in industry. During the past decade or more, it has been common to graduate top students from excellent institutes without a detailed understanding of more than the simplest kinetic models. A problem now emerges that these chemists are gradually moving into positions of responsibility and no longer have the detailed understanding of kinetics that a past generation might have taken for granted.

Not so long ago it was quite hard to study solution reactions quantitatively. Simple approaches such as looking at the development of spots on a TLC plate help show us how the reaction is progressing, but do not give quantitative results. Off-line monitoring by quenching a reaction mixture and then running HPLC provides some level of quantitative results, but for many reasons it is hard to get good kinetic models from this and is a slow and tedious method if one wants to study the progress of a reaction under a large number of different conditions. Some reactions can be studied in an NMR tube. Some second order reactions are often studied under pseudo first-order conditions by putting one reactant in a large excess, but in real industrial practice we cannot do this, for example we cannot use pseudo first-order conditions in a manufacturing process, simply for the sake of monitoring, if one of the reagents is expensive. If a second order process is to be optimized with a ratio of reactants of 2:1, then these are the conditions we want to test in the pilot plant and hopefully use in the manufacturing plant, not a ratio of 100:1. Of course the physical chemist trying to obtain rate information is not worried because he or she is not concerned about the economics of bulk manufacturing and uses elegant ways round various difficulties which are not available to an engineer in a manufacturing plant.

Using on-line spectroscopy we are able to obtain quantitative information of how a reaction progresses, but the raw data will be in the form of mixture spectra requiring chemometrics to resolve into individual components, and kinetics is one way of fitting a model to these data that tells a chemist what is happening. However, the aims of modern industrial kinetics are rather different to those envisaged by the pioneers. The original kineticists were mainly concerned in interpreting the data in terms of fundamental physical processes, often using quantum chemistry or statistical mechanics in the final analysis. The modern day applied chemist is often more interested in kinetics from an empirical point of view. Knowing kinetic information may tell us when a product is at a maximum, or when the reaction has gone to 95 % completion, the influence of factors such as pH or temperature on the rate constant, or even if a reaction is deviating from expected behaviour (important if a factory is malfunctioning).

Both physical and analytical chemistry are quantitative disciplines. Yet there remains a gulf between the way the two cultures think. This is well illustrated by how people from each area approach the modelling of reactions, and it is important to appreciate this when trying to employ chemometric type approaches. It is possible to employ both types of information.

The physical chemist will be mostly interested in fitting data to a kinetic model. A classic example is of a clean first-order reaction of the form A → B, modelled by:

$$[B_t] = [A_0] - [A_t] = [A_0](1 - e^{-kt})$$

The average physical chemist will find a way of observing the concentration of either A or B independently, and then plotting for example $\ln([A_t])$ versus t as in Figure 8.8. The gradient of the best fit straight line gives us the rate constant and so allows us to predict, for example, the half-life of the reaction or the time the reaction reaches 95 % completion. In order to obtain a good estimate of the concentration of one of the reactants (ideally A because B may degrade further) one has to devise a suitable measurement technique. A common approach might be to find a region of a spectrum where A has a unique absorbance, such as an IR or UV/visible band and follow the changes in peak area or peak height with time. For many first-order reactions this is quite easy, and because the rate is not dependent on concentration, we do not even need to calibrate the peak area to concentration.

But in real life most reactions are more complicated, a typical example being a second-order reaction of the form:

$$A + B \longrightarrow C$$

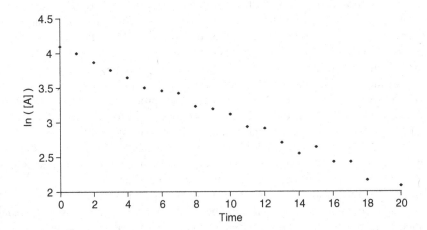

Figure 8.8 Typical kinetics plot for a first order reaction, as employed by physical chemists

Now there are three species. A common approach is to plot $\ln([A]_t / [B]_t)$ versus t giving a straight line of gradient $([A]_0 - [B]_0)k$ which enables us to determine the rate constant. The problems with this approach are several fold. First of all one needs accurate knowledge of the starting concentrations. A difficulty is that the estimate of the rate constant is dependent on the difference between two values. Small errors in these can have a major impact on the estimated rate constant. A particular experimental problem relates to dissolution, in that the starting materials may not dissolve at the same rate, especially when the reaction is warming up. Remember we are often dealing with real-world reactions that may occur in synthetic and process chemistry, and cannot, for example, necessarily have ideal conditions in a manufacturing plant. Second, it is necessary to estimate the concentrations of two components rather than one. In UV/visible spectroscopy there may no longer be selective regions in each spectrum for each of the chemical species, so it is increasingly difficult to quantify, especially if the reaction is complicated by degradation or side reactions which result in interferents that overlap with the existing spectra. Third, one needs to relate the peak heights to absolute concentrations and so requires good calibration. This is because a 2:1 peak area ratio will not usually correspond to a 2:1 concentration ratio as each compound will have different extinctions. In fact because standards need to be pure under exactly the conditions of the reaction this can be quite difficult as many reactions are run under conditions where one or more pure reactants is unstable, since, for example, pH, temperature and solvent, which are often factors that catalyse a reaction, can also influence the spectral characteristics of one or more components.

A physical chemist will probably concentrate on finding reactions and conditions under which he or she can obtain these pure signals and so can reliably measure the concentrations. This may take many months or years but an important aim of the physical chemist is to obtain hard models of reaction data, usually based upon kinetic theory, and try to fit the data to a known kinetic model. The advantage is that if we do know about the kinetics we can take this into account, and this allows us to use rate constants to predict, for example, reaction half lives. A problem is that if we do not know some information well, such as the starting concentrations, or if there are unknown side reactions, or the calibration is difficult, there can be serious difficulties in the estimation of these parameters. The traditional physical chemical approach is to improve the experimental procedure, until appropriate conditions are found to obtain this kinetic information. Although very elegant, this is often time consuming and means that reactions are studied under conditions that may not be suitable in real-life such as a manufacturing plant, and may not be optimized for a specific process.

The analytical chemist often approaches this problem entirely differently. He or she is not so concerned with the exact kinetics and may employ a spectroscopic probe (Section 8.2) such as NIR or UV/visible, where peaks overlap, to monitor the reaction. The main aim is to obtain concentration profiles for each reactant, rather than exact kinetics models, and the fact that no species exhibits a clean wavelength or spectral peak is not necessarily perceived as a problem.

Consider a reaction of the form $A \rightarrow B \rightarrow C$, for which we obtain UV/visible spectra of a mixture of components. There are various approaches to the deconvolution of such spectra, a simple one is MLR (Section 6.4) if one knows the spectra of all three pure components under reaction conditions because

$$C \approx X \cdot S^+$$

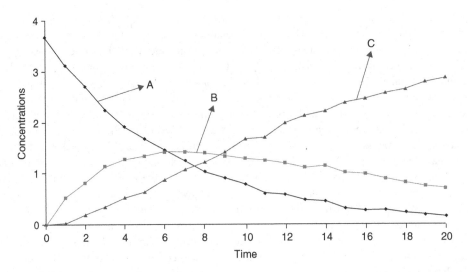

Figure 8.9 Result of resolving the components in a reaction of the form A → B → C

to give concentration profiles such as those of Figure 8.9. Even if one cannot obtain the spectra of the pure standards, this does not matter, one can calibrate the mixture spectra by using off-line and on-line methods. Consider quenching a number of samples, and performing off-line HPLC or another chromatographic technique to accurately determine the concentrations of the three compounds in the reaction mixture. Now the compounds are no longer unstable and we can take time to set up a calibration model between the HPLC estimates and the reaction spectra. Although we do not know the spectra of each individual component, approaches like PLS (Section 6.6) can be employed to predict concentrations using the equation (for each compound separately in the case of PLS1):

$$c \approx T.q$$

Now we can obtain profiles without knowledge of the pure spectra, or the kinetics, and, indeed, if there are unknown interferents it is even possible to perform PLS without knowing the characteristics of all the compounds in the mixture.

PLS can be viewed as a type of soft modelling, taking no account of the kinetics which in turn can be viewed as a type of hard modelling. It is really useful in that one does not need pure spectra and one does not need to make assumptions about initial concentrations. And one can measure reaction half-lives from these empirically obtained concentration curves. But the weakness is that it can lead to nonsensical physical results, for example, negative estimates of concentrations and in most cases is dependent on the quality of the training set for calibration, and if there is known kinetic information, why not use it? Whether PLS or other chemometrics models are preferred or kinetics can be incorporated depends a little on the process, if very little is known about a reaction, it may be unwise to assume kinetic models whereas if a reaction has been developed over several years and its mechanism is well known, incorporating such information is useful. In the next section we will look at further elaborations and how kinetics and multivariate information can be combined.

8.4 DEVELOPMENTS IN THE ANALYSIS OF REACTIONS USING ON-LINE SPECTROSCOPY

In Section 8.3 we discussed two contrasting approaches to the analysis of spectroscopic reaction data, one based on kinetics and one based on multivariate methods, below we describe some extensions that are increasingly in vogue.

8.4.1 Constraints and Combining Information

In Section 8.3 we described how two sorts of information can be obtained about a reaction, kinetic and multivariate. These two types of information can be combined, and we will look at how to simultaneously use kinetics and multivariate information, rather than treat each type separately.

There are a wide variety of ways to combine both types of information, but most involve using two alternating criteria. Consider knowing the pure spectra of the reactants and products, as well as the starting material. We can then 'guess' a value of k (the rate constant), which gives us the profiles of each compound in the mixture, often represented by a matrix C. We can then use this to 'guess' a value of the original data matrix X since $X \approx C.S$ or, in other words, the observed reaction spectra are the product of the concentration profiles and pure compound spectra. If our guessed rate constant is good, then the estimate of X is quite good. If this is not so good, we need to change our rate constant, change the profiles and continue until we get a good fit.

This theoretically simple approach takes into account both the 'hard' kinetic models used to obtain C and the 'soft' spectroscopic models used to see how well the estimated spectra equal the observed spectra. Life, though, is never so simple. We may not know the spectra of every compound perfectly, and certainly not in a quantitative way, we may be uncertain of the starting concentrations, there may be unknown impurities or small side reactions. So many sophisticated elaborations of these approaches have been proposed involving changing several parameters simultaneously including the initial concentrations.

PCA also can be incorporated into these methods via Principal Components Regression (Section 6.5). If we know the scores T of the data matrix, and guess C using our kinetics models, we can then find a transformation matrix $R = T^{+}.C$. Now we can see how well the scores are predicted, using matrix R and adjust k (and any other parameters) accordingly, and continue in an iterative way.

The most common techniques used are ALS and ITTFA which been introduced in Sections 7.10 and 8.1 and have a similar role in kinetics analysis, although several tailor made elaborations have been developed specifically for reaction monitoring.

Kinetics knowledge is not the only information available during a reaction, so the methods above are by no means restricted to kinetics. There are many other constraints, that can be incorporated. Simple ones are that spectra and concentration profiles are positive, but more elaborate ones are possible to incorporate, for example concentration profiles with time are likely to be smooth and unimodal, the profile of the starting materials will decrease with time, and all other compounds should increase and then sometimes decrease. These can all be incorporated into algorithms allowing the models to contain as many constraints as required, although the convergence of the algorithms can be quite tricky if a large number of factors are taken into consideration.

8.4.2 Data Merging

An important facility that is emerging is the ability to obtain several blocks of data simultaneously, which is especially important in spectroscopy, e.g. MIR and UV/visible spectra may provide complementary information about a reaction mixture. Instead of analysing each set of data separately, they can be merged. This means for example, if a reaction were monitored at 200 UV/visible wavelengths and 150 MIR wavenumbers, a data matrix of 350 columns can be obtained, the first 200 corresponding to the UV/visible and the next 150 to the MIR data. These can then be treated as one large dataset, but an important trick is to be able to scale these blocks correctly. Scaling depends in part on the number of data points each block, for example, if there is only one temperature monitored but 500 spectroscopic data points, we do not want the single temperature measurement to be overwhelmed by the spectroscopic variables. A second factor relates to the original size of the variables in each block, if the size is radically different, then this is usually accounted for by standardizing. Finally each block must be scaled by how important or useful the information is, sometimes some measurements are more helpful than others. The usual way for scaling is to weight each variable according to its importance.

After that, many of the existing approaches can be applied as outlined elsewhere in this chapter, but to a larger merged dataset.

8.4.3 Three-way Analysis

For first-order (or pseudo first-order) reactions, yet more sophisticated approaches are possible using three-way data analysis (Sections 5.11 and 6.8). Because of the novelty of these approaches, several chemometrics groups have raced to pioneer these methods. There are several variants but one, applicable to first order data, converts a two way dataset (consisting of a number of spectra recorded against time) to a three-way dataset by a simple trick of taking the first data points and dividing the time series into two parts each containing some overlap. For example, if we record 100 points in time, why not create two separate time series of 70 points each, the first from points 1 to 70, and the second from points 31 to 100? These can be combined into a three way dataset as in Figure 8.10. The trick is that the top and bottom planes differ by 30 points in time throughout. For first-order reactions the relative concentration of the starting material at two points in time is constant and depends only on k and the difference in time. Hence the ratio of concentrations between any point in the top and bottom plane of Figure 8.10 is the same. Clever chemometric algorithms can convert this information into rate constants. Because there are a huge number of points in

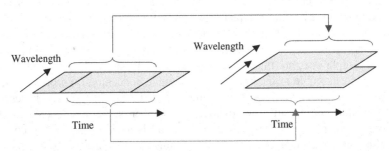

Figure 8.10 Creating three-way data from two-way data in reaction monitoring

each plane, there are a huge number of measures of k and so good estimates can be obtained. These represent multivariate enhancements to traditional approaches developed for single wavelength data many years back.

8.5 THE PROCESS ANALYTICAL TECHNOLOGY INITIATIVE

Process Analytical Technology (PAT) has been around as a concept for several years. However, although chemometrics as a discipline has been recognized for over 30 years, with coherent flagships such as dedicated named journals, textbooks and conferences available since the mid 1980s, it has been slow to be adopted by regulatory authorities. In the pharmaceutical industry, especially, regulation is important because processes have to be shown to result in products of given quality in order to be allowed on the market. Impurities and by-products can have unknown physiological effects which have not been assessed properly via toxicological trials and so there is a continuous need to monitor processes to ensure that product quality remains within well defined limits, which are often defined legally or via regulatory bodies such as the US Food and Drug Administration (US FDA).

Chemometricians will argue that multivariate methods provide important information as to the quality of a product. A problem is that multivariate statistical ideas are often difficult for legislators and harder to visualize compared with simpler measurements, even though these simple measurements may not necessarily paint such an accurate picture. For example, it is easier to state that a product's quality is passable if a peak height at a given wavelength of a specified sample size is above a predefined threshold than if its spectrum falls within a certain predefined range in a scores plot, yet the peak height at a given wavelength may be influenced by several factors and as such, although easier to measure, this peak height may not necessarily be such a good indicator of quality. The recent PAT document produced by the US FDA [3] contains many recommendations for the use of methods that are generally regarded as originating in chemometrics. A summary has been published [4]. This has caused a significant revolution in how the pharmaceutical industry in particular views chemometrics, as the PAT initiative incorporates many multivariate methods into the suggested approach for ensuring product quality and performance of processes. The ability to monitor reactions in real time using spectroscopic probes in combination with data analysis has been one of the driving forces for such an initiative. These recommendations are not prescriptive but represent what the FDA regards as good practice.

According to the FDA, there are four main tools in PAT. Below we list these and indicate how chemometrics methods have a potential role.

8.5.1 Multivariate Tools for Design, Data Acquisition and Analysis

These tools are all chemometric by origin and are achieved through the use of multivariate mathematical approaches, such as statistical design of experiments, response surface methodologies, process simulation, and pattern recognition, in conjunction with knowledge management systems. Especially important is the need to monitor and assess product quality and problems with processes. Using measurements such as spectra, the underlying signal is usually multivariate in nature both because different ranges and peaks in a spectrum are influenced by the nature of the product and because there will be several interlinked factors that influence the appearance of a spectrum. It is expected that a multivariate model has a chance to give a more reliable answer than a univariate one (e.g. a single physical measurement or a single chromatographic or spectroscopic peak). However, it is usually first necessary to develop such models using careful training or calibration sets, e.g. measuring spectra of products of varying quality, and these are then related using statistical methods to properties of a product. Of course, a key to the success is to have a sufficiently representative training set which often requires some formal experimental design. In addition records from long-standing databases can be used to refine the model and also to monitor instrumental problems which can interfere with the chemical signal. Some methods for experimental design are discussed in Chapters 2 and 9.

8.5.2 Process Analysers

Multivariate methodologies are often necessary to extract critical process knowledge for real time control and quality assurance. Comprehensive statistical and risk analyses of the process are generally necessary. Sensor-based measurements can provide a useful process signature. A key advance over the past few years is that the complexity of signal that can be monitored has increased from simple measurements (e.g. pH or temperature) to complex spectroscopic signatures. Hence the development of new PAT requires chemometrics to disentangle the signature.

8.5.3 Process Control Tools

These are slightly different to process analysers, because they require feedback often in real-time, whereas information from process analysers can be used post-event, e.g. to determine the quality of a batch or simply for passive monitoring. Process control requires on-line decisions. Chemometrics analysis is more common post-event, and certainly formal assessment of product quality often has to be performed on batches, but there is a developing use of on-line multivariate software. Most process control tools are rather simpler, using well known charts such as Shewhart charts of easily measured univariate parameters (Section 3.13), however, a few common methods such as PLS PCA, Q and D charts can be implemented on-line and be used to provide key indicators of the progress of a process: this is often called Multivariate Statistical Process Control and some of the techniques are discussed in Section 3.14.3. One issue involves the relative complexity of interpreting such parameters and whether it is easy for plant operators and managers who may have very limited knowledge of chemometrics, to make decisions based on multivariate parameters. Software development is crucial for the acceptance of chemometrics approaches on-line.

8.5.4 Continuous Improvement and Knowledge Management Tools

Chemometrics has a modest role to play here, but it is important to monitor, for example, how new instruments affect the multivariate signal, how day to day variation in instrumental performance influences spectroscopic parameters, and to build up knowledge of batches of products that differ in quality. Because there are not many public domain case studies of chemometrics being used over a period of several years to monitor products, due to the relative novelty of multivariate methods in process analytics, it will be interesting to see whether this aspect of chemometrics develops in the future, although it certainly will be important to determine whether a chemometric model developed, for example, in 2005 has relevance in 2008 or whether the full cycle of setting up training sets and determining models needs repeating regularly.

REFERENCES

1. R. Gargallo, R. and Tauler A. Izquierdo-Ridorsa, study of the protonation equilibria of a transfer ribonucleic acid by means of a multivariate curve resolution method. *Quimica Analitica*, 18 (1999), 117–120
2. E.R. Malinowski, *Factor Analysis in Chemistry*, 3rd Edn, John Wiley & Sons, Ltd, Chichester, 2002
3. US Department of Health and Human Services, FDA, *Guidance for Industry PAT – A Framework for Innovative Pharmaceutical Development, Manufacturing, and Quality Assurance*, September 2004 (http://www.fda.gov/cder/guidance/6419fnl.htm)
4. R.G. Brereton, Chemometrics and PAT, *PAT Journal*, 2 (2005), 8–11

9

Improving Yields and Processes Using Experimental Designs

9.1 INTRODUCTION

The average synthetic chemistry research student has very little interest in optimizing yields. Normally he or she will fish out a compound from the bottom of a flask, purify it and then send for nuclear magnetic resonance (NMR) and mass spectrometry (MS) analysis. In a few good cases crystals will be obtained for X-rays, but it is not necessary to optimize yields to report a new synthesis in the scientific literature or for a PhD.

In the pharmaceutical manufacturing industry, however, yields are very important. Normally only a very small proportion of candidate drugs ultimately see the light of day in terms of the marketplace, but it is these elite compounds that result in the profitability and viability of large international pharmaceutical giants.

In order to understand the importance of this one has to have some appreciation of how profits are obtained. Let us consider the economics of an imaginary drug that retails at £1000 per kg. How does this price relate to profits?

A major cost for all future looking pharmaceutical companies is research and development. If companies do not constantly look for new drugs, in the long run they will either go bust or be taken over. The reason for this is that patents have only a finite lifetime, so a particular lead product cannot be protected for ever. Additionally, there is competition and other companies may find alternatives to established products, and finally very occasionally health scares or doubts can result in a drug being withdrawn from the market. This ongoing research takes up a substantial portion of a company's budget, because the vast majority of candidate drugs evaluated are not ultimately marketed. Also, there is a long time delay between first discovering a new potential drug, passing clinical trials and being licensed for medical use. A typical company may spend £400 of the £1000 price tag in research mainly towards other potential future targets, leaving £600 towards manufacturing the product and profits.

Consider the situation where it costs £250 for raw materials to make 1 kg of product. However, one also needs to maintain a factory, paying employees, maintaining equipment, electricity charges and so on, perhaps £100 per kg relates to factory costs. And then there are analytical expenses. The quality of a product must be constantly checked, in case there are any problems with a batch. Both on-line and off-line methods are generally used, equipment

Applied Chemometrics for Scientists   R. G. Brereton
© 2007 John Wiley & Sons, Ltd

is expensive and employees have to be quite skilled, so assume that for every 1 kg of product, analysis costs £50, which may include some expensive methods such as NMR occasionally if the managers are suspicious as well as cheaper spectroscopic approaches such as near infrared (NIR). Most products have a specification and it is necessary to prove that the purity of each batch is within legal limits, often using quite elaborate protocols. The manufacturing, factory running costs and analysis reduce the profit from £600 to £200 in this case.

But there are other expenses. All serious products must be patented, and lawyers do not come cheap. Furthermore patents must be protected around the world. There are costs involved in packaging, distribution and marketing. Finally some other costs such as the salary of the Chief Executive and the cost of the Human Resources Office and Accountants must be paid. Assume that these come to £60 per kg (again these are imaginary figures and may well be on the cautious side). This means that the remaining profit is £140 per kg or 14 % of the cover price. Figure 9.1 illustrates this fictitious calculation.

What do we need the profit for? The shareholders definitely want a dividend, otherwise investors will be lost and the company will not have funds in the bank, and for many institutional investors such as pension funds and investment banks these dividends represent their survival. There may be bank loans for buildings that need to be paid off. Expansion might be on the agenda, why not construct a new building, or even take over a small company? Perhaps build a new factory? Or it might simply be useful to save up for a rainy data, the next year might not be so profitable but it is necessary to pay one's employees on a regular basis. So it is always important to be able to maintain profits.

One way is to increase the price of the drug, but it is not so simple. For example there may be competition in the market. Also the customers might be government health services who have strict budgets and simply will not spend more than a certain amount. Many drugs are to treat people who are very ill or aged, and adverse newspaper headlines of

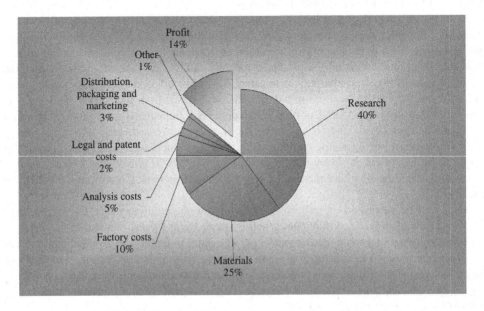

Figure 9.1 Imaginary breakdown of the cost of a drug

high drug prices whilst shareholders are getting good dividends, and the Chairman receives an enormous bonus, do not help the image of a company. Some sales are in Third World countries that cannot afford expensive medicines, and so the price must be limited, if not bootleg suppliers could enter the market and produce drugs that appear similar but do not actually have the correct physiological effect, an ethical as well as financial dilemma, especially in uncontrolled markets.

So an alternative way to improve profits is to increase the yield. Let us say we can improve the yield so that by spending £250 on raw materials we are now able to produce 1.05 kg, a 5 % increase in yield, and all other costs are approximately similar, perhaps add on £10 extra cost per kg to account for the fact that there will be small increases in distribution, factory running and analysis costs (this is well above a 5 % rise in these categories). Now the income is £1050 (as 5 % more product has been manufactured) but the cost has risen from £860 to £870. Our profit, however has risen from £140 to £180 or by nearly 30 %. This is a massive change. The shareholders will be happy. Maybe it really is possible to construct a new building or takeover a small company. And perhaps it is now possible to sell some cheap drugs to organizations in developing countries.

What has this got to do with chemometrics? The whole house of cards depends on the increase in yield of 5 %. In fact if a process consists of several reactions, this may correspond to only a very small improvement in yield for two or three of the steps in the chain. So it is worth investing effort into optimizing yields. The laboratory based academic chemist often does not feel this and has very little interest in whether a yield can be improved by 2 %. A research student is unlikely to want to spend a year increasing a yield on a single step of a synthetic reaction, he or she will get very bored and be more excited by devising new reactions and finding new compounds. But in industry this is a vital procedure, and chemometric experimental designs are employed with great effect for the optimizing of yields.

9.2 USE OF STATISTICAL DESIGNS FOR IMPROVING THE PERFORMANCE OF SYNTHETIC REACTIONS

It is particularly important in industry to be able to improve the performance of a reaction. Often the requirement is to find conditions that maximize yields, but also we might want to reduce impurities, speed up the reaction or even minimize the cost of reagents. The way to do this is to run reactions on a very small scale first in the laboratory under a variety of conditions and then choose those conditions that appear to give the best result. These can then be scaled up to be used in a manufacturing plant, although sometimes there can be slight differences in the optima when increasing the scale.

Traditionally, it has been conventional to use the method of one factor at a time for optimization. Consider, for example, a reaction whose yield depends on temperature and pH. Our first step might be to set a temperature, for example 50 °C and then gradually change the pH until an optimum is reached. Then, use this optimum pH and alter the temperature to change the pH until a new maximum is found. This approach can be extended to several factors such as stirring rate, solvent composition, ratios of reagents and so on.

What is the weakness? In statistical terms we have *interactions* between the various factors. This means that the influence of both pH and temperature are not independent. So, for example, the optimum pH at 40 °C is different to that at 60 °C. This means that we would

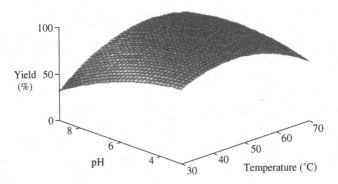

Figure 9.2 Imaginary response surface for a reaction as a function of temperature and pH

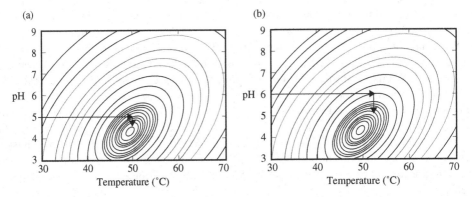

Figure 9.3 Two one factor at a time strategies for finding optima of the reaction profiled in Figure 9.2. (a) Strategy 1; (b) strategy 2

find a different apparent 'optimum' according to the experimental strategy employed. This problem has been discussed previously in Section 2.1.

In reality what we are trying to do is to explore a response surface, an example of which is illustrated in Figure 9.2. The observed response surface may be a little more bumpy and irreproducible because of experimental error, but ideally we would like to obtain the underlying model that relates our yield (or any other interesting response) to the experimental conditions, and then find the optimum, which is the maximum on the response surface.

The response surface is more clearly represented by a contour plot (Figure 9.3), and here we can understand the dilemma. Consider experimental strategy 1. We start first at pH 5, find the best temperature and then use this temperature to find the best pH. The optimum is a little bit away from the true one. Strategy 2, starting at pH 6 is even worse. In fact, unless we fortuitously choose the true optimum pH right at the start (which we clearly do not know in advance), we will not find the best conditions using the strategy of altering one factor at a time.

How could we improve the experimental approach? One way might be to perform a grid search. For example we could choose 10 pHs and 10 temperatures, and run the reaction at all hundred possible combinations, but this clearly is very time-consuming. It may not find the exact optimum, but a region in which we suspect contains the best conditions, so we can then continue this sort of experimentation (another grid with much smaller intervals in pH and temperature) until we get to the true optimum, perhaps in 200 experiments we will get a sensible result. For computational simulations, this is a feasible approach to optimization, but not when each experiment is time consuming and costly.

This is only the tip of the iceberg. It is unlikely that there are only two factors that influence the yield of a reaction. What about catalyst concentration, or stirring rate, or solvent composition or ratio of reactants? Maybe there are 10 factors that have potential significance. How can we perform a grid search experiment now? By running 10^{10} or 10 billion experiments? Clearly this is impossible. Because of interactions, we cannot rely on performing a smaller set of experiments on two or three factors, and using the optima we have discovered for these factors as conditions under which we try to optimize a further couple of factors.

Even more difficult is the fact that we might want to optimize other responses, not just the yield, for example how pure is the product and how fast can we obtain the product (this may relate to cost because the longer a factory is in use the more expensive the product, so a slightly lower yielding product produced much faster will sometimes be preferred dependent on the circumstances). We may wish to obtain several pieces of information from the experimental data. Some methods will home in on only one optimum, so we may need to repeat part or all of the experiment again to study another response.

Looking at the experimental response illustrated in Figure 9.3, it is obvious that some conditions are not likely to result in a good yield, on the periphery of the experimental region, so we could reduce the number of experiments in the grid search. However, we would have to be extremely careful especially if there are several factors as we could miss the optimum altogether or gather nonsensical data. So what would help is a set of rules that help find the optimum, saving time, and with minimum risk of making a mistake. These involve using statistical experimental designs. Below we will review strategies for optimizing reaction conditions employing chemometrics approaches. The methods in this chapter build on those introduced in Chapter 2.

9.3 SCREENING FOR FACTORS THAT INFLUENCE THE PERFORMANCE OF A REACTION

In Section 9.2 we looked at the motivations for developing and applying formal rules for experimental design in optimization of reaction conditions. In this section we will look at how such methods can be applied.

In many modern laboratories, it is possible to automate, using robotics, running many reactions simultaneously. These may involve performing the same overall reaction, but with subtle differences in conditions, such as when the catalyst is added, the proportion of reagents, the solvent composition, stirring rate, the pH and so on. The aim is to determine the best conditions to optimize the yield. Ten or twenty such reactions can be studied simultaneously.

How can we establish conditions that allow us best to home in on optimum conditions rapidly and with least risk of obtaining a false optimum? Here statistical experimental designs come into play. Normally the first step is one of screening, there are often a huge number of factors that could have an influence. Which ones are important?

The first step is to list down the possible factors. There may be a large number of such factors. They are of three fundamental types:

1. *quantitative*, such as pH or temperature;
2. *categorical*, e.g. is the reaction stirred or not;
3. *compositional*, such as the proportion of a solvent. These are limited to a number between 0 % and 100 % and all the solvents must add up to a constant total.

The second step is to determine sensible limits for each of these factors. As a rule initially one can treat all factors as having two levels, low and high, but different types of factors have to be handled in varying ways. Each factor at this stage must be capable of being varied independently of each other one.

1. For quantitative factors it is quite easy, for example, we may simply take pH, e.g. between 5 and 8, within which we expect our optimum. The question is whether pH has a significant influence over the optimum.
2. For categorical factors, either an absence or presence, for example the reaction could be performed in the light or the dark, or two completely different conditions such as different catalysts. These limits are easy to establish.
3. One needs to think a little harder about compositional factors. For example, we might want to run the reaction in a three solvent mixture of acetone, methanol and tetrahydrofuran (THF). Some people like to study these types of factors separately using mixture designs (Section 2.12), but it is relatively easy to use ratios of solvent composition, for example, we could vary the ratio of methanol to acetone between the extremes of 2 and 0.5, similarly the ratio of THF to acetone. It is easy to show that the proportion of acetone will be equal to $1/[1 + (\text{methanol/acetone}) + (\text{THF/acetone})]$, and then the other proportions can easily be calculated. The calculation is given in Table 9.1. See Section 9.6.1 for further examples.

One is then able to 'code' each factor, with $+$ standing for a high level and $-$ a low level. Table 9.2 is of a fictitious reaction of the form $A + B \rightarrow C$ where we add the second reactant to the first over a period of time under varying reaction conditions. We have discussed elsewhere another case study (Section 2.10). The levels of the factors have to be physically sensible of course, and must relate to what the chemist actually expects.

The third step is to produce a list of conditions under which the experiment is performed. There are a variety of designs, but a key requirement is that the columns in the experimental matrix are *orthogonal* to each other (see also Sections 2.6, 2.8 and 2.9), in other words independent. One rule of thumb is to choose the design according to the number of factors of interest and round up to the nearest multiple of 4 which must be at least one more than

Table 9.1 Upper and lower limits for the ratio of methanol to acetone and of tetrahydrofuran (THF) to acetone in a three-component mixture

Independent limits			Solvent composition		
Methanol/acetone	THF/acetone		Acetone	Methanol	THF
2	2		0.200	0.400	0.400
0.5	2	$\longrightarrow$	0.286	0.143	0.571
2	0.5		0.286	0.571	0.143
0.5	0.5		0.500	0.250	0.250

Table 9.2 Upper and lower limits for 13 factors in an imaginary reaction

Factor		Low	High
x_1	NaOH (%)	30	50
x_2	Temperature (°C)	30	60
x_3	Nature of catalyst	A	B
x_4	Stirring	Without	With
x_5	Reaction time (min)	80	240
x_6	Volume of solvent (ml)	100	200
x_7	Methanol: acetone (v/v)	0.5	2.0
x_8	THF: acetone (v/v)	0.5	2.0
x_9	Light	Dark	Light
x_{10}	Catalyst: substrate (mol/ml)	2×10^{-3}	4×10^{-3}
x_{11}	Time to add second reactant (min)	5	20
x_{12}	Ratio of reactants (mol/mol)	1	3
x_{13}	Concentration of first reactant (mM)	50	200

the number of factors and equal to the number of experiments that need to be performed. If you have 8, 9, 10 or 11 factors you want to study, round up to 12, if you have 12, 13, 14 or 15 factors round up to 16. Then choose the design. There are two main classes of design. If the multiple of 4 is a power of 2 (e.g. 8, 16 or 32), it is possible to use a fractional factorial design (Section 2.8), otherwise a Plackett–Burman design (Section 2.9). Of course there are other alternatives also. To construct a Plackett–Burman design start with what is called a generator, examples are listed in Table 9.3; there are only a certain number of generators that have the desired properties. This allows us to establish a table of conditions. In all the generators listed there is one more + level than − level. For the first experiment all factors are at the − (lowest) level providing that the generator contains one more '+' than '−'. The generator then gives the conditions for the second experiment, and each subsequent experiment is formed by shifting the generator by one column either to the right or left (so long as this is consistent, it does not matter). The condition in the end column cycles round to the first column in the next experiment. This is illustrated in Table 9.4, where the construction of the third row (or experiment) from the generator in the second row is shown. This table can then be converted to a set of experimental conditions. If the number of factors is less than 7, 15, 19 or 23, we use 'dummy' factors for the last columns. These are factors that have no significance on the outcome of the reaction, such as the technician that handed out the glassware, the colour of the label on the flask, etc. So if we want to study 13 real factors, we choose a design involving 15 factors in total, involving 16 experiments.

We will not discuss the construction of fractional factorial designs in this section (they have already been introduced in Section 2.8), but note that by changing around rows and columns they are equivalent (for 7, 15 and 31 factors) to Plackett–Burman designs.

The fourth step is to perform the experiment and obtain a response, usually a yield. However, it is possible to measure several different responses such as impurity levels, yield to cost ratio, and perhaps other desired physical properties of the product. Each may have a different optimum and be influenced by different factors.

Fifth and finally, we want to model the relationship between response and the value of each factor. We have already discussed how to produce a mathematical model (Section 2.4)

Table 9.3 Plackett–Burman generators

No. of experiments	Factors	Generator
8	7	+ + + − + − −
12	11	+ + − + + + − − − + −
16	15	+ + + + − + − + + − − + − − −
20	19	+ + − + + + + − + − + − − − − + + −
24	23	+ + + + + − + − + + − − + + − − + − + − − − −

Table 9.4 Constructing a Plackett–Burman design for seven factors

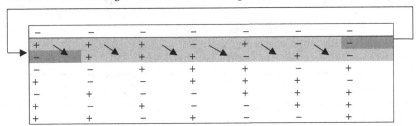

which may be performed in a big variety of commercial packages or in simple spreadsheets such as Excel. These provide coefficients which relate the response to the coded value of each factor. Since each factor has the same coded range, all coefficients are on the same scale. A large positive coefficient means that the response increases with increase in value of the factor, and vice versa. We would hope most factors to have little effect, and these will have a coefficient that is relatively small in magnitude. Ideally the coefficients can be presented graphically, as in Figure 9.4, in which three out of seven factors in an experiment are deemed to be significant. These are the ones we need to study in greater detail to improve the reaction conditions. In many practical cases, a large number of potential factors (15 to 20) can be narrowed down to a few (perhaps 4 or 5) that we think are worthy of detailed study.

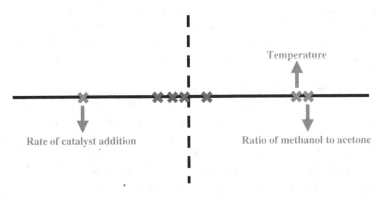

Figure 9.4 Graphical representation of the value of the coefficients for seven factors, three of which are significant

9.4 OPTIMIZING THE PROCESS VARIABLES

In Section 9.3 we reviewed some methods for screening the factors that are most important to improve the efficiency of a reaction. The next step is to change these factors until a true optimum has been obtained.

Normally after screening, we have achieved the following.

1. A large number of factors that have little or no significance on the outcome of the reaction have been eliminated.
2. A few quantitative factors remain, such as pH, which can be altered to take on almost any value. These are often called *process* variables.
3. Some important *mixture* variables, normally corresponding to solvent compositions, have been identified: methods for dealing with these types of variables will be described further in Sections 9.5 and 9.6.
4. Some factors that are *categorical* in nature, i.e. stirring or not, light or dark, remain. In such cases it is often obvious that the factors must be set at one level and cannot be further optimized.

The process variables are the most straightforward to handle. There are a variety of designs such as the central composite design (Section 2.11) that can be used to establish a series of conditions for performing the reaction. The basic design consists of $2^f + 2f + 1$ experiments for f factors. The $2f + 1$ experiments represent a so-called 'star' design, and the 2^f experiments a 'factorial' design. These two designs are superimposed, as illustrated in Figure 9.5, for two factors. The total number of experiments for between two and seven factors are presented in Table 9.5. Usually a few extra replicates in the centre are also performed, as this is a check on the reproducibility of the conditions, a typical number is 5.

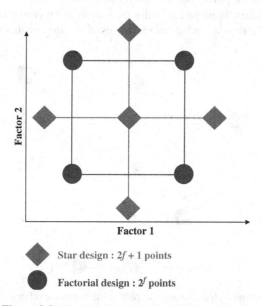

Figure 9.5 Central composite design for two factors

Table 9.5 Number of experiments required for full central composite designs of varying sizes (excluding replicates in the centre)

Factors	Experiments
2	9
3	15
4	25
5	43
6	77
7	143

However, one problem arises in that the number of experiments can become excessive if one has several factors, especially if it is required to replicate some points. A method of overcoming this is to use a fractional factorial design (Section 2.8). This involves removing some of the factorial experiments, and for a 3 factor design, we can reduce the total from 15 (or 7 star points and 8 factorial points) to 11 experiments (or 7 star points and 4 factorial points), as in Figure 9.6. It is inadvisable to reduce the factorial points for a 3 factor design drastically if one wants to produce a full model containing the interaction terms as there are not many degrees of freedom remaining to determine how well the model is obeyed using the lack-of-fit (see Section 2.2). However, when there are more factors, this is often a sensible idea because three factor or higher interactions are normally not very meaningful or even measurable, so for a five factor design, one can safely reduce the number of experiments from 43 to 27 simply by performing a half factorial, without a serious loss of information.

Once a response is obtained, it is necessary to model it, which normally involves fitting an equation connecting the response (such as a synthetic yield) to the variables as described in Section 2.5. This equation, however, is really the means to an end, and in itself is probably not very interesting. Generally coded values are used for this mathematical model but the

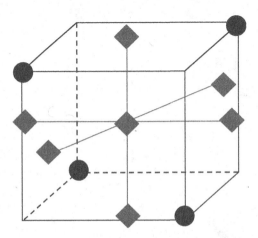

Figure 9.6 Using fractional factorial designs to reduce the overall number of experiments in a central composite design

optimum coded conditions can then be converted to true experimental conditions, either automatically by software or by using a spreadsheet or calculator. These models are not designed to give physical insights into the quantum chemistry or mechanism of the reaction but are primarily empirical in nature. The majority of synthetic and process chemists do not have the time or interest to understand complex reactions in terms of fundamental molecular parameters and are mainly interested in statistical modelling which can be used to predict an optimum.

One important aspect to remember is how the points in the design relate to the nature of the model:

- The star design is at three levels (including the centre point) and so allows squared terms. In reality these are used to model curvature in the response which is very common, and often a feature of an optimum.
- The factorial points allow us to estimate interaction terms. As we have seen the influence of the factors is not usually completely independent (Section 9.1) so we must have a model that permits these to be studied.
- Reducing the factorial points reduces the number of high level interactions that can be studied.

See also Section 2.11 for further discussion.

Normally it is sufficient to study two-level interactions. The middle column of Table 9.6 lists the number of such combinations for varying numbers of factors. For the enthusiasts, this number is given by $f!/(2!(f-2)!)$. In order to capture all these interactions, and the square and linear terms, one must have at least the number of experiments of the middle column plus $2f + 1$, as indicated in the right-hand column of Table 9.6.

If one wants to determine how well the model is fitted and also have some idea of replicate error a few extra experiments are required, but Table 9.6 provides a good idea as to how much one can reduce the size of the design according to the number of factors shown to be significant. What is important is to appreciate the relationship between the actual arrangements of experiments and the type of information we are able to obtain.

Obtaining the mathematical model is actually quite easy. Most packages can do this, but it is not too difficult to multiply matrices together using standard Multiple Linear Regression (Section 6.4) to relate the response to the design matrix (Section 2.5). Some chemists are terrified of mathematics, but the principles are extremely simple, although some people prefer

Table 9.6 Minimum number of experiments required for modelling all possible two factor interactions for different numbers of factors, plus squared and linear terms

Factors	Combinations	Minimum no. of experiments
2	1	6
3	3	10
4	6	15
5	10	21
6	15	28
7	21	36
8	28	45

to spend hundreds or even thousands of pounds on a black box that does this calculation rather than spend a few hours learning the basis of matrix algebra and manipulation. The concept of the pseudoinverse, introduced in Section 2.4, is helpful.

The quality of the model can be tested, for example by replicate analysis, but if we are satisfied it remains to predict the optimum. Many approaches are possible. The simplest are visual, e.g. using contour plots of the response, which is fine if there are two factors, just possible using sophisticated computer graphics for three factors but not feasible for more. If the model is not too complicated it is possible to use simple pencil and paper and differential calculus. If you are not up to this, or if there are several significant factors, there are numerous computational optimization programs which are part of most standard data analysis packages and also specific experimental design software. These will predict the best conditions.

Finally, try these conditions in the laboratory. They may not always give a true optimum, because the model itself is subject to experimental error, and it is sometimes useful to perform another optimization around the conditions the model predicts as the optimum, but using a far narrower range of the factors, perhaps a tenth the original. Sometimes experiments may be slightly irreproducible or the optimum may be very flat. When you are very close, often intuition and the individual observations of the scientist have an important role to play, so do not rely totally on the automated output of a package, although if the experiments have been designed correctly you should have a good idea of the best conditions by this stage.

9.5 HANDLING MIXTURE VARIABLES USING SIMPLEX DESIGNS

When optimizing a reaction, in addition to the process variables, there are usually also mixture variables to be taken care of. These are variables whose total adds up to 1, usually solvents, but they could be the proportions of ingredients or even the proportion of gases. In some cases we may even have more than one set of mixture variables that we need to optimize simultaneously.

Mixture designs are very common in food chemistry but also play a role in synthesis, one use being to determine the optimum mixture of solvents for improving the performance of a reaction. Mixtures are normally represented in a special space often called a simplex. We have already discussed some of these designs previously (Section 2.12) but will expand on the use of these methods in this section.

9.5.1 Simplex Centroid and Lattice Designs

An important approach to handling mixture variables is by using simplex centroid or simplex lattice designs.

Simplex centroid designs consist of all possible combinations of $1, 1/2, 1/3, \ldots 1/f$ where f is the number of factors. Note that for a 3 factor design, unitary combinations consist of points on the corner, binary combinations on the edges, ternary in the centre of a triangle. There will normally be $2^f - 1$ possible experiments (Section 2.12). However if the number of factors is large, the number of experiments can be reduced by removing some of the higher order terms, for example, if we have five factors, instead of performing $2^5 - 1$

Table 9.7   Cutting down the number of factors in a simplex centroid design

A	B	C	D	E
1	0	0	0	0
0	1	0	0	0
0	0	1	0	0
0	0	0	1	0
0	0	0	0	1
1/2	1/2	0	0	0
1/2	0	1/2	0	0
1/2	0	0	1/2	0
1/2	0	0	0	1/2
0	1/2	1/2	0	0
0	1/2	0	1/2	0
0	1/2	0	0	1/2
0	0	1/2	1/2	0
0	0	1/2	0	1/2
0	0	0	1/2	1/2

or 31 experiments we could cut this down to 15 experiments if we perform experiments only at the one and two factor blends as in Table 9.7. This new design omits three factor interactions but allows all possible two factor interaction terms to be estimated. If several mixture variables are combined with several process variables, even if there are autoreactors that allow many experiments to be performed in parallel, the amount of data can become excessive unless there is some way of reducing the number of experiments.

The simplex lattice design is a somewhat more elegant approach when one wants to reduce the number of experiments where there are several factors, and can be denoted by a $\{f, m\}$ simplex lattice design. This implies that there are f factors and all possible combinations of factors from $0, 1/m, 2/m \ldots m/m$ are studied. The value of m cannot be larger than f. The smaller m the lower the number of experiments. Figure 9.7 and Table 9.8 illustrate a $\{3, 3\}$ simplex lattice design. For the enthusiasts, the number of experiments that need to be performed equals $(f + m - 1)!/[(f - 1)!m!]$ as given in Table 9.9 for several different designs. Notice that a $\{f, 2\}$ simplex lattice design corresponds to a simplex centroid

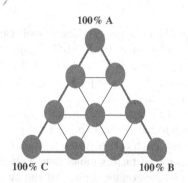

100% A

100% C   100% B

Figure 9.7 $\{3, 3\}$ Simplex lattice design

Table 9.8 {3, 3} Simplex
lattice design

A	B	C
1	0	0
0	1	0
0	0	1
2/3	1/3	0
2/3	0	1/3
1/3	2/3	0
0	2/3	1/3
1/3	0	2/3
0	1/3	2/3
1/3	1/3	1/3

Table 9.9 Size of various possible simplex lattice designs

Factors (f)	Maximum blend (m)				
	2	3	4	5	6
2	3				
3	6	10			
4	10	20	35		
5	15	35	70	126	
6	21	56	126	252	462

design that retains only binary blends, but when k is larger than 2 these two designs are quite different.

In some situations there may be 10 or more ingredients in blends and it is essential to consider approaches for reducing the number of experiments.

9.5.2 Constraints

One important experimental problem however is that in most cases it is not sensible to study the proportion of components between 100 % and 0 %. Imagine using a mixture of acetone, methanol and water as a solvent. Probably the reactants will not dissolve in 100 % of water, so a better approach is normally to vary the proportion of each component between fixed limits. The designs must now be altered to take this into account. This problem has been introduced briefly in Section 2.12.3 but several further solutions to this are described below.

A simple method is to keep to the original simplex designs but to change the upper and lower limits of each component. Each corner of the triangle (for a three factor design) will now correspond to a combination of factors rather than a single pure component. In order to do this, there are some restrictions on the limits that can be used, and it is necessary that the sum of the upper limit of each factor, and the low limits of all the remaining factors equals 1. Table 9.10 represents one such possibility for three factors. The design can then easily be generated. The new points in the design are given by:

$$\text{New} = \text{Lower limit} + (\text{Upper limit} - \text{Lower limit}) \times \text{Old}$$

where 'Old' represents the points for the simplex designs where each component varies between 0 and 1. The transformed simplex centroid design is given in Table 9.11, and is pictured diagrammatically in Figure 9.8. In fact any type of simplex design can now be generated if required.

However, there are several other possible ways to set the limits for each factor. A design is feasible providing the following two conditions are met:

Table 9.10 Possible upper and lower limits where the sum of each upper limit and other lower limits equals 1

Factor	A	B	C
Upper level	0.6	0.4	0.2
Lower level	0.5	0.3	0.1

Table 9.11 New conditions for constrained simplex centroid mixture design using the limits in Table 9.10

A	B	C
0.6	0.3	0.1
0.5	0.4	0.1
0.5	0.3	0.2
0.55	0.35	0.1
0.55	0.3	0.15
0.5	0.35	0.15
0.533	0.333	0.133

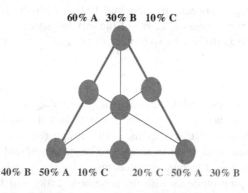

60% A 30% B 10% C

40% B 50% A 10% C   20% C 50% A 30% B

Figure 9.8 Graphical representation of the design in Table 9.11

Table 9.12 Impossible constraints for mixture designs

Impossible: $0.8 + 0.3 + 0.1 > 1$			
Factor	A	B	C
Upper level	0.8	0.4	0.2
Lower level	0.5	0.3	0.1
Impossible: $0.6 + 0.3 + 0.0 < 1$			
Factor	A	B	C
Upper level	0.6	0.4	0.2
Lower level	0.5	0.3	0.0

1. The sum of the upper limit of each individual factor and the lower limits of all the other factors does not exceed 1.
2. The sum of the lower limit of each individual factor and the upper limits of all other factors must at least equal 1.

Some impossible conditions are presented in Table 9.12.

There are now various approaches for setting up designs, and these designs are often represented as covering a polygonal region within the original mixture space (for example a region within a triangle or tetrahedron). Normally one performs experiments on the vertices, sides and in the centre of the region. The following rules are useful when there are three factors:

1. Determine how many vertices the polygon will contain, the maximum will be six if there are three factors. If the sum of the upper limit for one factor and the lower limits for the remaining factors equal 1, then the number of vertices is reduced by 1. The number of vertices also reduces if the sum of the lower limit of one factor and the upper bounds of the remaining factors equals 1. Call this number v.
2. Each vertex corresponds to the upper limit for one factor, the lower limit for a second factor, and the proportion of the final factor corresponds to the remainder, after subtracting from the level of the other two factors from 1.
3. Order the vertices so that the level of one factor remains constant between each vertex.
4. Double the number of experiments, by taking the average between each successive vertex (and also the average between the first and last), to provide $2v$ experiments. These correspond to experiments on the edges of the mixture space.
5. Finally it is usual to perform an experiment in the centre, which is simply the average of all the vertices.

A typical design with five vertices and so 11 experiments is illustrated in Table 9.13. These designs can be represented diagrammatically providing there are no more than four factors (Section 2.12). If each factor has the property that the sum of its upper limit and the lower limits of all other factors equals 1, the design reduces to one containing only three vertices and becomes identical in shape to a simplex centroid design, except that the vertices no longer correspond to pure components.

These types of design provide us with the conditions under which to perform the experiments. If there are several factors all with quite different constraints it can in fact be quite

Table 9.13 Setting up a constrained mixture design

	A	B	C
Lower	0.1	0.3	0
Upper	0.7	0.6	0.4

Step 1

- $0.7 + 0.3 + 0.0 = 1.0$
- $0.1 + 0.6 + 0.0 = 0.7$
- $0.1 + 0.3 + 0.4 = 0.8$

so $v = 5$

Steps 2 and 3 vertices

A	0.7	0.3	0.0
B	0.4	0.6	0.0
C	0.1	0.6	0.3
D	0.1	0.5	0.4
E	0.3	0.3	0.4

Steps 4 and 5 design

1	A	0.7	0.3	0.0
2	Average A and B	0.55	0.45	0.0
3	B	0.4	0.6	0.0
4	Average B and C	0.25	0.6	0.15
5	C	0.1	0.6	0.3
6	Average C and D	0.1	0.55	0.35
7	D	0.1	0.5	0.4
8	Average D and E	0.2	0.4	0.4
9	E	0.3	0.3	0.4
10	Average E and A	0.5	0.3	0.2
11	Centre	0.32	0.46	0.22

difficult to set up these designs. In areas such as food chemistry, materials and paints, there may be a large number of potential components. Quite specialist and theoretical mathematical papers have been written to handle these situations flexibly and there are also a number of software packages for experimental design under such situations. However, even if the designs appear complex, it is important to understand the main principles, and for three- or four-component mixture designs which tend to be the maximum in synthetic chemistry, the guidelines in this section will normally suffice.

9.6 MORE ABOUT MIXTURE VARIABLES

In Section 9.5 we looked at how to deal with mixture variables using simplex designs. There are, of course, several other approaches.

9.6.1 Ratios

One of the simplest methods for handling mixture variables has already been introduced previously (Section 9.3), in which we transform compositions to ratios. The ratios are

Table 9.14 Some possible mixture designs in which ratios are used as factors for a two factor full factorial design (four experiments)

A/B	A/C	A	B	C
Design 1				
0.5	0.5	0.2	0.4	0.4
0.5	2	0.286	0.571	0.143
2	0.5	0.286	0.143	0.571
2	2	0.5	0.25	0.25
Design 2				
0.25	0.25	0.111	0.444	0.444
0.25	4	0.190	0.762	0.048
4	0.25	0.190	0.048	0.762
4	4	0.667	0.167	0.167
Design 3				
0.5	0.25	0.143	0.286	0.571
0.5	0.5	0.2	0.4	0.4
2	0.25	0.182	0.091	0.727
2	0.5	0.286	0.143	0.571

completely independent (orthogonal) and we can use normal designs such as factorial (Section 2.6) or central composite (Section 2.11) designs rather than specific mixture designs. If there are three components in a mixture we can set up two independent ratios. Table 9.14 illustrates three such designs in the form of two level full factorials, involving four experiments (left-hand columns) and the corresponding proportion of each component (right-hand columns) in the actual reaction mixture. The experimental conditions can also be represented in three-component mixture space (Figure 9.9), and it is easy to see that they correspond to constrained mixture designs.

If one wants to use more than two levels, for example to set up a central composite design, when using ratios it is best that each level is coded logarithmically, for example, use the ratios for A/B of 0.5, 1 and 2 for the three levels rather than 0.5, 1.25 and 2. However using this approach all the normal designs for process variables can be employed, including central composite and factorial designs.

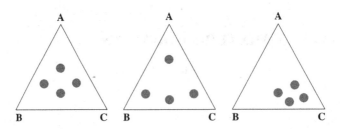

Figure 9.9 Experiments in Table 9.14 represented in mixture space

Table 9.15 A three-component mixture design, where A and B are minor components and C is a filler

A	B	C
0.05	0.05	0.9
0.05	0.1	0.85
0.05	0.15	0.8
0.1	0.05	0.85
0.1	0.1	0.8
0.1	0.15	0.75
0.15	0.05	0.8
0.15	0.1	0.75
0.15	0.15	0.7

9.6.2 Minor Constituents

Another approach can be employed if the majority of the mixture variables are minor components. For example, the main solvent may be acetone and we want to add various components at the level of a few per cent. In this case, providing the highest total percentage of all the minor factors is much less than 100 %, we can simply use conventional process designs for the minor factors and then set the proportion of the major factor to the remainder. Table 9.15 illustrates a design in which the factors A and B vary between 5 % and 15 % – this design can be considered as either a three level factorial design or a central composite design with the axial points at the same level as the factorial points.

9.6.3 Combining Mixture and Process Variables

In many cases we want to combine process and mixture variables, for example solvent composition, pH and temperature. This is quite a simple procedure; the two types of design can be superimposed on each other as in Figure 9.10. This particular design involves two process variables in the form of a two level full factorial and three mixture variables in the form of a simplex lattice design, resulting in 7×4 or 28 experiments in total. Naturally the rules for reducing the number of factorial points if there are several process variables, including extra points for modelling square terms in the process variables, and introducing constraints in the mixture variables, can all be applied.

9.6.4 Models

A final topic involves how to model the response obtained for the mixture experiments. Surprisingly, this is quite a difficult topic, primarily because the mixture variables are not orthogonal to each other, or in simple terms they depend on each other.

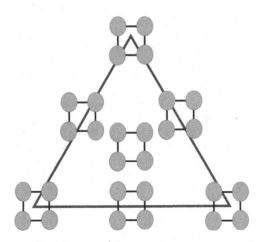

Figure 9.10   Combining mixture and process variables

There are several ways round this, the most common is to use what is called a Sheffé model (see also Section 2.12.1). This models all the one factor, two factor, up to f factor interactions (where f is the number of factors), so that for three factors we have a model of:

$$y = b_1 x_1 + b_2 x_2 + b_3 x_3 + b_{12} x_1 x_2 + b_{13} x_1 x_3 + b_{23} x_2 x_3 + b_{123} x_1 x_2 x_3$$

Note that for a three factor simplex centroid design there are exactly seven terms in the model and seven experiments. If we decide to miss out the ternary experiment in the middle of the triangle, it is simply necessary to remove the three factor interaction term. In contrast, removing higher order interaction terms in a model but retaining the corresponding higher order blends allows one degree of freedom for determination of how good the model is.

The equation above does not include the normal intercept or squared terms, but because the levels of each factor are related by $x_3 = (1 - x_1 - x_2)$ it is possible to re-express the equation differently to give what are called Cox models, with various squared, intercept, etc., terms included for $f - 1$ factors; since the final factor is completely dependent on the others this model takes all information into account.

Finally, if one sets the levels to be ratios, it is possible to use the logarithms of the ratios as terms for the independent model. If we use a three factor mixture design which is represented by a two level factorial we can model the response by:

$$y = b_0 + b_1 \log(x_1/x_2) + b_2 \log(x_1/x_3) + b_{12}[\log(x_1/x_2)][\log(x_1/x_3)]$$

The modelling of mixture designs is a big subject but is covered well in most experimental design and modelling software.

10

Biological and Medical Applications of Chemometrics

10.1 INTRODUCTION

In this chapter we will be examining some applications of chemometrics to biology and medicine. Biologists have a large number of spectroscopic techniques at their fingertips, examples being nuclear magnetic resonance (NMR), gas chromatography-mass spectrometry (GC-MS), fluorescence, infrared (IR) and pyrolysis, which are used to investigate complex mixtures. Chemometrics mainly comes into play when the aim is to relate complex instrumental analytical chemical data to biological properties.

10.1.1 Genomics, Proteomics and Metabolomics

With the human genome project has grown the triplet of genomics, proteomics and metabolomics (some people also distinguish metabonomics although the dividing line between the 'l' and 'n' varieties depends in part on who you are, different researchers promoting different viewpoints). Can we use fluorescence probes to determine gene sequences and can deconvolution be employed to sort out the spectroscopic data? Can we use mass spectrometry to determine protein structure and can pattern recognition be employed to interpret these complex data? Can NMR of biological extracts be used to determine the genetic make up of an extract – is it cheaper performing NMR than genetic profiling? In some cases the aim is to identify individual compounds or features, for example what biomarkers are used to determine the quality of a nutritional plant extract?

Quite different questions are asked for whole organism profiles, for example in metabolomics. In this case we may not directly be interested in the individual concentrations of compounds, but whether there are features in profiles such as obtained by NMR spectra that can be related to genotypes, strains, mutants or even environmental factors. Is the genetic composition of a crop over a large farm or a region of the country sufficiently similar?

In human or animal studies, there are a variety of secretions that can be analysed, varying from quite easy ones such as plasma, blood and urine, to quite hard but often chemically rich ones such as sweat and saliva. Chemometrics comes into play when the processes are

Applied Chemometrics for Scientists R. G. Brereton
© 2007 John Wiley & Sons, Ltd

often quite subtle, involving perhaps many compounds interacting at relatively low levels, as opposed to traditional methods which might search for large quantities of single substances in high amounts in, for example, urine. Animal metabolomics is a particularly challenging and fast growing area, which often poses huge challenges. Some people use the word systems biology to mean the combination of all these approaches. To the chemometrician, there is special interest in data arising from the application of analytical instrumentation to biological extracts as part of the overall strategy.

Plant metabolomics involves primarily analysing plant extracts, and whereas the biological signatures are still quite subtle, there is often greater ability to perform reproducible experiments and true analytical replicates. Plants can be grown under very controlled conditions, whereas humans, for example, cannot be controlled in this way, and in addition to their biology have a variety of personal habits that are hard to control.

10.1.2 Disease Diagnosis

Closely related to metabolomics is the desire to diagnose disease often from chemical signals. There has been a huge growth especially in the use of NMR and liquid chromatography-mass spectrometry (LC-MS) in this area over the past decade. Of especial importance is the possibility of early diagnosis of disease from looking at chemical signals in blood, plasma, urine, etc. This can be done by analysing extracts and using chemometrics in two ways. The first is predictive, classifying extracts into those from diseased and nondiseased patients, and the second is to determine which features in an extract are significant indicators of disease, often called biomarkers.

This is a very tough area, because the chemical signs of disease at an early stage may be very subtle, and also because the composition of an extract depends on many factors, not just whether the person is developing a disease but also their age, sex, what they ate, their lifestyle, genetics, ethnic group and so on. It is very hard to obtain a sufficiently large and representative sample with suitable controls. Often model studies are performed first on rats and mice, where it is possible to study the build up of a disease under controlled conditions, and then these are extended to much harder human studies. However, there is a huge amount of information that can be gained by spectroscopy and chromatography and a wealth of data that requires sophisticated chemometrics techniques and this is undoubtedly one of the fastest growing areas currently.

10.1.3 Chemical Taxonomy

Even before the 'omics' words were invented biologists have for many years used approaches such as pyrolysis mass spectrometry, e.g. to classify bacteria, and this has been a rich area for development of new chemometric techniques especially in microbiology. One problem is that some biological data are nonlinear, which means that conventional approaches for classification are often effectively supplemented by methods such as neural networks. The majority of chemometric algorithms assume that the response can be modelled by a linear combination of underlying factors [although nonlinear Partial Least Squares (PLS) and Principal Component Analysis (PCA) have been proposed but are not in very general use however they are easy to implement], however nonlinear methods have been developed in biology, such as neural networks, Support Vector Machines and Genetic Programming, to cope with these problems. There is no sharp division between what would be regarded as a chemometric approach and a pure computational approach, but most people might regard chemometrics encompassing reproducible methods, such as PLS or Support Vector Machines, and we will make this restriction in this text.

10.2 TAXONOMY

In traditional biology, taxonomists classify organisms into species according to their characteristics, which in early studies were often physical in nature, grouping animals and plants into different species, genera, families and so on. The classification of bacteria and other micro-organisms cannot always be performed using physical measurements but often various tests are performed for this purpose.

The results of these measurements are then used to see how organisms differ, and so group these often into different species, which are usually given distinct names. A conventional distinction between different species is that they cannot breed together, but even this is blurred, for example a mule is a cross between a donkey and a horse. In fact there are numerous ways of defining species. Within each species there are often distinct groups, e.g. strains or ecotypes or phenotypes. These subgroups may intermingle or can be quite distinct due to physical isolation, and can be the precursors of new species groups in the future.

Taxonomists often take several measurements, and one major early application of multivariate statistics, especially discriminant analysis, which aims to classify organisms into groups, involved conventional taxonomy. Fisher's iris data [1] is a much cited classical dataset. Four measurements, namely petal width, petal length, sepal width and sepal length were measured on 150 irises consisting of three species. No single measurement is an unambiguous indicator of class membership. The scores plot of the first two Principal Components

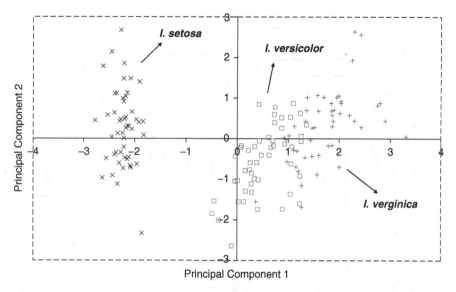

Figure 10.1 Scores of Principal Component 2 against Principal Component 1 for the Fisher iris dataset

(PCs) of the standardized data are presented in Figure 10.1 with each of the three species indicated by a different symbol. Whereas *Iris setosa* seems well discriminated, the other two species, *I. versicolor* and *I. verginica*, are not so well distinguished, and it would not be easy to unambiguously classify an unknown plant from its physical measurements into one of these two groups using just PC scores. However many other methods can be used for classification which we will look at in more detail later in this chapter, which allow class membership to be determined and go further than PCA in permitting a numerical indication of whether an unknown fits into one of the prior groups or even whether it is an outlier and belongs to no group currently studied.

In some cases one is interested in classifying a large number of species or strains. This subject is often called numerical taxonomy, and one method often employed is to use cluster analysis (Section 5.6) where the raw measurements are converted to what is often called a dendrogram, phylogram or tree diagram (note that there is a subtle difference between this and a cladogram in which the branches are of equal length : see Section 11.4 for more details). Figure 10.2 represents part of a dendrogram for angiosperms (flowering plants). Each group at the end of the tree may be further divided into species, strains and so on. The main relationships can now easily be visualized. The earliest botanists produced these diagrams by hand, as computers were not available. Numerical taxonomy, however, involves

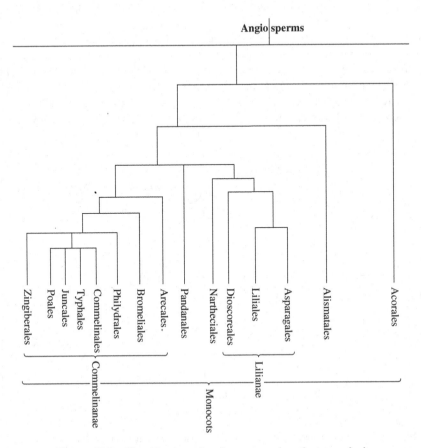

Figure 10.2 Classification of angiosperms using cluster analysis

computing the similarities between organisms, or samples, or even an average measurement on a series of organisms, to obtain diagrams such as dendrograms, computationally, from data obtained by biologists, most of which is quantitative in nature although categorical variables (which may take the form of 'yes' and 'no', or involve several answers such as colours) are also allowed. In fact there are a vast number of physical measurements that can be made on plants and animals. This can be extended, for example, to bacteriological tests with the results of these tests forming the original data matrix.

So what has this to do with chemometrics? In the modern day we are also likely to make chemical measurements on organisms (or extracts from organisms) in addition to physical measurements, so people have potentially a vast amount of chemical data available. This can be of numerous types. A simple approach involves using chromatography of extracts (e.g. plant extracts or tissue samples) to determine the relative concentration of a series of known compounds in a variety of samples, and then use this chemical information for classification and clustering. It is increasingly common to perform an analytical measurement such as pyrolysis mass spectrometry or NMR or GC-MS on a sample, and use the peak intensities for subsequent pattern recognition. In this case the chemical origins of the peaks are of peripheral interest, because it is not, for example, easy to assign the NMR peaks of whole extracts to any individual compound.

As it becomes more common to make analytical chemical measurements on biological samples, many of the ideas of traditional numerical taxonomy can be employed by the biological chemist.

10.3 DISCRIMINATION

Fisher's classic work on iris data is widely attributed as having catalysed many aspects of multivariate classification, especially discriminant analysis. In this section we will look at some of the approaches developed to assign objects into groups by biologists. Many of these methods are also used in chemistry, and build on principles already introduced in Section 5.8.

10.3.1 Discriminant Function

The key to Fisher's work is the concept of a discriminant function. In its simplest form, if we have data that belongs to two groups, the value of this function tells us which group a particular object belongs to. A high value may indicate group A and a low value group B. In mathematical notation, if the measurements of a given sample are denoted by a row vector x (for Fisher's iris data this consists of four measurements, namely petal width, petal length, sepal width and sepal length), a discriminant function between two groups can be defined as follows:

$$W_{AB} = (\bar{x}_A - \bar{x}_B).S^{-1}.x'$$

In above equation:

- $(\bar{x}_A - \bar{x}_B)$ is the difference between the mean measurements for groups A and B; the discriminant function as defined above is such that an object originating from group A will have a more positive value to one originating from group B.

- **S** is what is called a *pooled variance covariance matrix* of the variables. We have already defined the variance and covariance in Section 3.3. This matrix will be of dimensions 4×4 for the iris data as there are four variables. This matrix is an average of the variance covariance matrices for each class. If there are two classes, it becomes:

$$S = \frac{(N_A - 1)S_A + (N_B - 1)S_B}{(N_A + N_B - 2)}$$

We have encountered a similar idea in the context of the t-test (Section 3.7), but here we are dealing with matrices rather than individual numbers. The reason for this is that some variables may be correlated, and so provide very similar information, and we do not want these to overwhelm the analysis.
- x' is the transpose of the vector of measurements for any individual sample, for example a vector of length 4 consisting of the measurements for each iris.

This function can then be defined for every object in the dataset. If you are unfamiliar with matrix notation, see Section 2.4.

An important feature is that the Fisher score is scaled by the inverse of **S**. A simpler distance (the 'Euclidean') distance would involve computing $(\overline{x}_A - \overline{x}_B) \cdot x'$ instead, as discussed in Section 5.8 but this would omit information about how spread out each class is and also whether two variables are correlated or not. The motivation of introducing this extra term can be understood by reference to Figure 10.3, where we compare the size of two populations, one of elephants and one of mice. We can measure the 'average' size (e.g. weight or length) of the elephants and the mice. We then ask whether a baby elephant is best classified as an elephant or mouse by size. In fact the size of the baby elephant may be much closer to the average mouse than the average elephant, so using a simple Euclidean distance, we will misclassify the animal. However if we take into account the spread of in size of the two groups, the baby elephant should easily be classified as an elephant, and the variance covariance matrix makes this adjustment.

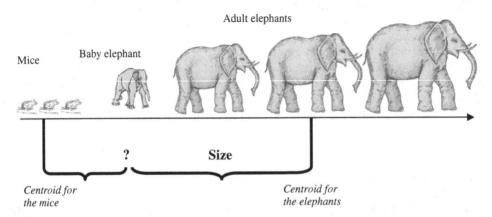

Is the baby elephant an elephant or a mouse?
Is the weight of the baby elephant closer to that of mice or adult elephants?

Figure 10.3 The size of a baby elephant is closer to the centroid of a population of mice to that of a population of adult elephants

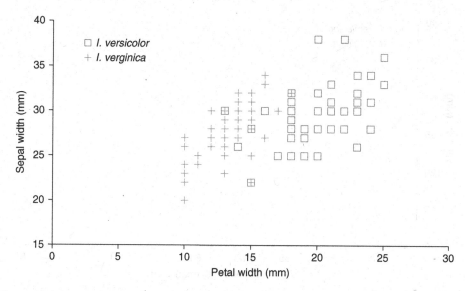

Figure 10.4 Graph of sepal width versus petal width for two of the species of the Fisher iris data

10.3.2 Combining Parameters

Why do we need to perform this elaborate procedure? Figure 10.4 is a graph of sepal width against petal width for two of the species from Fisher's data. We can say that 'on the whole' a low sepal width indicates *I. verginica* whereas a high one indicates *I. versicolor* but one cannot really classify the plants uniquely just using this measurement. Petal width is more promising, and if one discriminates the species between 16 mm and 17 mm, there will be only seven misclassifications or 7 %. However, we were lucky and chose the best discrim- inating parameter, change petal width to sepal length and we get the graph of Figure 10.5 which is harder to interpret. In many situations it is difficult to find a single parameter that be used to unambiguously distinguish between groups.

A combination of the four parameters describes best the data, and discriminant analy- sis allows us to use all the information together. Table 10.1 shows the main steps of the discriminant analysis for *I. verginica* and *I. versicolor* (each class has 50 samples and so is of the same size in this example). Figure 10.6(a) is of the discriminant function against sample number, it can be seen that a divisor can be drawn and now only three samples are misclassified. Notice however, that all the numbers in Figure 10.6(a) are negative.

It is sometimes useful to change the scale so that a positive value indicates membership of class A and a negative value class B, so that:

$$W_{AB} = (\bar{x}_A - \bar{x}_B).S^{-1}.x' - 1/2(\bar{x}_A - \bar{x}_B).S^{-1}.(\bar{x}_A + \bar{x}_B)'$$

The second term simply involves subtracting the average of the means of each class. This has the effect of shifting the scale [Figure 10.6(b)].

10.3.3 Several Classes

What happens if there are more than two classes? The method above can quite easily be extended. If we have three classes there will be three possible discriminant functions W_{AB},

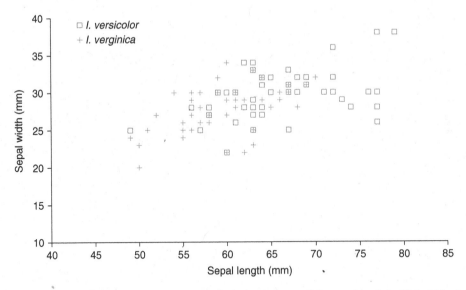

Figure 10.5 Graph of sepal width versus sepal length for two of the species of the Fisher iris data

Table 10.1 Determining the main matrices and vectors required for the discriminant function between *I. versicolor* and *I. verginica*

	Petal width	Petal length	Sepal width	Sepal length
S_A				
Petal width	3.909	6.516	4.111	5.554
Petal length	6.516	28.531	7.423	20.076
Sepal width	4.111	7.423	9.419	8.242
Sepal length	5.554	20.076	8.242	26.247
S_B				
Petal width	8.256	5.489	4.616	4.587
Petal length	5.489	29.850	6.995	29.722
Sepal width	4.616	6.995	10.192	9.189
Sepal length	4.587	29.722	9.189	39.626
S				
Petal width	6.083	6.002	4.363	5.071
Petal length	6.002	29.190	7.209	24.899
Sepal width	4.363	7.209	9.806	8.715
Sepal length	5.071	24.899	8.715	32.936
S^{-1}				
	0.268	−0.050	−0.104	0.024
	−0.050	0.106	0.011	−0.076
	−0.104	0.011	0.174	−0.039
	0.024	−0.076	−0.039	0.094
$(\bar{x}_A - \bar{x}_B)$				
	−6.8	−12.3	−2.1	−6.52

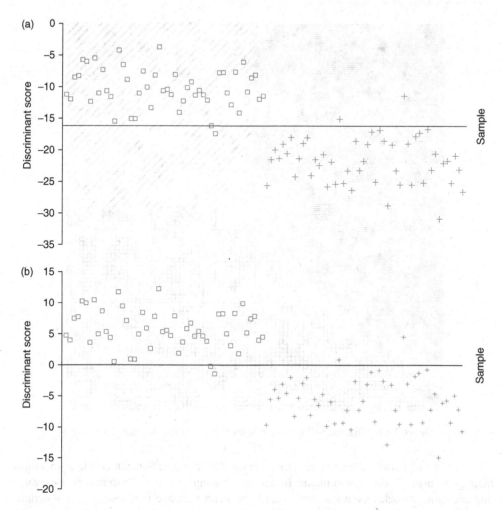

Figure 10.6 Fisher discriminant score. (a) Raw data; (b) Adjusted for means between *I. versicolor* and *I. verginica*

W_{AC} and W_{BC} and for four classes six possible functions, W_{AB}, W_{AC}, W_{AD}, W_{BC}, W_{BD} and W_{CD} and so on. It is theoretically possible to discriminate between any N classes by picking $N - 1$ of these functions. So if one has has three classes, it is necessary only to calculate any two of the three possible functions, and for four classes any three of the six possible functions. How can these be used to tell us which class a sample belongs to? If we have three classes and choose to use W_{AB} and W_{AC} as the functions:

- an object belongs to class A if W_{AB} and W_{AC} are both positive;
- an object belongs to class B if W_{AB} is negative and W_{AC} is greater than W_{AB};
- an object belongs to class C if W_{AC} is negative and W_{AB} is greater than W_{AC}.

If we were to plot the values of W_{AB} and W_{AC} for all the objects in a graph, it will be divided into three segments as in Figure 10.7, according to the class membership. Notice that the bottom left-hand region represents two classes.

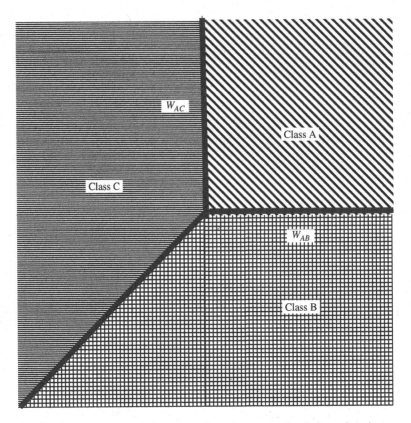

Figure 10.7   Discriminating three classes using two discriminant functions

There are of course a variety of other approaches for deciding which class an object belongs to using Fisher discriminant functions, although the one illustrated is a classical one and quite straightforward to understand, the main message being that these functions can be extended to use in multiclass problems.

10.3.4 Limitations

One major limitation of these methods is that for the matrix S to have an inverse the number of measurements must be less than or equal to the number of samples. In classical biology this condition is almost always met, for example in Fisher's data there are four measurements and 150 samples. These matrices are often referred to as long, thin, matrices. Chemists often have to deal with short, fat matrices, where the number of measurements greatly exceeds the number of samples, for example one may record intensities at 1000 spectroscopic wavelengths for a training set of 50 samples. Thus, chemometricians have needed to modify these methods, the first proponents of multivariate statistics found measurements hard and expensive and so could not have conceived the situations of rapid, computerized, data collection that are routine tools of the modern chemist. A common modification is to use PC scores (Section 5.2) rather than raw measurements as the variables, although the matrix S then becomes a diagonal matrix, because the scores are orthogonal. An alternative is feature selection, e.g. just choose the most useful spectral wavelengths or chromatographic peaks. Finally it may be possible to lump measurements together in blocks, for example averaging the intensity regions of a spectrum.

By the means described above, it is possible to calculate discriminant functions that are used to group objects. The methods can be considerably extended, as we will discuss later, for example is it really best to place a dividing line half way between the class centres, and what provision is there for objects that belong to neither group?

Note that in many applications it is important to validate the model as discussed in Section 5.7, although we will concentrate primarily on describing different methods in this chapter. However, the reader must appreciate that this validation step is often quite important.

10.4 MAHALANOBIS DISTANCE

In Section 10.3 we looked at the main ideas behind discriminant analysis, a technique common in biology, that is often used also in biological applications of chemistry. These principles can be extended to determination of class distances.

The idea behind this is that for each class one can determine a centroid, which is the centre of the dataset. We have already come across this concept before (Section 5.8.2) but we did not discuss the distance measure in detail. A simple approach would be to use the Euclidean distance, which is simply the distance to the geometric mean or centroid of each class. The closer an object to this, the more likely it is to belong to a group. Table 10.2 presents some data recorded using two variables, with a graph displayed in Figure 10.8. At least two of the samples of class B (numbers 12 and 18) are hard to classify; also sample 6 of class A is not so easy to distinguish. If we did not know the grouping in advance we might get into trouble. In vectors, the class distance for an object from the centre of class A is defined by:

$$d_A = \sqrt{(x - \overline{x}_A).(x - \overline{x}_A)'}$$

These are calculated in Table 10.2 for each sample. A sample is classified into the class whose centre it is closest to. We can see that samples 12 and 18 are incorrectly classified, as they are nearer the centre of class A than class B.

However, one piece of information that has been neglected so far is that class B is much more diffuse than class A, i.e. it is spread out more as can be seen from Figure 10.8. This can often occur, for example one species may have more variability than another. Even size matters, the variance of the body length of elephants will be much larger than that of mice, so some sort of scaling of the range of elephants' body lengths would be advisable : a linear model may confuse a mouse with an elephant, as shown in Figure 10.3 and discussed in Section 10.3.1, similar principles applying as for the Fisher discriminant function. Since the measurements for class B are more spread out in our example, a large distance from the centre of class B is not as significant as a large distance from the centre of class A. Another problem that occurs, when there are several measurements (in our case there are of course only two) is that there can be correlations between these. For example we may make five measurements on the wing of a bird. If the wings are roughly the same shape but differ mainly in size, these fundamentally measure the same thing, and these five variables may unfairly dominate the analysis in comparison with other measurements of a bird's body, so these five parameters really measure approximately the same property. By analogy, if we were measuring the concentrations of metabolites it might be that the primary metabolites

Table 10.2 Example for calculation of Euclidean and Mahalanobis distance

True class	Sample	x_1	x_2	Euclidean Distance			Mahalanobis distance		
				Distance to class A	Distance to class B	Prediction	Distance to class A	Distance to class B	Prediction
A	1	79	150	27.267	71.297	A	1.598	3.951	A
A	2	77	123	0.497	50.353	A	0.037	2.646	A
A	3	97	123	19.557	39.185	A	1.217	2.177	A
A	4	113	139	39.081	52.210	A	1.916	2.688	A
A	5	50	88	44.302	62.012	A	1.914	1.915	A
A	6	85	105	19.317	32.561	A	1.378	1.573	A
A	7	76	134	11.315	59.362	A	0.729	3.215	A
A	8	65	140	21.248	70.988	A	1.588	3.779	A
A	9	55	103	29.915	59.257	A	1.296	2.325	A
B	10	128	92	59.187	16.824	B	4.416	0.391	B
B	11	65	35	88.656	69.945	B	4.951	2.169	B
B	12	72	86	37.179	40.008	A	2.069	1.187	B
B	13	214	109	137.249	104.388	B	8.999	2.579	B
B	14	93	84	41.781	19.205	B	2.963	0.503	B
B	15	118	68	68.157	19.734	B	5.101	1.092	B
B	16	140	130	62.971	51.480	B	3.709	1.866	B
B	17	98	74	52.932	18.969	B	3.773	0.535	B
B	18	81	97	26.022	32.635	A	1.683	1.323	B
B	19	111	93	44.863	6.280	B	3.395	0.340	B

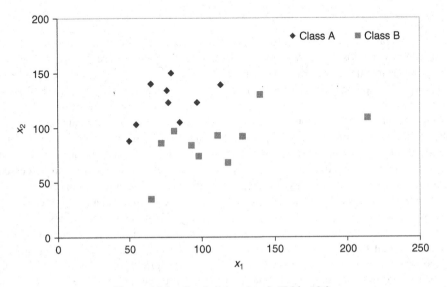

Figure 10.8 Graph using data in Table 10.2

dominate the dataset but have little discriminatory power and a small selective number of markers contain the best information about the difference between organisms.

The Mahalanobis distance overcomes this difficulty. Most chemometricians define the distance of an object to class A by:

$$d_A = \sqrt{(x - \overline{x}_A).S_A^{-1}.(x - \overline{x}_A)'}$$

where S_A is the variance covariance matrix of the measurements on the training set samples in class A (see Section 10.3). Each object has a specific distance from the centroid of each class. Notice that some classical statisticians like to group all the training samples together and so define the distance by:

$$d_A = \sqrt{(x - \overline{x}_A).S^{-1}.(x - \overline{x}_A)'}$$

where S is the pooled variance covariance matrix over all training set samples. This latter approach has the disadvantage in that the calculations have to be repeated if additional classes are added, but is simpler computationally. We will adopt the first definition. The Mahalanobis distances are calculated in Table 10.2, and we find that all the samples are correctly classified, as the bigger spread of class B has now been taken into account.

We can, of course, represent these distances graphically, as in Figure 10.9, the dividing line being where the two distances are equal. It is now possible to unambiguously classify the samples. We could alternatively have used the discriminant function described in Section 10.3 but this uses a single pooled variance and so will not always give the same answer, however, objects with a value close to zero could be regarded as of uncertain origin and are a little ambiguous. Figure 10.10 is of the Fisher discriminant function [adjusted

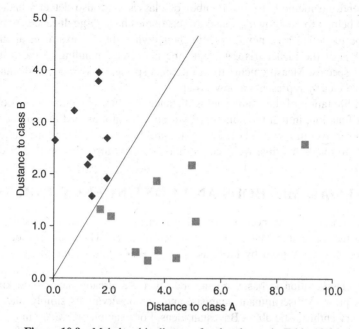

Figure 10.9 Mahalanobis distance for the classes in Table 10.2

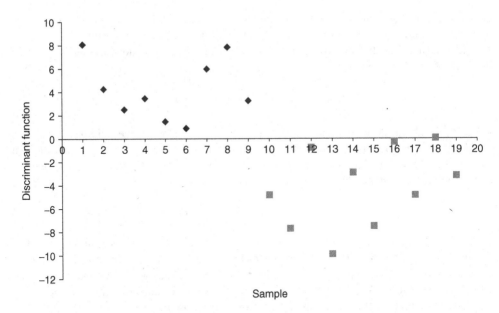

Figure 10.10   Fisher discriminant function for the data in Table 10.2

for the mean distance between the classes as in Figure 10.6(b)], and sample 18 is actually misclassified but has a score close to zero, and we may be suspicious of this point.

There are several more elaborate statistics, which allow for determining whether samples are ambiguous or are outliers, and which can be used to attach a probability of class membership rather than unambiguously assign a sample to a specific group. Of course one can extend the methods to any number of classes, and also detect whether there are samples that belong to no known groups: these samples have large distances to both classes. They may be modelled as a new group in their own right, for example, another species. Figure 10.11 is of the Fisher iris data, with the axes corresponding to the class distances to two of the species. Measurements from the third species, *I. serosa*, are distant from both classes and so clearly represents a new group.

The main limitation of the Mahalanobis distance measure is similar to that of the Fisher discriminant function, in that the number of variables cannot exceed the number of samples in the group (or overall dataset) because of the problem of calculating the variance covariance matrix. To overcome this, we need similar tricks to those discussed in Section 10.3.4.

10.5 BAYESIAN METHODS AND CONTINGENCY TABLES

Above, we described two important methods for classification, namely Fisher discriminant analysis and the associated Mahalanobis distance measure. There are in fact a huge number of approaches many used by biologists, all of which can be modified using Bayesian techniques.

Bayesian classification is based on the idea that we already have some knowledge of probabilities prior to discriminant analysis, and our observations simply alter this. Many people get very enthusiastic about Bayesian ideas. For example, if we are in the North Pole, the chances are the temperature will be quite low, relative to the Equator. This knowledge

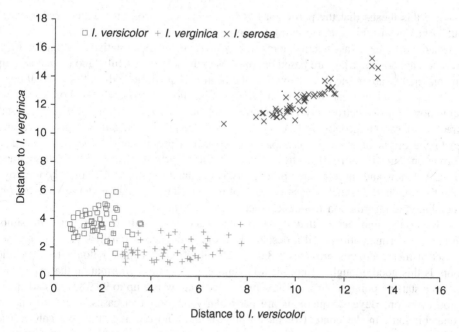

Figure 10.11 Mahalanobis distance of the Fisher iris data to *I. versicolor* and *I. Verginica*

can be part of our predictions of weather and is called the *prior* knowledge. In fact our prior knowledge may be quite elaborate, it may consist of historic weather data for different times of year, so we already know historically what the temperature in July, for example, usually is. Then we have some observation, of the current weather patterns and use this to change our predictions. These observations simply modify what we expect, and lead to a *posterior* probability. Bayesian methods have been illustrated in the context of signal processing in Section 4.5.

There are huge debates in the literature, especially medical statistics, about the applicability of Bayesian approaches, which can have a radical influence on predictions. Let us say a patient is asked to fill in a lifestyle questionnaire in which he or she admits to heavy smoking for 30 years. If the patient is undergoing tests for lung cancer, we know before even the tests have been performed that there is a good chance, so we can add this evidence to other evidence obtained from chemical or biological tests, which basically are used either to increase or decrease the probability of cancer. Mathematically one way this can be incorporated is into classification algorithms, for the Mahalanobis distance we write, for the distance to class A:

$$d_A = \sqrt{(x - \overline{x}_A).S_A^{-1}.(x - \overline{x}_A)' + 2 \ln(p_A)}$$

where p_A is the prior probability of class membership, however the approaches below are intuitively simpler and can allow quite easily for multiple tests or classification techniques. Bayesian methods can have a significant influence over our predicted probabilities of class membership. For example, we may know that 99 % of samples belong to one species, we only expect a small number from an unusual species, why not include this information? But of course this means that the particular test has to be very strong to suggest that a sample actually belongs to this specific minor group.

Because different classification methods are used in medical statistics, some of which diagnoses and clinical trials are based on, one has to think very carefully about what assumptions one makes about the data. However, the beauty of such methods is that several pieces of sequential evidence can be used to improve and modify probabilities. For example, it is common to use spectroscopy for screening, we may use near infrared (NIR) to classify extracts into various groups. NIR is unlikely to be confirmatory, but will give us a good idea of the origin of a specific sample. So we have a preliminary probability. Using the information gained from NIR, why not now try high performance liquid chromatography (HPLC)? A new and more refined probability is obtained, finally NMR could be employed as the final method. Instead of treating each of these techniques independently, it is possible to combine the information from each successive experiment.

A Bayesian might argue that if one is interested in classifying samples, one simply continues experimentation until a desired level of confidence is reached. Let us suppose a first set of measurements provides a 85 % probability that a sample belongs to a specified group. Is this good enough? If not, do some more, but use this information that we already have as a starting point, adding the next bit of evidence we are up to 95 %. Is this adequate? If not, carry on. Bayesian methods are particularly important in cases where no one test is unambiguous. In the context of the Fisher iris data, no one measurement can be used for perfect classification of samples, a cut-off can be determined which may for example be able to classify most but not all samples correctly and so on. In clinical diagnosis or forensic evidence this approach is very common as a combination of symptoms or tests are required to be sure of a diagnosis. Bayes' theorem states, in terms of medical diagnosis:

New probability after a test

= (Previous probability × Probability would test positive if the disease was present)/

(Previous probability × Probability would test positive if the disease was present

+ (1 − Previous probability) × Probability would test positive if the disease was absent)

This is probably best understood via a simple example, see Table 10.3 which is commonly called a *contingency table*. This illustrates the results of a classical test for anaemia on a number of patients which involves measuring the amount of serum ferritin in blood, if it exceeds a certain level the test is deemed positive, but otherwise negative. The test is not perfect, ideally we would like the off-diagonal elements to be 0, but in many situations this is not possible. Analogously this can be applied to multivariate methods for classification, the prediction ability of a disease by, for example, NMR of plasma extracts, may not be

Table 10.3 Contingency table for illustrating Bayes' theorem

		Disorder		
		Present	Absent	Total
Test result	+ve (<65 mM)	731	270	1001
(serum ferritin)	−ve (>65 mM)	78	1500	1578
Total		809	1770	

perfect, but would contribute to evidence that a disease is present. Using two or three tests would allow the building up of more certain evidence.

The table shows that:

- for 270 out of 1770 patients who do not have the disease, the test result is positive, i.e. there is a probability of 0.153 that there will be a positive result if the disease is present;
- for 731 out of 809 patients who do have the disease, the test result is positive, i.e. there is a probability of 0.903 that there will be a positive result if the disease is present.

We now can calculate posterior probabilities, i.e. probabilities that the disease is present if the test is positive. As for the prior probabilities this depends on what other evidence we wish to use. The simplest (and classical) assumption is that there is a 0.5 chance that a patient has the disease before evidence is obtained, using equation on page 301, this makes the chance of disease 0.856 after a positive diagnosis. However, let us say this disease is found in only 1 person in 20 of the population, it may be a genetic disorder that is expressing itself. Starting with this prior probability of 0.05 of positive result yields a posterior probability of only 0.237. Finally it may be that we are not asked to perform this diagnosis unless the doctor is pretty sure, perhaps 0.95 sure, this makes the posterior probability 0.991, which is quite a difference. Hence including this extra information can modify the overall calculated chance that a disease is present dramatically.

Bayesian methods are particularly important where no single piece of evidence in itself results in 100 % perfect predictions, as is common in many areas of biology. The problem is where to start from, i.e. what to use as the first prior probability, and this usually can only be determined from an understanding of the problem in hand. In terms of the contingency table of Table 10.3, this is equivalent to weighting each column, according to the relative (prior) probabilities, the difference between classical and Bayesian statistics is that in the former each column is weighted equally whereas in the latter the weights can be varied.

There is a lot of terminology associated with contingency tables and implicitly Bayesian methods for assigning probabilities of class membership (the two are often interlinked):

- *Likelihood ratios.* A positive likelihood ratio is the relative odds that a test is positive when a sample is a member of the corresponding group to the odds that the test is positive when the sample is not. A negative likelihood ratio is the reverse.
- *True positives* and *true negatives* represent the diagonal elements of the contingency table, whereas *false positives* and *false negatives* the off-diagonal elements.
- *Specificity* is defined as the number of true positives over the number of true positives plus false negatives, and *sensitivity* as the number of true negatives over the number of true negatives and false positives.

Note that the use of this terminology is not restricted to Bayesian analysis.

Most Bayesian methods can be combined with all the classification approaches described in this chapter.

10.6 SUPPORT VECTOR MACHINES

Support Vector Machines (SVMs) are a relatively new approach for classification, first proposed in the early 1990s, and are gaining popularity due to their promising empirical performance, especially in biology.

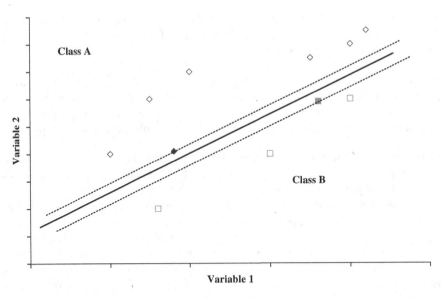

Figure 10.12 Finding samples that are on the boundary between two classes or Support Vectors

SVMs are distinguished from the methods above in various ways. The first and most straightforward property is that they are boundary methods. In simple terms, they do not try to model a class as a whole but to find the boundaries between two classes. In their simplest and most classical implementation they are implemented as two class classifiers, that is they try to distinguish between two groups: at the end of this section we will describe extensions to multiclass problems. Figure 10.12 illustrates this, in which samples that are on the boundary between the classes, or Support Vectors, are indicated with filled symbols. In order to model classes only the boundary samples or Support Vectors are required, and the trick of SVM methods is to identify these samples. An unknown is then classified according to which side of the boundary it belongs to. This can as usual be done on training sets via cross-validation or bootstrapping or on test sets as outlined in Section 5.7. The best boundary is defined as the one where there is maximum empty space between the two classes, and the algorithm searches for this.

So far, so good, as this does not appear radically difficult or different from other approaches, but the importance of SVMs lies in the way a boundary can be established. Many biological problems are nonlinear, that is a nice clean line cannot be drawn between two groups of samples. This differs from many problems in traditional analytical or physical chemistry which tend to be quite linear. Figure 10.13 represents two classes that are nonlinearly separable. SVMs try to obtain this boundary.

This can conceptually be done by finding higher dimensions (or feature space) that separate out the groups. This is illustrated in Figure 10.14. The input space represents the original variable space, each axis corresponding to a measurement (e.g. a chromatographic peak area or a spectral intensity at a given wavelength). Feature space involves adding an extra dimension that separates out the classes with a linear plane (or hyperplane). This concept may appear complex but is not too difficult to understand. Consider two groups separated by a parabola, one group above it and one below, adding an extra dimension of

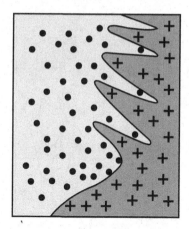

Figure 10.13 Two nonlinearly separable classes

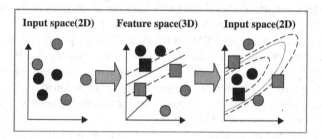

Figure 10.14 Finding a higher dimension that linearly separates two groups (the Support Vectors are illustrated by squares)

x^2 will make the two groups linearly separable. Finally this plane can be projected back into the original data space, and will appear curved or nonlinear.

A special computational method, called the 'kernel trick' makes this process of adding dimensions easier, and is the equivalent of adding these dimensions, but instead the boundaries are in the original input or data space. There are several popular kernels that are employed in SVMs, these are the Radial Basis Function, Polynomial Functions, and Sigmoidal Functions, dependent on the number of adjustable parameters and the complexity of the boundaries required. The Radial Basis Function is probably the most popular due to its relative simplicity. In simple terms the role of these functions is to adjust the boundaries accordingly.

A key problem about SVMs is that it is possible to obtain boundaries of almost indefinite complexity, so that ultimately 100 % of training samples are classified correctly, the ultimate boundary enclosing every sample or group of samples, see Figure 10.15, posing a dilemma. In SVMs there are several important theories about complexity, but these are related to the idea of Structural Risk Minimization. The more complex the boundary, the better the training set is predicted, but the higher the risk of over-fitting on the test set. Validation is essential, but an important guiding principle of SVMs is that the more samples in the dataset, the more complex a boundary can be obtained for a given risk of misclassification

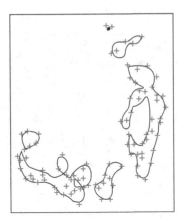

Figure 10.15 Complex boundaries

of test samples. Hence complexity relates to the number of samples in the training set, and, also, of course, how linearly separable or otherwise the data.

Optimizing the performance of SVMs is tricky and there are not yet rules that are set in stone, but a common strategy is to take a function such as the Radial Basis Function and change the parameters in a 'grid search' usually using cross-validation or a bootstrap on a training set, until the minimum number of misclassification is found. Note that the training and test set must have much the same composition for this to be successful, otherwise there is a risk of over-complexity in the SVM model.

SVMs are an alternative to older approaches such as neural networks and have the advantage that they are reproducible providing the kernel is known, so it is possible to return to data in many months or years later and obtain an identical answer.

SVMs were not originally designed for multiclass problems where there are several groups, but can be applied in such situations in a variety of ways. The simplest is to perform a 'one against all' computation in which a boundary is placed between each class and all other samples, for example if there are four classes, four such computations are performed. Often these classes are called the 'target' and 'outlier' class and it is hoped that each sample is only once assigned to a target class, but there are elaborations to decide what is the membership of a sample if it appears in a target class more than once (or never), there are various statistics that can be obtained about how close a sample is to a boundary, if required. An alternative, but more computationally intensive is to perform several 'one against one' classifications, in which all samples are assigned to one of two of the original classes (the model can be developed just using samples of two classes in each training set), and the final membership is determined by a majority vote.

There is a burgeoning literature on SVMs, but the text by Christianini and Shawe-Taylor [2] is a classic. A review of SVMs in chemometrics has recently been published [3]. Over the next few years we expect to see much more about SVMs and kernel methods in chemometrics, especially with applications to biology developing rapidly.

10.7 DISCRIMINANT PARTIAL LEAST SQUARES

Another technique that can be employed for classification is discriminant PLS (D-PLS). We have already discussed PLS as a calibration method (Section 6.6). We showed how PLS

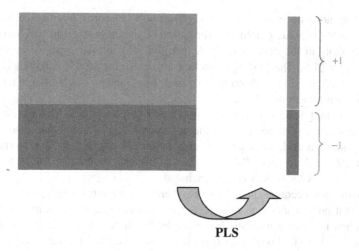

Figure 10.16 Discriminant PLS

can be used to relate one block of data (e.g. 'x') such as a spectrum to another (e.g. 'c') such as the concentration of one or more components in a spectrum.

PLS can also be used for discriminant analysis or classification. The simplest approach is when there are only two classes. In such case one sets up a 'c' vector which consists of a number according to which class an object belongs to, there are two common variations, using '1' and '0' or '+1' and '−1' for classes A and B. We will stick to the latter convention although there is no fundamental difference between either approach. PLS can be used to form a calibration model between the 'x' block which may, for example, be a set of spectroscopic measurements, and the 'c' block which may provide information as to the biological origins of a sample. This is illustrated in Figure 10.16. The results of calibration are a number. A simple rule for interpreting this number may be to assign an object to class A if it is positive, and class B if it is negative. Of course there are elaborations, for example a large positive or negative number much greater than ±1 might indicate an outlier, and there may be a region around 0 where we feel that the classification is ambiguous, so it is possible to choose different decision thresholds. If the two classes of interest are, for example, whether a patient is diseased or not, and it is possible to adjust the decision threshold to reduce either the number of false positives or false negatives (Section 10.5) according to which is most important for the specific application. Some people incorporate more elaborate statistical tests such as the F-test, to determine class membership, so interpreting the result of calibration in the form of a probability.

An important aspect of PLS involves choosing the number of significant components for the model, often using cross-validation or the bootstrap (Section 6.7) usually by maximizing the per cent correctly classified (Section 5.7) although the predicted values of the 'c' block can be employed as an alternative criterion. One important consideration relates to model validation – if the quality of model is to be tested this should be a separate operation from choosing the number of significant components. The former involves optimizing the number of significant components, the latter validation, and usually an independent test set is required, which contains samples that have not been used in model building via cross-validation. It is a mistake to use the optimum cross-validated prediction error as an assessment of how good the PLS model is.

In addition an advantage of D-PLS is that there will be scores, loadings and weights and these can be used to provide graphical information on the class structure. We have discussed these types of plots in the context of PCA (Section 5.3), similar types of information can be obtained using PLS. The loadings can help us determine which variables (for example chemicals) are most useful for characterizing each class, and so could help in a real context, e.g. in finding potential biomarkers.

Another important feature is the ability to add extra variables. Many biological problems, in particular, are nonlinear, unlike the most common situations that arise in analytical chemistry so one can add, for example, squared terms. If there are 100 variables in the 'x' block, simply add another 100 variables that are the squares of the original ones. In additional several blocks of data can be included.

Where the method becomes quite tricky is when there are more than two classes involved. One method that only works well in specialized cases is to calibrate against a 'c' vector in which each class is given a number: for example, for three classes we may assign '+1' to A, '0' to B and '−1' to C. Sometimes there is a reason why the middle class is intermediate in nature to the extremes one being in genotyping as we possess two chromosomes for each gene, so class A may correspond to an individual with two alleles 1, class B with mixed allele 1 and 2 and class C both alleles 2. However, these examples are quite specific, and normally there is no particular sequential relationship between the classes. An alternative approach is to set up a 'c' matrix of $N - 1$ columns, where N is the number of classes. For example, if there are three classes, set up the first column with '+1' for class A and '−1' for either classes B or C, and the second column with '+1' for class B and '−1' for classes A or C. It is not necessary to have a third column. Obviously there are several (in this case three) ways of doing this. It can be a bit tricky to interpret the results of classification, but one can use an approach similar to that used in discriminant analysis (Section 10.3.3) when there are more than two classes.

A more common approach is to use a 'c' matrix where each column relates to membership of a single class, so for three classes we use three columns, with +1 for members and −1 for nonmembers. We still have to be a bit careful how the result of calibration is interpreted. For example, a negative number means that an object does not belong to a specific class. What happens when all the three values from the calibration are negative? Do we say that the object is an outlier and belongs to no class? Or do we find the value closest to '+1' and assign an object to that specific class? There are a variety of ways of making choices and interpreting the results.

If the 'c' block consists of more than one column it is possible to use PLS1 on each column separately for classification or PLS2 to classify all the objects simultaneously. There are no general guidelines as to the most appropriate approach. However, D-PLS is quite popular and allows all the techniques of validation, looking for false positives/negatives, contingency tables and so on to be applied and the algorithms are quite straightforward, so suited for repetitive computation, for example when it is required to choose test sets or perform bootstraps in an iterative way, and have the advantage over many more sophisticated methods that they are easy to optimize (just choose the right number of components).

10.8 MICRO-ORGANISMS

Some of the earliest work in the development of chemometrics pattern recognition as applied to biology was in the area of micro-organisms, probably because they are easy to study in the laboratory, and obtain controlled extracts from. It is important to be able to characterize micro-organisms, especially bacteria, for a variety of reasons. In the control of diseases we need to know if an isolated microbial culture matches a known species or strain. The features of organisms in reference culture collections need to be monitored. New strains may be identified in different cultures grown in different laboratories.

Conventionally biologists use morphological measurements such as features determined by inspecting plates under the microscope or the results of specific tests. In certain specific cases people use marker compounds which are characteristic (often secondary metabolites) for specific organisms.

10.8.1 Mid Infrared Spectroscopy

The problem with traditional morphological approaches is that they are laborious and often require elaborate methods for sampling of cultures and preparation of slides. Alternative approaches are based on molecular spectroscopy where the culture is analysed, normally during growth, by a variety of techniques. Many of studies involve using Fourier transform mid infrared (MIR) spectroscopy, generally recorded between 800 and 4000 cm^{-1}. Unlike in conventional organic chemistry where spectra are usually of pure compounds, the biological spectra arise from signals originating from complex mixtures, such as proteins, nucleic acids, polysaccharides and lipids. It is often difficult to assign any specific peak to an individual compound, but a spectrum can be divided into regions and the intensity of the signal in each region obtained, generally automatically, without manual intervention.

The advantage of this approach is that the methods are noninvasive, quick and do not require huge amounts of sample. Ideally a series of spectra can be recorded, these can be grouped by PCA (Section 5.2) or cluster analysis (Section 5.6). For specific species or strains one can apply a variety of methods for supervised pattern recognition including several of the approaches discussed above such as SVMs, Mahalanobis distance and D-PLS, as well as neural networks and genetic programming (which in themselves are a big subject). The usual tests such as cross-validation, the bootstrap, and using a test set can be employed to see how well the classes are modelled, and then unknowns can be classified. Ideally this should be done automatically by a machine, for example in which a database of known organisms are available and the classification of an unknown is matched to this database.

Of course life is not always so simple. There are practical instrumental imperfections that need to be overcome. A common difficulty in IR spectroscopy involves baseline problems. Often quite elaborate approaches to baseline correction are used. There are many factors some of which are a little irreproducible and so it can be difficult if spectra are recorded on different days and under varying conditions. One simple and common approach for attempting to overcome this is to use derivatives, the Savitzky–Golay method (Section 4.8) increases resolution as well as reducing baseline problems. This is illustrated in Figure 10.17 for three simulated noise free spectra. By the time of the second derivative they are essentially indistinguishable : of course if there is substantial noise there may be limitations to the use of derivatives.

Other problems relate to small peak shifts in different spectra, there are ways of overcoming this, such as dividing the spectrum into windows and summing intensities within these windows, rather than trying to identify the exact frequency of each peak in the spectrum. Another common consequence is that only certain regions of a spectrum are informative. If

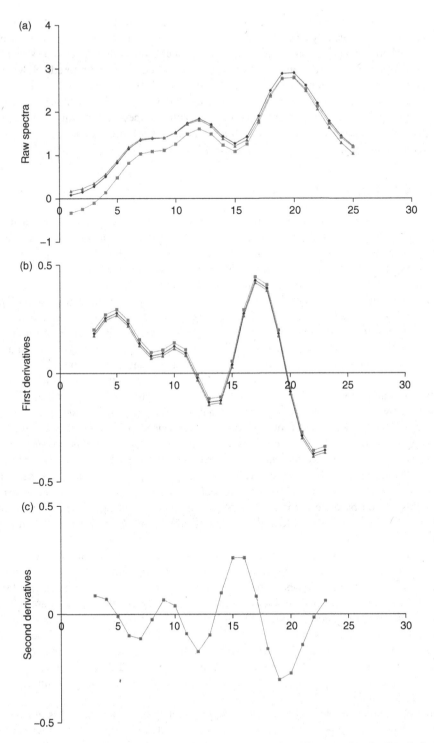

Figure 10.17 Use of derivatives to reduce baseline problems. (a) Raw spectra; (b) first derivatives; (c) second derivatives

one uses some methods for scaling the data, such as standardization, attaching equal impor-tance to each wavelength or region of the spectrum, it is possible to get poor results, as each region, even those containing no useful information has equal influence on the data. What one wants is to see the results of supervised pattern recognition on different regions of the spectrum. Some people do this fairly simply, just by looking at the per cent correctly classi-fied choosing any specific window of wavelengths, but others use quite elaborate approaches to variable selection or feature reduction such as genetic algorithms. One may not end up with a single contiguous region of the spectra for pattern recognition but various different wavelength ranges.

10.8.2 Growth Curves

Spectroscopy is not the only consideration when making measurements on micro-organisms, their growth cycle must also be taken into account. Most cultures of micro-organisms have several phases in their growth as illustrated in Figure 10.18. The lag phase is when they are settling in, the log phase is when they are multiplying exponentially. By the time the stationary phase is reached there is an equal balance between organisms multiplying and dying. The death phase is when nutrients are exhausted and the population decreases. There can be differences in spectroscopic characteristics at each stage in the cycle, so either it is necessary to sample during one stage (the stationary phase), or else to have an average spectrum, for example four or five times during the culture.

10.8.3 Further Measurements

Providing care is taken both in the biology and spectroscopy, chemometrics methods for pattern recognition have a major role to play in the automated classification of micro-organisms. Although much of the classic work has been performed using MIR, there are important advances in the use of Raman spectroscopy and NIR, both techniques having

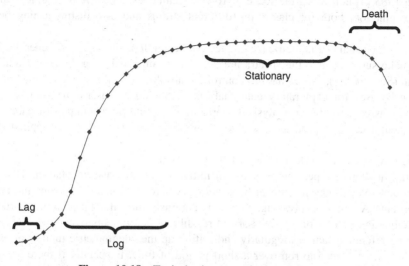

Figure 10.18 Typical micro-organism growth curve

been effectively employed for microbial classification. For the enthusiasts it is also possible to incorporate other types of data, such as morphological data or any conventional biological measurement and test; these can be modelled as extra variables in addition to the spectroscopic wavelengths. Various tricks allow specific scaling, for example, if there are 100 spectral measurements but only five biological measurements, rather than allowing the spectroscopy to swamp the biology, it is possible to scale the biological measurements to be of equal importance.

The application of sophisticated pattern recognition techniques to the classification of microbes using molecular spectroscopy has been and continues to be a method of fertile research with important real-world applications.

10.8.4 Pyrolysis Mass Spectrometry

A classic application of chemometrics has been in the area of pyrolysis mass spectrometry, and this was one of the original driving forces for the development of multivariate classification techniques in biological analytical chemistry.

The English newspaper *The Independent* has stated the pyrolysis mass spectrometric method is 'so simple a chimpanzee could be trained to do it'. Samples are heated to break down into fragments; for bacteria, conditions that specifically cleave proteins and polysaccharides are used. The trick is not to cleave into moieties that are too small and simple as they would not be very diagnostic, but to cleave into medium sized moieties (often called the pyrolysate). After heating the material is then bombarded with electrons to produce both molecular and fragment ions. Typically the m/z range between about 51 and 200 is then studied, resulting in 150 mass measurements. The relative intensities in the mass spectra at each m/z number are then used to characterize the samples. Because the patterns are quite complex and depend on many different factors, multivariate chemometric techniques are normally required for interpretation of these complex datasets.

Some of the most important pioneering reported work in the 1980s was performed in the FOM Institute (Netherlands) by Windig and colleagues. Their well regarded paper published in 1983 [4] helped set the scene. Pyrolysis techniques appear to have unique abilities for discrimination, both for classifying different strains and also distinguishing between species.

Many of the usual chemometric techniques can be applied, although it is generally important to make sure first that the spectra are all on the same scale, e.g. by row scaling each spectrum (Section 5.5), as it is hard to control the absolute quantities of biological material. PCA can be used for exploratory data analysis, following by discriminant analysis if we know the classes, and cluster analysis. Computationally intense techniques of data analysis such as neural networks and genetic algorithms have more recently been applied to such data.

A big advantage of the technique is that it is relatively fast and cost effective once set up, costing perhaps £1 per sample after an initial outlay for instrumentation. Thus it has an important role to play in clinical bacteriology where cost effective taxonomic analyses are important. A major disadvantage though is that instrumental signals are not particularly reproducible and so it is often necessary to recalibrate instruments over time, for example by running reference samples regularly and setting up the chemometric models afresh, and a batch of samples need to run over a short period of time. Especially if there are several different users of the same instrument, all having slightly different needs, this can be quite a serious limitation. Pyrolysis is not like a technique such as NMR where chemical shift and intensity information are reasonably stable from one day to another.

However pyrolysis mass spectrometry is a major success story of chemometrics, as multivariate data analysis is the key to interpreting information which would be very difficult to understand using manual methods, unlike many conventional spectroscopies, and it has an important role to play especially in biology. The community is very active and the *Journal of Analytical and Applied Pyrolysis* is a major source of leading edge research in this area, a surprisingly large number of papers employing chemometric methods to interpret the results.

10.9 MEDICAL DIAGNOSIS USING SPECTROSCOPY

In Section 10.8 we have looked at how multivariate methods combined with spectroscopy can be employed for the classification of organisms. The potential applications of chemometrics for classification are much broader and have significant potential in clinical science.

Conventional medical diagnoses rely on a variety of well established approaches including histopathology, radiology and clinical chemistry. Although these have a very important role to play, over the past few years new spectroscopic approaches have been developed, primarily using IR spectroscopy.

Normally a tissue or blood sample is obtained from the patient and then analysed spectroscopically. There are a number of traditional methods. These may involve reagent specific colorimetric approaches which test for individual compounds or even individual inspection. However, most biological material is quite complex in nature and statistical approaches can be used to look at the spectra of extracts which consist of a mixture of compounds.

Most common chemometrics approaches can be applied. PLS may be employed for calibration to determine the concentrations of certain diagnostic compounds. PCA is very common especially to separate out various groups. Cluster analysis and supervised pattern recognition have important roles to play. The range of applications is surprisingly large, and a few early papers in this ever expanding area are listed below, to give a feeling for the diversity.

Eysel and coworkers developed methods for diagnosing arthritis using synovial fluid [5] [for the nonexperts, synovial fluid is described as ' a viscous (thick), straw coloured substance found in small amounts in joints, bursae, and tendon sheaths']. Four classes of tissue were diagnosed by other means (control, rheumatoid arthritis, osteoarthritis and spondyloarthropathy), the aim being to see whether spectroscopic approaches provide a sensible and cheap alternative. Linear discriminant analysis was the main method of choice. Work was performed finding the most diagnostic parts of the spectrum for the best and most robust discrimination. Over 95 % of samples were correctly classified by this approach. Although the medical establishment is fairly conservative and slow to take up these methods, preferring conventional methods, in fact spectroscopy can provide very efficient and cost effective classification. The diagnosis of Alzheimer's disease is another important application, both using normal IR [6] and fluorescence IR [7] spectroscopy combined with pattern recognition. The methods described in the paper by Harlon *et al.* [7], involve taking tissue specimens of both Alzheimer's and non-Alzheimer's subjects on autopsy (after death) and subjecting them to spectroscopy, using PCR models to classify into groups. The ultimate aim is to develop a noninvasive approach. The use of fluorescence, combined with PCA, has been extended further [8]. Cluster analysis of derivative IR spectroscopy can be used for the detection of leukaemia [9]. Dendrograms separate out two groups of the normal lymphocyte cells and three groups for the leukaemic cells, which can be interpreted in terms of DNA and lipid content. The analysis was performed independently on two spectral regions. The differentiation between leukaemic and normal cells is consistently good. Fourier transform IR microspectroscopy combined with chemometrics has also been applied successfully for detection of cervical cancer [10], and has potential for automated screening in a clinical environment.

The literature of combined spectroscopy and chemometrics for medical diagnosis is an expanding one, and the studies described here are only the tip of what is probably quite a large iceberg. An excellent review of applications in IR and Raman spectroscopy has been published recently [11]. Possibly the main barriers are to get these approaches accepted in mainstream medicine as acceptable diagnosis, especially since there may be legal implications if patients are misdiagnosed. Another common problem relates to obtaining sufficiently large and representative training sets, which can cause difficulties in terms of obtaining models that are sufficiently representative of a population.

10.10 METABOLOMICS USING COUPLED CHROMATOGRAPHY AND NUCLEAR MAGNETIC RESONANCE

Possibly metabolomics (or some say metabonomics – there are claims that these words have subtly different meanings but this is by no means universally accepted) has been one of the fastest areas to take off over the past few years, and chemometric strategies for pattern recognition are continually developing in this area. All the usual methods for medical and biological pattern recognition can be performed on these types of data, such as preprocessing, variable selection and discrimination, which are all integral parts of the data analytical protocol but special methods for preparing the information have been developed, and this section will focus exclusively on these techniques. The methods in this chapter and Chapter 5 can then be employed to further assess the data.

10.10.1 Coupled Chromatography

Many measurements are made by coupled chromatography, using LC-MS or GC-MS of extracts, most commonly of plasma and urine. The aim is to obtain a matrix for pattern recognition in which the rows are samples and the columns represent variables obtained from the chromatography. The difficult part is this first step, after which all the standard pattern recognition methods (as discussed above) can be applied. The usual wide range of diseases can be studied in humans and animals as well as other factors such as genetics, or even for plants, the environment under which the plant was grown, and so the potential productivity of a plant (for certain crops this may be several years or seasons away, e.g. trees and so it could be economically important to predict this from young saplings).

There are two fundamentally different approaches for handling the data.

10.10.1.1 Peak tables

The first involves obtaining a peak table, which is a list of peaks found in each sample together with their intensities. Each column in the data matrix consists of the intensities

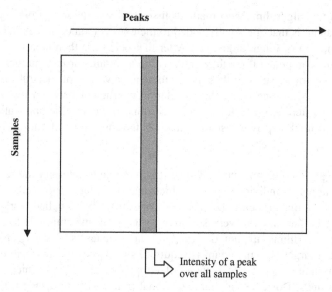

Figure 10.19 Peak table

(usually baseline corrected and integrated areas although sometimes peak heights) of a specific peak, hopefully corresponding to a single compound, over all chromatograms as illustrated in Figure 10.19.

This seemingly easy process is in fact much harder than it may appear. First of all many metabolomic datasets are quite large. A dataset consisting of 1000 chromatograms and 500 peaks per chromatogram involves half a million peaks in total, too much for manual identification of alignment, so approaches need automating. There are normally several steps:

1. The first is to find how many and where peaks are in each chromatogram, usually involving identifying their positions, start and end points. Many methods are available often involving derivatives although there are a legion of papers describing new approaches. There are several common difficulties. The first involves low signal to noise ratios, and when to judge whether a 'bump' above a baseline is actually a peak or not. In some cases peak picking is done on all the variables (in GC-MS this is called the Total Ion Current) but in other cases a judicial choice of variables (e.g. *m/z* numbers that appear significant) can improve the signal to noise ratio, and there are many ways of choosing these a common one being the Component Detection Algorithm (CODA) algorithm [12]. The second problem involves coping with overlapping peaks. This can be a serious issue as a two peak cluster can appear to originate from a different compound from its pure components as the two spectra (e.g. in the mass spectral direction) are mixed up and their separate origins may not be recognized. This can be overcome by using deconvolution as discussed in Sections 7.8 and 7.9 or for mass spectrometry detectors looking at selective ions as discussed in Section 7.2.2.

2. The second step, often called peak matching, is to recognize which peaks in different chromatograms arise from the same compound. This is a very sensitive step, because if not done correctly the results of pattern recognition may simply reflect problems with the

peak recognition algorithm. Most methods first involve some sort of prealignment, e.g. to reference standards that are usually added to the extract (sometimes called landmarks) to provide some sort of rough alignment. After that peaks are then identified as originating from the same compound if (a) they elute within a certain chromatographic time window and (b) the spectra are sufficiently similar : there are a variety of spectral similarity indices available, a common one being the correlation coefficient between two spectra. These two functions often need careful tuning according to the specific application.

3. The third and final step is to usually integrate the intensities of each peak, to obtain a peak table.

The advantage of using peak tables is that it is possible to interpret the results of pattern recognition in terms of individual peaks, which may be related to specific compounds by interpretation of the mass spectra, the diagnostic ones often being called marker compounds. A disadvantage is that they are very sensitive to the peak matching algorithms and many public domain algorithms are not designed for large datasets. For example, when peak matching often a target chromatogram is required, so we try to answer the question which peaks from chromatogram 2 match those of chromatogram 1 and which originate from different compounds. However, in a large data matrix it is hard to decide which is the most appropriate target chromatogram, especially when there are a lot of presence/absence of key peaks, and no one chromatogram contains a significant portion of all the peaks found in all the samples. In addition there may be problems with peak detection in cases where some peaks are heavily overlapping.

10.10.1.2 Total Chromatographic Profile

The alternative approach is to use the summed intensity of the chromatogram, or Total Ion Chromatogram (TIC) for mass spectral based approaches. There is no requirement to detect or match peaks, although the interpretive power is less, the data matrix consists of the chromatographic intensities at each elution time for each sample rather than peak areas.

Using summed intensity methods poses new problems, however. The most serious is one of alignment, this is critical. In peak table based approaches a small misalignment does not matter as common peaks are identified within a window. There are several ways of overcoming this, the first is to use methods for alignment to known chromatographic landmarks, these could be internal standards, or peaks that are present in most of the samples. In some cases such as plant metabolomics there are often compounds that are common to most extracts. The regions where the peaks occur are aligned exactly and the regions between are warped, which may involve stretching or contracting regions between these landmarks. Methods such as Dynamic Time Warping or Correlated Optimized Warping have been developed for this purpose [13], although the success involves several factors. First there must be a sufficient number of reliable landmarks for the alignment. Second there must be a good way of recognizing that these landmarks are common, often involving a second spectral dimension (although this is not strictly necessary). Third, the choice of target chromatogram can be critical, often more so than for peak table methods.

Once aligned, it is necessary to baseline correct the chromatograms. This is much more critical than for peak table methods. After this pattern recognition can be performed on each individual data point in the aligned and baseline corrected chromatogram, although the size of the data matrix is often quite large, sometimes involving 10000 or more variables, each representing a single scan number. In order to overcome this, and to cope with small alignment problems, it is quite common to divide the chromatogram into the summed intensities of regions, each consisting of several (typically 10 to 100) sequential data points in the aligned chromatograms, and use these combined variables for pattern recognition.

10.10.2 Nuclear Magnetic Resonance

A popular alternative to coupled chromatography for metabolomic studies is NMR. One advantage of NMR is that the chemical shift scale is reasonably constant, there can be some small changes between samples, but nothing of the magnitude of the elution time shift in chromatography, allowing a method that is relatively reproducible. With high field instrumentation (currently up to 1000 MHz) both good signal to noise ratios and resolution are possible. Peaks also have well defined shapes. Finally NMR can be performed in solution. Spectra are fast to acquire, unlike good chromatograms than can take an hour or more to obtain, and so there is good potential for high throughput analysis. For interesting spectra it is possible to perform many further experiments such as pulse sequences to further probe the structures of compounds in a mixture.

A disadvantage is that the NMR spectra are usually only obtained in one dimension, for the purpose of pattern recognition meaning that there is no secondary spectral information (for example spectral similarity) for correct alignment of spectra. Often this is overcome by bucketing spectra, that is summing up intensities in a specific region of each spectrum, a typical procedure may, for example, involve dividing an original spectrum acquired over 16K data points into 100 regions each of around 164 data points each, so that shifts of a few data points are no longer influential, and performing pattern recognition on this new data array.

All the normal methods for pattern recognition can be performed on these data, although one disadvantage is that the variables, unlike in chromatography, do not correspond to biomarkers, or individual compounds, but to regions of the spectrum, so there can be problems interpreting the data in terms of underlying chemistry. However for the purpose of classification this may not matter too much: we may be interested in which samples are part of which group (e.g. whether a patient has a disease) but not necessarily why. In addition certain regions of the NMR spectra may be known to contain resonances arising from specific compound classes and these can provide general information as to what sort of components in the mixture are responsible for differences, and so allow further investigation.

REFERENCES

1. The dataset is available at http://www.math.tntech.edu/ISR/Statistical_Methods/Data_and_Story_Library/thispage/newnode17.html
2. N. Cristianini and J. Shawe-Taylor, *An Introduction to Support Vector Machines*, Cambridge University Press, Cambridge, 2000
3. Y. Xu, S. Zomer, R.G. Brereton, Support Vector Machines: A Recent Method for Classification in Chemometrics, *Critical Reviews in Analytical Chemistry*, 36 (2006), 177–188
4. W. Windig, J. Haverkamp and P.G. Kistemaker, Interpretation of sets of pyrolysis mass-spectra by discriminant-analysis and graphical rotation Analytical Chemistry, 55 (1983) 81–88
5. H.H. Eysel, M. Jackson, A. Nikulin, R.L. Somorjai, G.T.D. Thomson and H.H. Mantsch, A novel diagnostic test for arthritis: multivariate analysis of infrared spectra of synovial fluid, *Biospectroscopy*, 3 (1998), 161–167

6. L.P. Choo, J.R. Mansfield, N. Pizzi, R.I. Somorjai, M. Jackson, W.C. Halliday and H.H. Mantsch, Infrared-spectra of human central-nervous-system tissue – diagnosis of Alzheimers-disease By multivariate analyses, *Biospectroscopy*, 1 (1995), 141–148

7. E.B. Hanlon, I. Itzkan, R.R. Dasari, M.S. Feld, R.J. Ferrante, A.C. McKee, D. Lathi and N.W. Kowall, Near-infrared fluorescence spectroscopy detects Alzheimer's disease in vitro, *Photochemistry Photobiology*, 70 (1999), 236–242

8. J.N.Y. Qu, H.P. Chang and S.M. Xiong, Fluorescence spectral imaging for characterization of tissue based on multivariate statistical analysis, *Journal of Optic Society of America – A. Optics Image Science and Vision*, 19 (2002), 1823–1831

9. C.P. Schultz, K.Z. Liu, J.B. Johnston and H.H. Mantsch, Study of chronic lymphocytic leukemia Cells by FT-IR spectroscopy and cluster analysis, *Leukemia Research*, 20 (1996), 649–655

10. B.R. Wood, M.A. Quinn, F.R. Burden and D. McNaughton, An investigation into FTIR spectroscopy as a biodiagnostic tool for cervical cancer, *Biospectroscopy*, 2 (1996), 143–153

11. D.I. Ellis and R. Goodacre, Metabolic fingerprinting in disease diagnosis : biomedical applications of infrared and Raman spectroscopy, *Analyst*, 131 (2006), 875–885

12. W. Windig, J.M. Phalp and A.W. Payne, A noise and background reduction method for component detection in liquid chromatography/mass spectrometry, *Analytical Chemistry,* 68 (1996) 3602–3606

13. G. Tomasi, F. van den Berg and C. Andersson, Correlation optimised warping and dynamic time warping as preprocessing methods for chromatographic data, *Journal of Chemometrics*, 18 (2004), 231–241

11

Biological Macromolecules

11.1 INTRODUCTION

In Chapter 10, we discussed several applications of chemometric methods in biology and medicine that use analytical chemical information such as infrared (IR), pyrolysis, chromatography and nuclear magnetic resonance (NMR) data. In these studies, we determined similarities between samples and groups of samples by a variety of approaches, such as principal components analysis (PCA), discriminant analysis and cluster analysis. This can allow us to predict the answer to a variety of questions ranging from whether an organism is a mutant, whether a bacterial sample from a patient corresponds to a known species, or whether tissue is carcinogenic.

Analytical chemical data are not the only information available, a major source of information comes from gene and protein sequences which also provide information about the similarities between organisms. We obtain a list of bases or amino acids from comparable DNA strands or proteins from each organism and use these to obtain a similarity measure which can be employed for subsequent pattern recognition. One aim is to often look at evolutionary or genetic differences between organisms as characterized by their DNA or proteins. We may be interested in seeing how species are related (much of the earliest work, for example in comparing haemoglobin sequences was in this area), but also in more subtle differences between two individuals of a similar species (even two humans) possibly for a specific genetic trait, in detecting new strains, and in economic assessments, for example, in agriculture to find out whether a particular new source such as a supplier from a different part of the world of new plant or animal products, is producing sufficiently similar material to an existing and well established source. A major step is to be able to compare two sequences between two organisms, and ask 'how similar are they?'

In this chapter we will concentrate primarily on pattern recognition aspects of bioinformatics, and not stray from this scope. It is often possible to obtain several blocks of data from organisms, for example, spectra and chromatograms of extracts and corresponding protein sequences and try to relate these types of information, so a chemometrician working in the area of biology should be aware of the nature of genetic and protein data, which can be calibrated or combined with chemical information.

For further reading there are several excellent books on bioinformatics [1,2].

Applied Chemometrics for Scientists   R. G. Brereton
© 2007 John Wiley & Sons, Ltd

11.2 SEQUENCE ALIGNMENT AND SCORING MATCHES

An important first step in handling such data involves performing *sequence alignment*. This involves taking two strands and lining them up so that they correspond as closely as possible, before comparing their similarities. A numerical score of similarity can then be produced between these two strands. The more similar they are the closer the evolutionary relationship is hypothesized to be.

In the simplest case two strands of DNA are compared and the more bases they have in common, the more similar they are. However, correctly aligning to DNA sequences is harder than proteins, because there are only four bases (adenine, A; cytosine, C; thymine, T; guanine, G) so the chances of finding pairs that are identical in two strands are reasonably high even by chance. One additional evolutionary feature that can be employed to advantage is that new bases (usually three at a time as three base pairs code for a single amino acid) can get inserted into strands of DNA, so two closely related sequences may differ length by multiples of three. Hence, an insertion of three bases into a DNA strand which is otherwise identical to the original does not necessarily imply there is a huge difference, meaning that the objective is not only to align two strands by shifting each strand until the maximum number of bases are identical, but one also has to permit gaps. Figure 11.1 is of two small strands in which 12 bases are identical. An additional three bases (AAT) have been inserted in the top strand, which probably in biological terms reflects a simple evolutionary change in sequence. When aligning sequences, this gap should be taken into account. Alignment algorithms can then be used to score the similarity between two strands from different sources, the more bases in common, the higher the similarity. However, one problem is that the more gaps allowed by the algorithm, the easier it is to align two strands well, but there may not be too much significance if the number of gaps is large. So in addition to scoring the number of identical bases in any two strands, after taking into account the gaps, the higher the better the match, a 'penalty' can be calculated according to the nature of the gaps. This penalty may vary according to method, for example a single long gap is probably less serious than a large number of small gaps. This is subtracted from the overall score of the number of matching bases. At the end, however, two strands of DNA are aligned as well as is possible, taking gaps into account, and a similarity measure is obtained.

Aligning proteins is a somewhat different affair, because in contrast to there being only four bases in DNA, there are 20 common amino acids. Thus the probability of matching is much lower and so it is possible to be more specific in detecting what constitutes a good match. It is, of course, also possible to allow for gaps, but unlike the case of DNA where a gap consists of three base pairs, in proteins gaps of single amino acids are common. Figure 11.2 represents two amino acid sequences (represented by single letters as

```
A  T  C  C  C  G  A  A  T  C  G  T  A  T  G  G  A  C
|  |  |     |  |  |           |  |  |  |     |  |  |  |
A  T  G  C  C  G  -  -  -  T  G  T  A  T  C  G  A  C
```

Figure 11.1 Two DNA strands that are fairly similar but of different lengths; one contains three base pairs that have been inserted

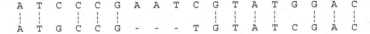

Figure 11.2 Two amino acid sequences

Table 11.1 Letter notation
for amino acids

Alanine	A
Arginine	R
Asparagine	N
Aspartic acid	D
Cysteine	C
Glutamic acid	E
Glutamine	Q
Glycine	G
Histidine	H
Isoleucine	I
Leucine	L
Lysine	K
Methionine	M
Phenylalanine	F
Proline	P
Serine	S
Threonine	T
Tryptophan	W
Tyrosine	Y
Valine	V

Table 11.2 Scoring matrix for amino acid alignment

	A	R	N	D	C	Q	E	G	H	I	L	K	M	F	P	S	T	W	Y	V
A	2																			
R	−2	6																		
N	0	0	2																	
D	0	−1	2	4																
C	−2	−4	−4	−5	4															
Q	0	1	1	2	−5	4														
E	0	−1	1	3	−5	2	4													
G	1	−3	0	1	−3	−1	0	5												
H	−1	2	2	1	−3	3	1	−2	6											
I	−1	−2	−2	−2	−2	−2	−2	−3	−2	5										
L	−2	−3	−3	−4	−6	−2	−3	−4	−2	2	6									
K	−1	3	1	0	−5	1	0	−2	0	−2	−3	5								
M	−1	0	−2	−3	−5	−1	−2	−3	−2	2	4	0	6							
F	−4	−4	−4	−6	−4	−5	−5	−5	−2	1	2	−5	0	9						
P	1	0	−1	−1	−3	0	−1	−1	0	−2	−3	−1	−2	−5	6					
S	1	0	1	0	0	−1	0	1	−1	−1	−3	0	−2	−3	1	3				
T	1	−1	0	0	−2	−1	0	0	−1	0	−2	0	−1	−2	0	1	−3			
W	−6	2	−4	−7	−8	−5	−7	−7	−3	−5	−2	−3	−4	0	−6	−2	−5	17		
Y	−3	−4	−2	−4	0	−4	−4	−5	0	−1	−1	−4	−2	7	−5	−3	−3	0	10	
V	0	−2	−2	−2	−2	−2	−2	−1	−2	4	2	−2	−2	−1	−1	−1	0	−6	−2	4

listed in Table 11.1). In the example, there are nine hits between sequences which represents a similarity measure, which can sometimes be expressed as the percentage of aligned amino acids that are identical. However, some people prefer to use more complex criteria for determining protein similarity. Unlike the four bases, which occur pretty much in the same frequency in nature, amino acids have different distributions, some being very common, and others rare. In addition some amino acids are very similar in function and replacing them by each other is not particularly significant, for example, I (isoleucine) and L (leucine). So people have developed matrices, which score alignment corresponding to the evolutionary significance of the matching. One such method is called the PAM250 matrix (Table 11.2). Notice that, for example, the score for matching I and L (2) is the same as that for S (serine) matching itself, this is because S is very common so matched pairs could easily arise by accident: W (tryptophan), in contrast, is very rare and often has a quite specific structural role so finding two matched Ws is scored as 17 and is very significant. The overall score is determined by aligning two sequences as well as possible, taking gaps into account, looking at which amino acids correspond to each other, then adding their scores. If there is some doubt as to the exact alignment and where gaps should be it is often acceptable to optimize the overall score by changing the position of gaps until as high a match is obtained as possible. A penalty is then subtracted from the overall score for each gap. Using the matrix of Table 11.2, and a gap penalty of -1 (which perhaps is rather small), we reach a total alignment score S of 46 for the two sequences of Figure 11.2.

Although the raw scores can be used as similarity measures, they can be transformed as discussed below, and act also as an aid for alignment of sequences, to find two strands that match as closely as possible.

11.3 SEQUENCE SIMILARITY

In Section 11.2 we discussed how to align two sequences of either nucleic acids or proteins, and described some simple scoring systems. The next step is to produce a numerical similarity or dissimilarity value, based on these scores.

Many biologists use cluster analysis (see Section 5.6 for an introduction) for converting these data into a way of visualizing the relationship between organisms based on the similarity between their DNA or protein sequences. Although alternative approaches such as PCA or discriminant analysis may often appear to be more appropriate, an important advantage of cluster analysis is that it can work from pairwise similarity matrices. This means that the starting point involves taking matches between two of the sequences rather than all the sequences at the same time. This has significant computational advantages because it is very tricky to align several sequences simultaneously, as opposed to performing alignment independently on pairs.

In classical analytical chemistry, similarity matrices are based on mathematical measurements, such as correlation coefficients, Euclidean distances and Mahalanobis distances. In macromolecular sequencing we use quite different criteria: the raw data are no longer numerical measurements but match functions.

What we want to do is to have a value that relates to how closely related the sequences are. The simplest is just the percentage matches, e.g. if 16 out of 20 amino acids are identical for two protein sequences, this number is 80 %, the higher, the better. Of course some adjustments are required if there are gaps, and this can also be expressed as a dissimilarity

Table 11.3 Typical pairwise similarity matrix between five peptides (A–E) using a percentage matching criterion, the higher the more similar

	A	B	C	D	E
A	100				
B	75	100			
C	38	56	100		
D	70	24	82	100	
E	52	30	64	68	100

(the number of mismatches). In this way a matrix such as that of Table 11.3 (for five sequences) can be obtained.

In Section 11.2 we saw that simple hit and miss matching criteria are not always sufficient to build up a suitable picture of sequence similarity, especially for proteins. For example, we introduced a score (S) based on Table 11.2 that takes into account that certain matches will be more common than others, just by chance occurrence. Building on this idea, there are a number of other possible statistical measures [3] that can be used to convert this score into probabilities or more similarity measures. The basis is that the chance that some amino acids occur randomly is different. This can be measured through databases where the relative abundance of different amino acids can be obtained. Consider an example where there are only two possible types of amino acid, **1** and **2**, but the chance of finding type **1** is 0.9, whereas that of finding type **2** is 0.1. If one sequence consists of two amino acids, **1 1**, what are the chances of one or more matches if the second sequence were entirely randomly generated?

- The chances of both sequences being the same equals the chance that the second sequence is **1 1**, or $0.9 \times 0.9 = 0.81$.
- The chances of only one amino acid in the second sequence matches equals the chances that the second sequence is **1 2** or **2 1** $= 2 \times (0.9 \times 0.1) = 0.18$.
- The chances that both amino acids in the second sequence are different is **2 2** or $0.1 \times 0.1 = 0.01$.

Hence if there is only one or no match this is a pretty convincing proof that the two sequences really are fundamentally different, but having two matches is not all that significant, and could have arisen by chance. In contrast, if one sequence consists of two amino acids, **2 2**, then the chances of two matches arising by chance is only 0.01. So we read much greater significance of two sequences of type **2 2** being identical than two sequences of type **1 1**. Naturally for real samples consisting of 20 amino acids, also taking into account certain likely matches such as I and L, and gaps, the calculations become significantly more complex, so much attention is placed on finding indicators of similarity from these matches.

An indicator function E can be defined by $E = Kmn \exp(-\lambda S)$, where m and n are the length of the two sequences to be compared, and K and λ are constants obtained from the similarity scores, usually obtained by mining a large database. This value is the expected number of sequences that have a score of at least S. The higher this is, the less likely the matching arose by chance and so the more significant the similarity. The value of S can be converted to a so-called 'bit' score given by:

$$S' = \frac{\lambda S - \ln K}{\ln 2}$$

that put various different ways of measuring S on a common scale (Table 11.2 represents only one of several approaches for similarity measures). Finally the scores can be converted to a probability:

$$p = 1 - \exp(-E)$$

The lower the probability the more significant the matching.

These measurements, and other elaborations devised in the literature, are the chemical biologist's equivalent to the analytical chemist's (dis)similarity measures and can be used in subsequent steps for clustering to obtain diagrams as described in Section 11.4, although it is also possible to use straightforward similarity measures from the matching frequency if desired.

11.4 TREE DIAGRAMS

Biologists classically have used tree diagrams to illustrate the relationship between organisms in classical taxonomy. Chemical biologists have borrowed some of these ideas when interpreting the output of clustering algorithms to represent the similarity between nucleic acid and protein sequences. Analytical chemists also use these but much less frequently and do not always use the full variety of possibilities available.

11.4.1 Diagrammatic Representations

A simple diagram can be represented by a series of lines, each connected by nodes. These nodes are the points at which each line joins another. Tree diagrams are special figures where there are no circuits, that is each branch originates from a parent branch and is not connected to any other higher level branch. Of the three diagrams in Figure 11.3, only Figure 11.3(a) and (b) are tree diagrams. Each tree diagram also has a series of nodes, which are either points at the ends of the branches or where new branches join the main tree. Figure 11.3(a) has nine nodes, whereas Figure 11.3(b) has six nodes. The number of nodes is always one more than the number of branches, as illustrated in Figure 11.4 for one of the tree diagrams.

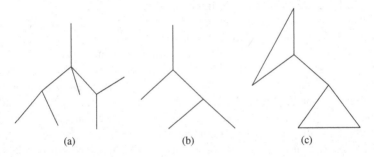

(a)   (b)   (c)

Figure 11.3   Three diagrams. (a, b) Tree diagrams; (b) is also a dendrogram

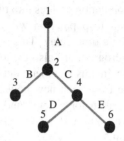

Figure 11.4 Tree diagram consisting of six nodes (1–6) and five branches (A–E)

Tree diagrams are widely used especially in statistics and decision theory, these for example, could represent probabilities: we may wish to know the outcome of tossing a die twice, there are thirty-six possible outcomes, which can be represented as terminal nodes in a tree, the first level branches relating to the first toss of a die and the second level to the second toss.

11.4.2 Dendrograms

In biology, there are special types of tree diagrams called dendrograms. These are also called binary tree diagrams and are distinguished by the fact that only two new branches emerge from each node. Figure 11.3(b) is also a dendrogram, whereas Figure 11.3(a) is not. Different authors use varying terminology for the features of a dendrogram; the alternative names are indicated in Figure 11.5. Some dendrograms are represented with a root, but this is often omitted. Note that chemists do not use dendrograms as much, so the terminology in the chemical literature tends to be much less well developed, and some of the subtleties below are overlooked.

11.4.3 Evolutionary Theory and Cladistics

Dendrograms visually relate to a hypothesized evolutionary representation of relationships between organisms in that each nonterminal branch always has two daughter branches. This particular and, in some circles, rather controversial assumption, in evolutionary terms, is

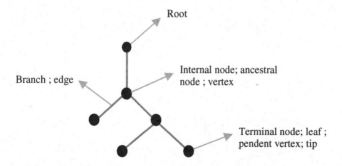

Figure 11.5 Terminologies for dendrograms

based on the hypothesis that each population divides into two groups over time. The principle is that it is very unlikely that a single population will divide into three or more groups simultaneously, there will always be a time delay. This is one assumption of what is called 'cladistics' and many types of dendrogram are also cladogram. Much molecular biological information is represented visually in this manner, and most cluster analysis algorithms allow only two branches to emerge from each new node. This important assumption about evolutionary splitting of groups is taken for granted and incorporated into most clustering and pattern recognition algorithms used by biologists, although some would argue that there is insufficient evidence to be certain.

11.4.4 Phylograms

A final refinement is the phylogram. In cladograms, the length of branches of the tree diagram have no significance and are usually of equal length. In a phylogram, they are related to the extent of evolutionary similarity, or to the supposed time at which different groups became distinct. In Figure 11.6, we modify the tree diagram to have different lengths of branches, labelling the three terminal nodes. It can be seen that objects B and C are quite closely related, and had a common ancestor to object A probably some time in the past. It is now a simple step to visualize A, B and C as strains represented by different proteins or organisms.

Quite often, phylograms are represented as in Figure 11.7, using straight rather than slanting lines, but both ways of viewing the data are equally meaningful. The vertical axis, however, can now represent a numerical index of similarity measures. Thus we can use sequence similarities to show how closely different organisms are related, and so draw

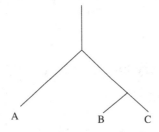

Figure 11.6   Phylogram

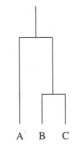

Figure 11.7   Alternative representation of a phylogram

a diagram that uses molecular biological information to derive a visual representation of proposed evolutionary relationships.

11.5 PHYLOGENETIC TREES

Previously we have discussed (Sections 11.2 and 11.3) how to measure similarities between nucleotide and protein sequences. We have also explored tree diagrams (Section 11.4) which are the most common visual representation of these similarities. Many molecular biologists try to convert molecular biological sequence information into dendrograms or phylogenetic trees.

It is important to recognize that although dendrograms are widely used within biology, they represent only one way of handling information, based on specific evolutionary theory. In other areas of chemometrics, cluster analysis (or unsupervised pattern recognition) is much less commonly employed, and it is arguable that a principal component plot based on molecular similarity information could give an equally useful viewpoint of the differences between organisms based on their sequence information. However, conventions are hard to challenge, although some might argue that one of the basic foundations of molecular bioinformatics is built on sand rather than rock.

The procedure of cluster analysis has been briefly introduced earlier (Section 5.6) primarily in the context of analytical chemistry. In mainstream chemistry, we use similarity measures based on statistical indicators such as correlation coefficients or Euclidean distances, normally between spectra, chromatograms or molecular fingerprints. In molecular biology, the similarity matrix arises from quite a different basis, not from pure statistical functions but from sequence information. Nevertheless it is possible to obtain a matrix, an example is in Table 11.3 where there are five objects (which may represent five protein sequences) and a similarity measure of 100 which indicates an exact match or a lower number which suggests a poor match. Note that the matrix is symmetric above the diagonal, and we miss out the top half of the information as it is the same as the bottom half.

The next step is to perform hierarchical agglomerative clustering. This involves first finding the pair of sequences that are most similar and combining them. In our example, this pair is C and D with a similarity measure of 82.

The next stage involves creating a new similarity measure between this group and the other remaining groups (A, B and E). In evolutionary terms we are saying that C and D had a common ancestor that is closest in time to the present day, and the combined group C and D represents this ancestor, how is this primitive species related to modern day A, B and E? There are numerous approaches to the calculation of this new similarity and the newcomer to the field should look very carefully at the definitions employed in any particular software package and paper.

The UPGMA (unweighted pair group method using arithmetic mean) is one of the most common, and is illustrated in Table 11.4. Once a cluster is joined, take the average of the distance between each of the other clusters, and use these as new similarity measures. So after cluster CD is created, its distance between A is $(38 + 70)/2$ or 54, its distance between D is $(56 + 24)/2$ or 40 and between E is $(64 + 68)/2$ or 66. Then in the next matrix search for the highest similarity, which is 75 and continue. The new similarity measures are indicated in italics and the best similarity at each step highlighted. The resultant dendrogram is presented in Figure 11.8. It is possible to assign distances to the branches. For example the point at which C and D join the cluster is 18 (equal to $100 - 82$), and A and B is

Table 11.4 UPGMA method for clustering. New coefficients are indicated in italics and best matches highlighted

	A	B	C	D	E
A	100				
B	75	100			
C	38	56	100		
D	70	24	82	100	
E	52	30	64	68	100

	A	B	CD	E
A	100			
B	75	100		
CD	54	40	100	
E	52	30	66	100

	AB	CD	E
AB	100		
CD	47	100	
E	41	66	100

	AB	CDE
AB	100	
CDE	44	100

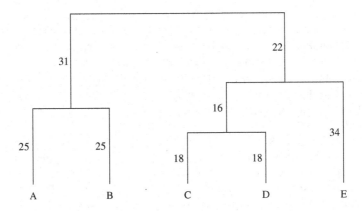

Figure 11.8 Result of clustering from Table 11.4

25 (equal to $100 - 75$). That between object C/D (the common ancestor of species C and D) and object E, is more tricky to calculate but equals 16, or ($100 - 66 = 34$) (which is the similarity point at which E joins the cluster) minus 18 (the point at which C and D join the cluster). This additive property is often called 'ultrametric'. The distances on the diagram are loosely related to evolutionary time, so we could replace these by an imaginary timescale. An inherent assumption when interpreting such diagrams is that evolution is at an equal rate in all branches (i.e. the chances of mutation and population divergence is similar), although this assumption is not necessarily proven.

Named methods of cluster analysis in molecular biology often differ from those in mainstream statistics and chemometrics, for example, the method of calculating similarities is denoted as 'weighted' rather than 'unweighted' in conventional analytical chemistry terminology, but this depends on what we imply by a weight. If we are joining a cluster of four objects to one of one object and we use the mean similarity measure between the objects. Is this really unweighted or is it biased towards the object in the smaller cluster? An evolutionary biologist will say it is not weighted because we are dealing with two single population groups back in time, and the fact that some have diversified into more species than others later on, is irrelevant, therefore each ancestral species is of equal importance. A spectroscopist may have a quite different perspective and may feel that information arising from, for example, an average of four spectra is more significant than information from a single spectrum so should be weighted accordingly. As in all science, to understand algorithms one must first think about the underlying assumptions and suitability of the method prior to applying it.

When using cluster analysis to interpret sequence information it is always important to appreciate the underlying philosophy behind the methods rather than simply plug data into a computer program, and in fact evolutionary theory is quite deeply buried into how molecular biological information is represented.

REFERENCES

1. D.R. Westhead, J.H. Parish and R.M. Twyman, *Bioinformatics – Instant Notes*, BIOS Science Publishers, Oxford, 2002
2. A.M. Lesk, *Introduction to Bioinformatics*, Oxford University Press, Oxford, 2002
3. See http://www.ncbi.nlm.nih.gov/BLAST/tutorial/Altschul-1.html#head2

12

Multivariate Image Analysis

12.1 INTRODUCTION

Multivariate Image Analysis (MIA) involves the application of multivariate chemometric methods to enhance the quality of images that have usually been obtained using spectra recorded over a range of wavelengths. The principle is that within an image there are different sources of information, which have different characteristic spectra.

For chemists, image analysis is important because it can be used to indicate the distribution of compounds or elements within a material. We will focus specifically on this aspect. Normally a microscopic image is taken of some material and it is the aim of chemical image analysis to determine spatial distributions. The signals due to each constituent are mixed together in the original image, the aim of multivariate techniques is to deconvolute this information so as to determine the distribution and, in many cases, spectrum, of each component.

Applications are varied. An important one is in the analysis of pharmaceutical tablets. These generally consist of an active drug plus something else that binds the tablet together, often called an excipient. Image analysis can be used to determine the relative distribution of the excipient and active ingredients. Why might we wish to do this? This distribution might influence properties such as solubility, and so ultimately must be important and checked for each batch. In addition we want a reasonably even distribution, because if we produce tablets, it is important that the dosage in each tablet is approximately similar, and corresponds to the overall proportions in the batch. Finally this can give clues to the method of manufacture and may be important in patent work. Often infrared (IR) or Raman microscopy is used and then various chemometrics approaches are developed for interpreting the resultant image. It is important to recognize that deblurring is not the only (or even main) reason for chemical image analysis, as, unlike in conventional photography, we want to distinguish the main components, either by separating into different images, or by using what are called false colour techniques to indicate concentrations of different compounds, where each compound is represented by a colour in the resultant image.

Another important application is looking at materials such as alloys or semiconductors. Often X-ray or near infrared (NIR) images are taken. The properties of a material depend on the distribution of elements. Image analysis can provide insights into the distribution of the material during the manufacture process. A particular distribution may relate to the properties and so one has a better insight into product quality. In addition it allows the possibility of

Applied Chemometrics for Scientists   R. G. Brereton
© 2007 John Wiley & Sons, Ltd

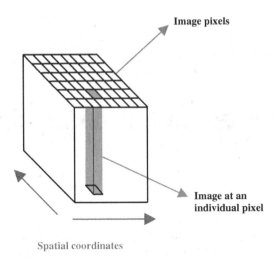

Figure 12.1   Computer representation of an image

controlling the production process. It is important not only to know the composition but the distribution of the elements in the material during the manufacturing process.

An image can be represented in a computer in various ways. One approach is to view it as a 'box' or 'tensor'. Figure 12.1 is a representation of the image. There are two spatial coordinates which may be geographical (for example an EW and NS axis) or related to the surface of a material. A digitized image is usually composed of pixels, which are small 'dots' on the screen. Each pixel contains some information presented normally as a colour of a given intensity on the photographic or microscopic image. A typical image may have 500 pixels along each axes, and so can be represented by a 500×500 matrix. At each pixel is a series of intensity readings. These may, for example be a spectrum recorded over several hundred wavelengths. In photography, it is common to reduce the data to several frequency bands (seven or eight) each of which is represented by an intensity. In the latter case, the image could be viewed as seven or eight superimposed photographs each over a different wavelength range.

We could treat the image simply as several independent images at different spectral frequencies, as in Figure 12.2. This means that we could provide several different photographic images. Chemometrics methods for MIA allow more sophisticated treatment of the data.

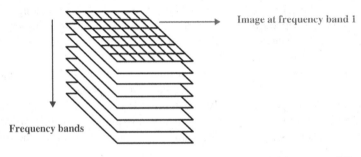

Figure 12.2   A photograph can be considered as several superimposed images at different frequency bands

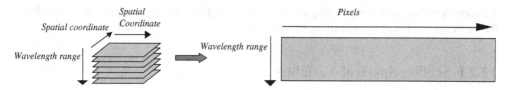

Figure 12.3   Unfolding an image

The image can be unfolded as in Figure 12.3, and so treated as a long data matrix, with one-dimension corresponding to all the pixels in the image, and the other to the wavelength ranges or spectra. Generally this matrix is then transposed, so that the pixels represent rows and the columns spectral frequencies. After this we can perform a variety of chemometric transformations as discussed below, and then the image reconstructed or folded back again.

The main steps and procedures in MIA are outlined below.

12.2 SCALING IMAGES

As discussed above, an image may be represented as a three way data matrix, sometimes called a tensor, two of the dimensions usually being spatial and the third spectroscopic. Some images are even more complex, for example it may be possible to build a four-dimensional array by including a third spatial coordinate, and even to add a fifth-dimension if we record a sequence of images with time (see Section 12.7 for further discussion). This section will discuss methods for scaling the data prior to application of multivariate methods.

Previously (Section 5.5) we have introduced the idea of data scaling as a form of preprocessing when handling two-way multivariate data matrices. In the case of image analysis, the possibilities are quite large as there are extra dimensions. In MIA, we normally aim to represent the image pixels by a statistic such as a Principal Component (PC) scores, a factor score or a class distance, rather than as raw spectroscopic data, so preprocessing will destroy the original image. This does not really matter, what is important is that the resultant chemometric variables can be interpreted. The aim of MIA is not primarily deblurring but to detect factors, such as the distribution of elements or sources within a map or picture.

For a two-dimensional (see Section 12.7) image, which is the most common type, the data can be represented in the form of a box (Figure 12.4). Each plane corresponds to the spatial distribution at one, or a series, of wavelengths, for example, varying from the violet to red [in some cases it may be possible to widen the range for example by using IR or ultraviolet (UV) spectroscopy, and although we cannot visualize this directly, because the colours in chemical image analysis do not have to correspond to real colours there is no

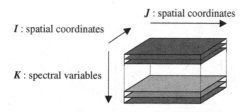

Figure 12.4   Dimensions of an image

problem handling this image computationally]. Each plane represents one of K variables. There are several ways of handling these data.

12.2.1 Scaling Spectral Variables

One of the simplest approaches for preprocessing the image is to scale each spectral variable independently. This can be done by centring each variable prior to further multivariate analysis. An alternative approach is standardizing: an advantage of this is that each variable is now equally significant, but some variables may consist of noise, so this approach is not always successful.

Sometimes we may want to combine two images, for example, X-ray and NMR or two types of spectroscopy, such as in Figure 12.5. It is possible that there will be far more measurements for one technique than another, so a further approach could be to scale each block to be equally significant, for example, after standardizing each independently, divide the values of each block by the number of variables in the block, so if one set of measurements were represented by 100 variables, divide the standardized data by 100. Otherwise one technique may heavily dominate the image if the number of measurements is larger.

12.2.2 Scaling Spatial Variables

There are, however, a variety of other approaches for preparing the data. It is not necessary to scale individual spectral variables, alternatively one can also transform single pixels in the image onto a similar scale, for example, by normalizing the spectrum at each pixel so that the sum of squares equals one. In this way the image at each point in space becomes equally significant, so that the overall intensity everywhere is comparable. Such scaled data will mainly look at contrasts or composition throughout the object. If a material is approximately equally dense then we may be more interested in whether there are clusters of compounds or elements in different parts, and how these are arranged, rather than absolute concentrations.

12.2.3 Multiway Image Preprocessing

The image may also properly be regarded as a multimode (three-way) dataset and there are, in fact, a large number of other ways to scale such types of data, as obtained in multivariate image analysis, often involving elaborate centring and unfolding. However, not all have an obviously interpretable physical meaning in image analysis and are more appropriate to other fields. Many of the principles are described in the text by Smilde *et al.* [1]. In

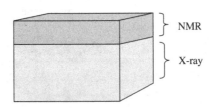

Figure 12.5　Combining two image blocks

addition, Geladi and Grahn have written an excellent book [2] which introduces some of the basic concepts in preprocessing of images. The interested reader should refer to one of these sources of reference which are goldmines of information in image analysis and the handling of multiway data from well established pioneers of the subject.

12.3 FILTERING AND SMOOTHING THE IMAGE

In addition to scaling (as discussed in Section 12.2) it is often important to preprocess the image by filtering and smoothing. This is usually performed prior to scaling, but it is always important to consider a sensible combination and order of operations, and there are no general guidelines.

It is importance to recognize that most images are digitized. A typical image may occupy an area of 512×512 pixels, which means that 262144 spatial points are used to represent the information. Higher resolution results in better definition but slower representation in a computer, more storage and memory space, and also algorithms are much slower. Sometimes raw data are available in analogue form and has to be converted to a digitized image. In addition each spatial image has a depth. In spectroscopic image analysis this equals the number of wavelength ranges used to describe each pixel. These may either be spectra centred on a chosen number of wavelengths or an average spectrum over a specific range. The more the wavelengths or wavelength ranges employed the bigger the image is to store and handle. If 10 wavelengths are chosen, and if the image is 512×512 pixels in resolution, over 2.5 million pieces of information are needed to describe the image.

In order to reduce storage space it is often quite common to represent each piece of information by a one byte number, i.e. an integer between 0 and 255. The higher the value of this integer, the greater the intensity. This implies that the image of the size above will require 2.5 Mbyte for storage. One problem is that many scientific packages use 4 bytes for representation of numerical data (in technical computing jargon this is called 'double precision'), so the image, when processed becomes four times larger. Performing operations such as Principal Components Analysis (PCA) on such a large dataset can be noticeably slow, due partly to memory problems and also due to the way computers perform high precision arithmetic. Reducing the digital representation of these numerical data helps sort out problems of storage, memory and speed, but it is harder to deal with contrasts as may occur where some features are overwhelmingly more intense than others. A way round this dilemma may involve transforming the data, e.g. logarithmically as described below.

It is often useful to apply simple spatial filters to improve the quality of the image and remove noise. There are numerous ways of doing this. The principle is that an image is continuous and so the intensity at each pixel can be replaced by the average of the surrounding pixels without significant loss of information. Figure 12.6 illustrates a 3×3 grid, with one pixel in the centre that displays high intensity, the others having low or zero intensity. The pixel in the middle could be replaced by the average of all nine pixels in the grid which will smooth out bumps. In real life we expect objects to be quite smooth and not contain sudden discontinuities, so this makes sense. We illustrate such an averaging technique on the data in Table 12.1, which could represent a 9×9 image recorded at a single wavelength. Assuming that pixels outside the area of the image have an intensity of zero (this will only affect the outermost image points), the change in appearance using this simple filter, involving replacing the intensity of each pixel by the average intensity of itself and the surrounding eight pixels in a 3×3 spatial grid, is given in Figure 12.7.

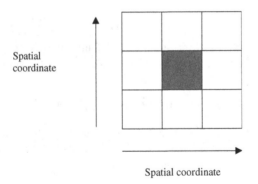

Spatial
coordinate

Spatial coordinate

Figure 12.6 A pixel as the centre of a 3×3 grid

Table 12.1 Numerical representation of the intensities of pixels in a 9×9 image

0.0811	0.0306	0.0333	0.1245	−0.0786	0.2072	0.0859	−0.1549	0.0124
−0.0200	0.0388	0.0452	0.0615	0.1305	0.1788	0.1199	0.0051	0.1144
0.1948	0.0878	0.2290	0.0372	0.3539	0.0557	0.0922	−0.0207	−0.0320
0.0919	0.2394	0.6955	0.7251	0.6106	0.7494	0.2675	0.0788	0.1098
0.1447	0.0046	0.3674	0.9563	0.9111	0.9471	0.4724	0.3726	0.2384
0.1511	0.2763	0.5211	0.7979	0.8337	0.5760	0.2602	0.3536	0.0433
0.2076	0.1488	0.0910	0.1621	0.2287	0.3997	0.3577	−0.0531	0.1215
−0.0869	0.1629	0.0336	−0.0622	−0.0133	−0.0026	−0.0129	−0.1859	−0.2326
0.1637	0.0184	0.0024	0.1546	−0.0495	−0.1775	0.0233	−0.0322	−0.0339

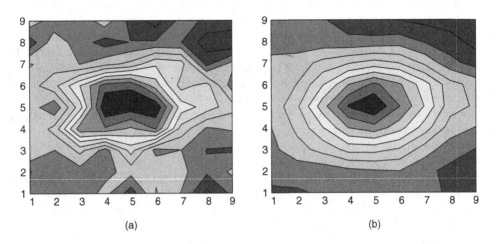

(a) (b)

Figure 12.7 Filtering the data in Table 12.1. (a) Raw; (b) filtered

It can be seen that there now is a clear object in the centre of the image which appears approximately circular. This procedure can then be performed separately for every variable (e.g. every spectroscopic wavelength range), so if there are 10 wavelength ranges, this is repeated 10 times for each pixel. Additional methods include median smoothing, and

Savitzky–Golay filters (Section 4.3) or derivatives (Section 4.8). It is also possible (but computationally intense) to take the first PC of a small region and use the score of this local PC as a smoothed estimate of the centre point, PCA being employed as a method for noise reduction in this case.

Occasionally individual pixels are missing from the image, often due to problems with the data acquisition. This can easily be overcome by using spatial filters, but excluding the missing pixel. For example if we used the average of a 3×3 grid, then if the centre point is missing we simply replace it by the average of the surrounding eight points, ignoring the missing central point.

In addition, it is also possible to filter in the spectral dimension, providing a reasonable number of wavelengths have been recorded. The normal range of methods such as moving average or Savitzky–Golay filters, and even Fourier and wavelet transforms are available as discussed in Chapter 4, but usually the number of spectral wavelengths is very large (except for so-called 'hyperspectral imaging'), limiting the range of methods that are sensible to apply. In addition it may be possible to perform spectral baseline correction, or perhaps subtract the image from a blank image. This can be important if there are instrumental problems, for example on the borders of the image or at various orientations, to ensure that the resultant data relate to the underlying objects rather than the measurement process itself.

Sometimes it is desirable to transform the raw data in a nonlinear fashion. There are several reasons for this. The first may be that the raw data have been recorded in transmittance mode (common in optical spectroscopy) but we want to build up a picture of concentrations of compounds in an object. To convert from transmittance to absorbance it is usual to logarithmically transform the data. Note that this may not always produce a nicer looking image, but intensities will be proportional (providing the spectral absorbance is within the Beer–Lambert range) to concentration, which may be important under certain circumstances. A different reason for logarithmic transformation may simply be because there are big contrasts in the image, some very intense objects in the presence of smaller ones, and the larger objects would otherwise dominate the image, effectively casting a shadow or virtual darkness over everything else. We may be quite interested in looking at trace concentrations, for example we may be looking at a tablet but really not be very interested in the main excipient, primarily the distribution of the chemically active compounds which may be present in small quantities in the material.

12.4 PRINCIPAL COMPONENTS FOR THE ENHANCEMENT OF IMAGES

PCA (Section 5.2) has a major role to play in chemical image analysis. The principle is that each object, for example, each element or each compound in a material, has its own spectrum, therefore we can use this different response to distinguish various types of object which might overlap in a conventional image.

In order to demonstrate the application of PCA, we will represent images as contour plots. This technique is often called a false colour technique, the colours representing intensity rather than absorbance at a specific wavelength. Our example will consist of a simple image of 9×9 pixels, of which there are two objects that are approximately circular in nature. We will represent the intensity using contours at 10 levels, with a greyscale gradation from light (high) to dark (low). For users of Matlab, these are the default greyscale contouring routines. Figure 12.8 represents the underlying positions of these objects; Figure 12.8(a) and

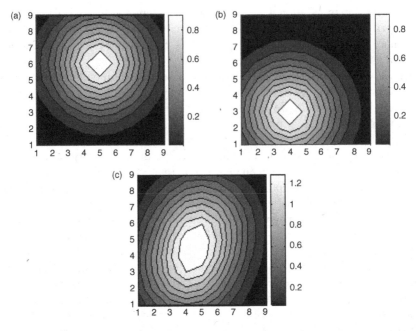

Figure 12.8 (a, b) Two objects and (c) their combined image

(b) are of the raw objects and Figure 12.8(c) is their combined image. From the summation it is hard to see that there are two objects, and also to determine details such as their position.

Each of the objects may have a unique, and in this case, different, spectrum if, for example, they have a different chemical composition. Let us suppose that an image is recorded at six different wavelength ranges, and there is a little noise as well. The data can be presented as a cube, each plate consisting of an image at a single wavelength (or spectral range), as shown in Figure 12.9. The underlying image is the product of the objects by their spectra plus associated measurement errors. We can look at each plane in the cube, so could present the information in the form of six different images, each at a single wavelength, for the data we will use in this example, these are shown in Figure 12.10. No single wavelength range is ideal. Some will be better at detecting one object than another, and some will be more dominated by noise. For example, the top left image best picks up a different object to the bottom right one. Naturally it is possible to play with the individual

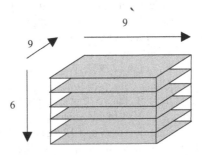

Figure 12.9 Organization of data for PCA image analysis example

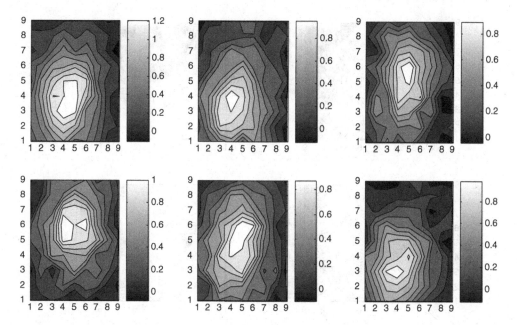

Figure 12.10 Presentation of the data for the PCA image analysis example at six different wavelength ranges

images by selecting different contouring options, and an enthusiast will probably develop his or her own favourite method, for example to look at the image using different regions of the spectrum to identify objects of different nature, but no single image makes it clear that there are two objects and where they are. By peering at these images one may predict that there is something strange and that the combined image probably corresponds to more than one object, but it is by no means obvious, and would take time and patience. In reality, for example looking at pharmaceutical tablets, there may be several hundred objects in a typical image and manual inspection would be near impossible.

When there are several objects, often of different kinds, it is nice to be able to look at where they are in the image simultaneously. PCA takes advantage of the fact that the spectrum of each type of object is different. A common approach, discussed previously (Section 12.1), is to unfold the image and then perform PCA on what is, in the current example, now a 81 pixel × 6 wavelength matrix. The scores are then calculated for each pixel. There will be several scores matrices, one for each PC, but these can be refolded back into up to six matrices (each corresponding to a PC), each of dimensions 9 × 9.

In Figure 12.11 we show the image of the first four PCs. Apart from the scale, the map of the first PC is very similar to the overall sum of images due to the two objects, and is not very useful for discrimination. However, when looking at the second PC we can now see that the two objects are easy to distinguish. The object in the lower left-hand corner is represented by a positive score, whereas that in the upper right-hand corner by a negative score. The third and fourth PCs represent largely noise and so do not show a very useful pattern.

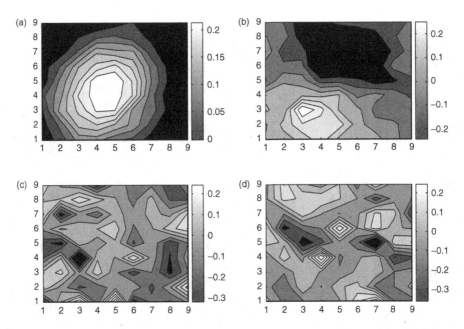

Figure 12.11   PCA of the data in Figure 12.10. (a) PC1; (b) PC2; (c) PC3; (d) PC4

If there were several objects of two different spectroscopic types we could easily distinguish them in a composite image, as in the second PC of Figure 12.11. Once however the number of different types of object increases, the interpretation becomes more complicated, and it is usually necessary to employ more than two PCs. Calibration and factor analysis can further help decompose the image into different sources, as will be discussed subsequently.

12.5 REGRESSION OF IMAGES

There are many reasons for regression between images which involve relating two different pictures.

The simplest involves comparing two spatial images, such as two greyscale photographs. Each photograph can be described as a series of pixels, each of different intensities, and so represented by a matrix. For example, one photograph may have been taken under different filters or with different apparatus to the other.

Why may we want to do this? Perhaps we are familiar with an old type of image and wish to change our recording mechanism, but still want to be able to return to the old type of picture that may have been available for several years, so as not to destroy historic continuity. More interestingly we may want to determine, for example, how a spectroscopic image can relate to physical features known from other measurements, e.g. the presence of a tumour in the spectrum of a brain, or to the concentration of a chemical.

It is in theory quite a simple operation. We take two images, for example Figure 12.12(a) and (b), and ask whether we can relate them mathematically. This can be done by unfolding these into vectors. For example if each image is 100×50 pixels in size, unfold to

(a)
(b)

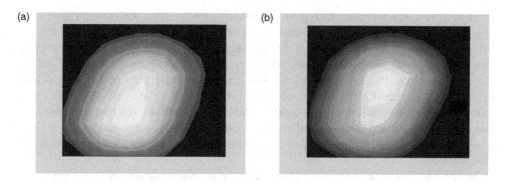

Figure 12.12 Relating two images

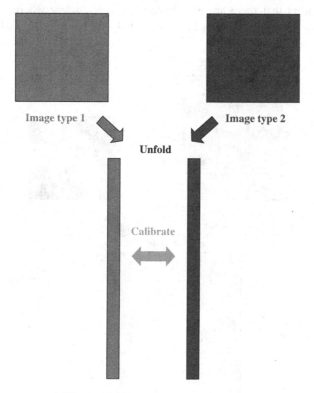

Figure 12.13 Calibrating two images

vectors of 5000 data points. Then try univariate regression (Section 6.2) to calibrate the new image to the old one and see if we can relate the two pictures. This process is illustrated in Figure 12.13. In addition to the linear terms it is possible to add squared and higher order terms to the calibration. In practice this procedure cannot provide dramatic improvements but it can emphasize features that perhaps are not clear on one image, and is quite simple.

More commonly we want to regress something such as a known concentration of an object or onto an image that is composed of a series of spectra, for example, if the spatial image is 100×50 pixels in size and one records a spectrum at each pixel, which in turn is divided into six spectral bands, the image could be represented by a 'box' or 'tensor' of dimensions $100 \times 50 \times 6$. A reference image may have been obtained by another method (or the dimensions of the embedded objects may be known from the sample preparation), and this information is then used for calibration. Both the reference and observed image are unfolded and the unfolded matrix is calibrated to the reference image, as illustrated in Figure 12.14, normally using an approach such as Partial Least Squares (PLS) (Section 6.6).

In order to illustrate the principles we will use a smaller image, of 9×9 pixels and six spectral ranges, similar in size to the example in Section 12.4. Figure 12.15 represents six images each taken using different spectral ranges, according to the wavelengths employed. It appears that there are two objects but it is unclear where these objects are located. The sum of these six images is presented in Figure 12.16 and it is now less

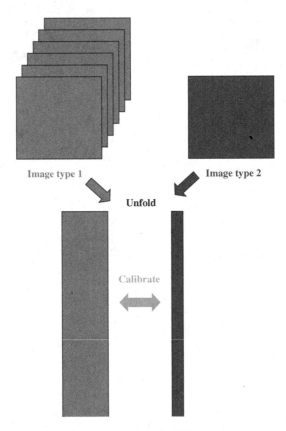

Figure 12.14 Calibrating a multivariate image to known properties, e.g. positions of objects of specific types (image type 2)

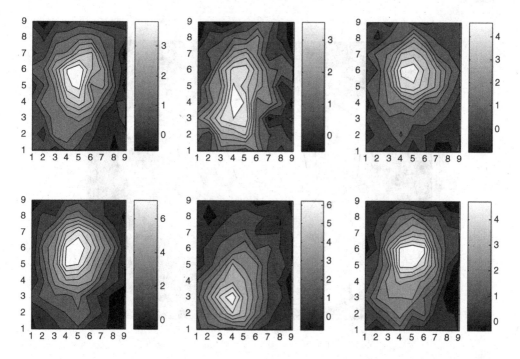

Figure 12.15 An image of size 9 × 9 pixels recorded at six different wavelengths

obvious that there are two objects. However, we may know from independent evidence the positions of the objects, for example the material may have been manufactured in a very precise way that we know about or we may have a separate method for observing the position for each type of object. One of the objects represented in the image is illustrated in Figure 12.17. All that is necessary is to perform calibration between the image in Figure 12.17 and the dataset represented in Figure 12.5. Using centred PLS and keeping two significant components results in the predicted image of Figure 12.18. We can see that we have been able to identify the object quite well. Note that different types of calibration models can be produced for different types of object, which takes into account that each type of object is likely to have a different spectroscopic response: for example in a tablet there may be several different components, each of which has a characteristic response.

We can, if desired, test out the quality of calibration using all the usual methods, example cross-validation, which involves removing some pixels from the overall image, calculating the model using the remainder of information and seeing how well the missing pixels are predicted. It is possible to compute residuals between predicted and actual images, and on occasion this may also result in finding new unsuspected features. In image analysis it is often possible to visualize residuals which can provide insights, e.g. into unsuspected impurities or structure.

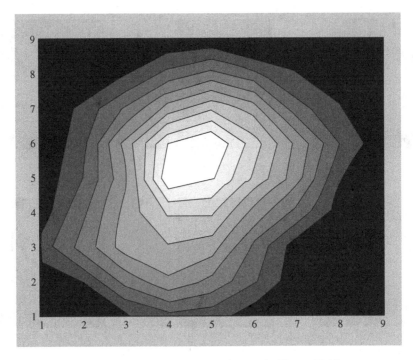

Figure 12.16 Sum of the six images in Figure 12.15

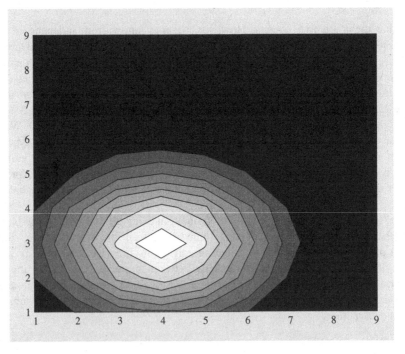

Figure 12.17 Underlying position of one of the objects in the images of Figures 12.15 and 12.16: this may be known from an independent method

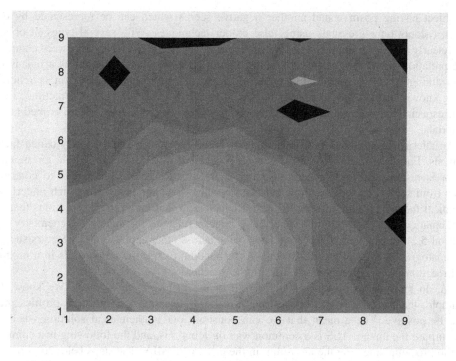

Figure 12.18 Calibrating Figures 12.15 to 12.17 using PLS: this is the PLS estimate of the object's shape

In some situations only part of the information is available from an image, for example the reference image may be known just in certain regions, calibration can be performed using these common areas, the remaining part of the map being used to predict features in the rest of the image. Ultimately the aim of calibration is to be able to dispense with the reference image and so use the real images to determine underlying features in the material. Chemists commonly use techniques such as IR or Raman microscopy to look at materials. Regression, especially PLS, plays a major role in chemical image analysis.

12.6 ALTERNATING LEAST SQUARES AS EMPLOYED IN IMAGE ANALYSIS

An important emerging technique in chemometrics, is called Alternating Least Squares (ALS) which has been introduced in Sections 7.10 and 8.1. One other common use is in the area of image analysis, especially when we are looking at several different chemical features of an image.

In Sections 12.4 and 12.5, we have described the use of PCA and PLS as applied to multivariate image analysis. The former technique has great potential in defining features in images, but if there are several different chemicals (or elements in an alloy, for example), it may often be hard to interpret each PC individually in terms of the underlying chemical composition. The example illustrated in Figure 12.11 consisted of only two types of object, and we were able to distinguish their positions using the second PC, with one type of object having positive and another negative scores which can be represented by different colours on a greyscale using false colour techniques. With several types of object, this clearly is not so simple. PCA images can be very effective but the direct chemical interpretation sometimes can be difficult. PLS, on the other hand, can act as a means for determining positions of different types of chemical features but calibration is required using known standards or else an independent image is required. This is sometimes difficult, especially if spectral characteristics change when chemical species are embedded in materials.

An alternative approach is to use methods based on what is often loosely called factor analysis. Each chemical or element can be considered as a factor. In simple terms, each factor is responsible for a specific spectral characteristic. The image is due to contributions from several factors, but how can we say which features belong to which underlying chemical factor? PCs are also called abstract factors in that they provide solutions that are mathematical in nature, but do not normally directly relate to the underlying chemistry (see Section 5.2). In Chapters 7 and 8 we have come across other applications of factors such as chromatographic elution profiles/spectra corresponding to specific compounds in a mixture and reaction profiles/spectra corresponding to reactants.

We do know something about the nature of the factors in most cases. We know, for example, that spectra should be positive. We know that the concentration profiles must also be positive. These mean that we can constrain our mathematical solution when we decompose the image. ALS is a common way of doing this, and the following is a common method for implementing the algorithm in the context of MIA. The steps below outline one way of using ALS, although the method is relatively flexible and can be adapted according to the specific situation.

1. Perform PCA on the unfolded data (Section 12.4) after suitable preprocessing and smoothing has been performed. It is necessary to decide how many PCs to keep, which should ideally relate to the number of chemical factors: several methods have been introduced in Section 5.10. The unfolded image is then represented by a product of scores and loadings: $X \approx T \cdot P$.

2. In the next step we take one of either the scores or loadings matrices and make all numbers in the matrix positive. For example, we can take the loadings matrix, P, and set any negative values to 0; the remaining numbers are the original loadings. The principle behind this is that each row of the loadings matrix represents a pure spectrum and relates to the spectral matrix.

3. Then use this new matrix which we will now rename S to predict a concentration matrix, C using regression. We use the pseudoinverse (Section 2.4) (denoted by $^+$) so that $C \approx X \cdot S^+$.

4. The new matrix C may contain negative values, so change this to replace all negative values by 0, as we know that concentrations will all be positive.

5. Then use this new value of C to predict and new value of S again by $S \approx C^+ \cdot X$.

6. This cycle continues from step 2, alternating between making C and S positive and then predicting the other one until convergence, which means there is not much change in either the predicted spectral or concentration matrices: see Section 7.10 for a further discussion of the general principles.

At this point we take the columns of C and refold each column separately. Each column should represent the concentration of a single chemical species, and the rows of S should represent the corresponding spectra. So, if we chose to calculate six PCs, column 3 of C and row 3 of S will ideally correspond to one of the chemically distinct species in our image.

This procedure works well providing the number of PCs is well chosen. If there is an inappropriate choice, some of the later factors may represent noise and trying to force the solution to be positive can be a little disastrous. However, one can easily visualize the solutions, and it should be normally obvious by eye whether we have found a sensible one or not and it is simply necessary to repeat the calculations with a different number of components. Each of the spectra can be interpreted and in turn allow us to assign each of the factor images to a chemical species in the mixture.

12.7 MULTIWAY METHODS IN IMAGE ANALYSIS

Above we have concentrated on using classical, two-way, approaches such as PCA and PLS for the interpretation of multivariate images.

Over the past decade, so-called multiway approaches, such as PARAFAC and Tucker3 (Section 5.11) have gradually gained acceptance in the chemometrics community.

In order to understand why we may want to use more sophisticated methods of image analysis we need to be able to understand why there are different sorts of images available. A recent paper has introduced an *O* (object) and *V* (variable) notation [3]. Most three way images are of three types.

1. *OOO*. In this sort of image, each pixel normally represent either a point in time or space. The axes may for example simply represent directions in a three-dimensional object. We record one reading at each point in space, e.g. a shade of grey. Alternatively two axes may be spatial and the third could represent time, for example we may have a number of greyscale images recorded at different times.
2. *OOV*. This is the most common type of image, and has been the basis for the discussion in Sections 12.2–12.6. We may have a two-dimensional picture and at each pixel record a spectrum or series of variables.
3. *OVV*. These types of images do occur in some forms of analytical chemistry. We may be looking at a single set of objects, for example a series of samples, but record a two-dimensional spectrum such as fluorescence excitation emission, for each object. Hence, two of the three-dimensions are spectral.

Even more complex data can arise when four types of related information are recorded, for example consider obtaining a two-dimensional picture consisting of a series of fluorescence excitation emission spectra, so we have an *OOVV* image, two variables being spatial and two spectral. An *OOOV* image could, for example, be formed by recording a series of two-dimensional images using several wavelength ranges over time, if we were to record fluorescence excitation information, this becomes even more complex, an *OOOVV* image. Conventional approaches such as PCA start to break down as it is incredibly complex to

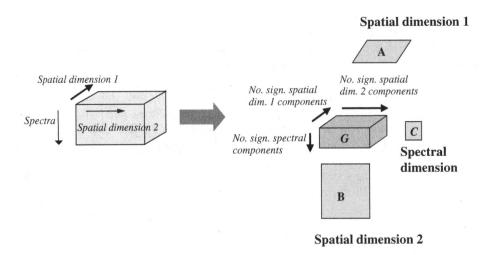

Figure 12.19 Tucker3 decomposition of an *OOV* image

define how to unfold and scale such large and complex datasets, so multimode methods play an important role for the interpretation of these datasets. Some of the techniques are still in the realms of advanced research, but for the simpler cases, such as *OOV* images, a well established literature is developing.

Figure 12.19 illustrates what is commonly called a Tucker3 decomposition of an *OOV* image. The original image is transformed into a core array *G* consisting of three-dimensions each corresponding to the original type of measurement. An important feature of this decomposition, is that each dimension of the core array can be different in size. To put it in simpler terms, we may find that there are four significant components for the wavelengths or variables, but 20 for each of the spatial variables. Hence the dimensions of a matrix *C* (corresponding to the variables) are equal to four times the number of original variables. This distribution is important because often the number of spectral measurements is limited compared with the number of spatial variables, so it may be possible to construct an image using, for example, 20 components for each spatial variable but only four spectral components. Alternatively different models can be obtained for the first, second, third, etc., spectral component, retaining all the spatial components. There is a great deal of flexibility.

PARAFAC is probably more widespread but has the limitation that each of the three types of matrix must have a common dimension, and a core matrix is not used. Each PARAFAC 'component' corresponds to a column (or row) through three matrices, two corresponding to the spatial and one to the spectral dimension in the case of an *OOV* image, analogous to PCs which have corresponding scores and loadings matrices. PARAFAC sometimes works well but the sizes of the matrices are limited by that of the smallest dimension, usually the spectral variable (hyperspectral imaging is as yet less commonly employed in chemistry but involves using a large number of wavelengths, typically 250 to 500, rather than a few 'slices', and whereas more flexible, poses other problems primarily in the area of computational speed).

Multiway methods are an area of research interest to chemometricians. There is the potential of obtaining significant datasets and this is the subject of cutting edge research by a select number of well regarded investigators worldwide. However, the mathematics of many of the methods is quite complex and so will probably take some years to filter into general use.

REFERENCES

1. A.K. Smilde, R. Bro and P. Geladi, *Multi-way Analysis: Applications in the Chemical Sciences*, John Wiley & Sons, Ltd, Chichester, 2004
2. P. Geladi and H. Grahn, *Multivariate Image Analysis*, John Wiley & Sons, Ltd, Chichester, 1996
3. J. Huang, H. Wium, K.B. Qvist and K.H. Esbensen, Multi-way methods in image analysis – relationships and applications, *Chemometrics Intelligent Laboratory Systems*, 66 (2003), 141–158

13

Food

13.1 INTRODUCTION

One of the biggest application areas, and an important original driving force of chemometrics, has been the food industry. There are many historic reasons for this, one being that many early pioneers worked in Scandinavia where there is a significant agricultural sector. In the UK, many of the earliest chemometricians were applied statisticians by training where there was tradition of development of multivariate statistics within highly regarded agricultural research institutes. In Denmark, over the past decade, there has been substantial drive within agricultural institutes. European initiatives in food safely and regulation have also helped this drive in analogy to the US Food and Drug Administrator's influence on the acceptance of multivariate methods within the pharmaceutical industry, for example the Process Analytical Technology (PAT) initiative (Section 8.5).

Why should chemometrics be so important to the food industry?

13.1.1 Adulteration

One key area involves the detection of adulteration. Often people water down food products. This can be quite hard to detect. An example is orange juice, which is often stored freeze dried, exported and then diluted again in the country of origin. How much is the orange juice diluted by? Is it mixed with a cheaper orange juice and passed off as the genuine product? Sometimes there are quite subtle differences between the adulterated and genuine article, but these can often be detected by spectroscopy such as near infrared (NIR) combined with multivariate data analysis. Often countries that export orange juices have less rigorous regulations to the countries they are imported to, and given that the orange juice market is measured in billions of dollars per year, small bending of rules (which could be entirely legal in the country of origin) can result in significant increases in profit margins and unfair competition. More sophisticated questions can be answered, for example is a turkey burger mixed with cheaper chicken meat? This allows the manufacturers to sell a product at a lower price, or else make higher profits. In the brewing industry often highly expensive products such as whiskeys can be watered down with cheaper blends.

Applied Chemometrics for Scientists R. G. Brereton
© 2007 John Wiley & Sons, Ltd

13.1.2 Ingredients

Measuring the amount of ingredients in food is also important. We trust the label on a packet. Some people may be concerned, for example, about the amount of protein, or fat, or carbohydrates in a given product. The reasons may be dietary or even medical. How can we be sure of the constituents? Supplies of raw material can vary in quality according to time of year, how the food was grown, temperature and so on, therefore, the proportions will change a little. This needs to be monitored. Traditional approaches such as chromatography are rather time-consuming, and a faster alternative involves combining chemometric and spectroscopic measurements. This can be used either for labelling and regulatory reasons, or to reject batches of materials that are unsuitable. Sometimes less good batches can be used for other purposes, for example, for animal rather than human consumption.

13.1.3 Sensory Studies

Sensory studies are very important to the food industry. Chemical composition will influence taste, texture, appearance and so on. How can we relate, for example, chromatographic peak heights to taste? Often a taste panel is set up and their results are calibrated to analytical chemical measurements. Hence from analysing a batch, we can make predictions about the market acceptance. Niche markets are also important. For example cat foods might appeal to specific breeds or ages of cats, and a very targeted product can fill a gap. Can one identify a blend that Siamese cat kittens specifically like?

13.1.4 Product Quality

It is also possible to determine what factors influence product quality, by seeing which variables are most influential in multivariate calibration. So-called sensometrics and qualimetrics are related to chemometrics, and multiway calibration, for example, where a product is judged on several criteria by several judges in a panel, is of increasing interest. Process control is important in food chemistry, for example in the manufacturing of soft drinks. The quality may change slightly according to the manufacturing conditions and the raw materials, can this be monitored, on-line, in real time, to determine if a batch is outside the accepted limits of quality?

13.1.5 Image Analysis

Finally, imaging of food also provides important information about constitution and quality. Multivariate methods using spectroscopic information such as Raman are regularly employed, and can show the distribution of different chemicals in fruits or meat. If there is a problem, this distribution may vary from the norm, and we are alerted. Imaging approaches have been discussed in Chapter 12 and will not be introduced again in this chapter, but the reader is reminded that multivariate image analysis has a significant role in the food industry.

13.2 HOW TO DETERMINE THE ORIGIN OF A FOOD PRODUCT USING CHROMATOGRAPHY

A very important application of chemometrics is the determination of the origin of a food. Some of the earliest applications of chemometrics in the 1970s and 1980s were in the brewing industry, where it is important to check the quality of alcoholic beverages. There are many reasons, one being to see whether an expensive brand name has been watered down or adulterated, another to check the quality of a product before it reaches the market. In this section we will illustrate the methods by a classic example relating to whiskey, more details being available in a well known paper published in 1978 [1].

An important problem involves checking the authenticity of expensive brand names. How can we be sure that a whiskey selling at £10 is really the true thing? Of course taste is an important factor, but it could be argued is not conclusive. What happens if someone opens up a bottle of expensive whiskey, tastes it, believes it is a little strange and then reports this, or someone orders a drink in a bar and feels that the whiskey is not quite the usual quality? This is unlikely to be used as conclusive evidence. Maybe the customer has already had several drinks and his senses are blurred or maybe he has opened a bottle and wants another one free and is trying to achieve his objective by complaining or has a cold or just is not good at judging the quality of the drink? How do customs officers test whiskey? Clearly it would be impracticable for them to spend all day drinking samples of every type of whiskey bottle that passes through their hands. So, some analytical technique needs to be used.

Gas chromatography (GC) is ideal for whiskey because the drink can be injected straight onto a column and the volatiles are then separated. Also gas chromatographs are portable and it is not necessary to transport samples into specialized laboratories. One important additional feature of analytical pattern recognition methods is that it is not necessary to identify and assign every peak to a chemical constituent before coming to a conclusion. A customs officer may not really be very interested in the molecular basis of the differences between whiskeys, mainly what the origin is of the whiskey, not why the whiskey is strange but whether it is strange. There is some complexity in the analysis because whiskey can change in nature once a bottle is opened. So we are not only interested in analysing new bottles, but ones that have been opened already. An experienced whiskey taster has first to open the bottle before he can check its flavour. Sometimes a bottle may have been open for a few days, for example in a bar, before the adulteration is suspected.

In order to develop a method for determining whether a whiskey is adulterated, the first step is to perform GC on a series of whiskeys. In the case reported in the paper, the aim was to distinguish an expensive whiskey called Chivas from other less expensive brands. Three groups were examined: 34 non-Chivas (NC) samples coming from a variety of cheaper sources; 24 Chivas (C) samples from new bottles; and 9 from used bottles of Chivas (U). The aim is to be able to distinguish NC from the C and U groups.

How can this be done? GC identifies 17 peaks common to all 67 samples. After first calculating peak areas, row scaling each chromatogram, and standardizing these 17 variables, it is possible to perform Principal Components Analysis (PCA) (Section 5.2), and visual separation can be obtained between the two groups NC and C and U. However the line separating the two classes using a Principal Component (PC) plot is arbitrary and although promising, a somewhat more quantitative approach is required if we want to predict the origins of an unknown sample.

The next step is to obtain a model that can be used to discriminate between the groups. There are numerous approaches, but one method is to first determine which of the 17 peaks are going to be good at separating the groups. For each peak, a Fisher weight can be calculated, that tells how well a feature separates two groups (Section 10.3), which basically involves calculating the ratio between the square of the difference of the means of two groups to their pooled variances. The reason why the squared difference between the means is not used alone is that variables may be recorded on different scales and so this technique allows the possibility to identify quite small but very diagnostic features that may not be obvious at first but are good at discriminating. In this particular study we would like to find features that have high weights for separating both NC from C, and NC from U, but low weights separating U from C. In the published work, six main features were identified for this purpose.

The final step is to perform discriminant analysis for these features, using approaches such as SIMCA (Section 5.8.3) and K Nearest Neighbours (Section 5.9). More elaborate methods such as Support Vector Machines (Section 10.6) could also be employed, although most problems in analytical chemistry of food are fairly straightforward. As is usual the models can be validated in the normal way by using cross-validation, the bootstrap and test sets as appropriate (Section 5.7), and an automated approach then developed to assess whether the GC of a whiskey indicates it is authentic or not. These types of approaches have been continually refined over the years and are used in the brewing industry as well as in other applications of food authentication. Pattern recognition techniques are still not yet commonly employed as conclusive evidence in court, but can very effectively identify adulteration, allowing a comprehensive and more expensive investigation of a small number of products. Chemometrics combined with chromatography is very effective in narrowing down the number of suspect samples.

13.3 NEAR INFRARED SPECTROSCOPY

NIR spectroscopy has had a specially important role in the history of the application of chemometrics to the analysis of food, having been an economic driving force for around 20 years. Chemometrics methods are routinely available in software for NIR analysers. The number of applications reported in the literature is huge, and in this section we only have room for a summary of NIR applications.

13.3.1 Calibration

A major application of chemometrics together with NIR involves calibrating the spectral features of extracts of a food to their chemical composition. A well established classical example involves estimating the protein content of wheat [2]. It is important to measure this as the higher the protein content, the higher the quality: this influences the amount growers are paid and allows bakers to determine the quality of their flour. Traditionally a test called the Kjeldalh-N method is employed. This test is rather slow and also uses noxious chemicals. An alternative is to shine NIR onto the samples and measure the intensity of the reflected light.

A common method for determining the calibration model between the traditional measure of protein content and the NIR spectra is to use Partial Least Squares (PLS) (Section 6.6), validated as usual (Section 6.7), in order to be able to predict the protein content of new samples from their spectra. An important consideration is experimental design, to ensure a sufficiently broad range of wheat samples are measured for the purpose. The wheat samples must have a sufficiently wide range of protein content, and also there must be a suitable range of phenotypes and genotypes to make the model applicable over all possible sample ranges in the training set. If a new variety of wheat from a different source is expected in the future, it is advisable to recalculate the calibration model, including extra training set samples from this new variety.

This classic example is regularly cited in the literature in various guises, but there are innumerable reports in the literature of using NIR calibration for the determination of food content, in a wide variety of animal and plant food stuffs. As well as protein content, it is very common to employ spectroscopy to measure moisture in food, in addition the proportion of all sorts of other fractions such as acids, energy content and lipids can be predicted. The main procedure in most cases involves setting up a suitable training set, developing calibration models usually via PLS and validating these models.

13.3.2 Classification

There are many reasons why we might want to classify food samples into groups, some of which are as follows. The first is to determine the origin (supplier) of a food. Another is to determine how a food has been processed. Adulteration is a third reason, the normal and adulterated products being used to define two classes. Finally we may want to classify food into acceptable or not. In all these cases the aim is to set up class models from known groups as discussed in Chapters 5 and 10, to see how well they perform and subsequently use this as a predictive model on an unknown sample. This can be valuable both for customs and for quality control.

An example from the literature is where 10 different types of oils (soybean, sunflower, peanut, olive, lard, corn, cod liver, coconut, canola and butter) are distinguished by a variety of pattern recognition techniques including Linear Discriminant Analysis (LDA) [3] (see Sections 5.8.2, 10.3 and 10.4). Very high discriminant ability (around 95 % dependent on method) is obtained on the test set. It is important to recognize that there must be an adequately sized test set for this purpose to ensure that the validation statistics are useful.

Another slightly different application is where it is desired to see which variables or portions of a spectrum are most useful for classification. We may be less interested in whether we can group samples but what features have most discriminant power. There are various tricks (Section 5.8) to look at the modelling power of variables.

An interesting observation is that in food chemistry linear methods such as PLS, LDA and SIMCA are normally adequate for predictive models. Unlike in many areas of biology, nonlinear approaches such as Support Vector Machines (Section 10.6) are not commonly employed.

13.3.3 Exploratory Methods

Exploratory methods are also useful in visualizing spectroscopic data from food. The normal stages of data pretreatment are of course essential to consider prior to performing exploratory data analysis, usually PCA. This step is sometimes used as a precursor to classification as discussed above, but can, for example, be used to display geographical origins of foods, and to see if there are any unexpected trends such as suppliers, time of year, etc. It can however be employed as an end it itself, for example to determine geographical origins of products.

Enhancements such as cluster analysis using dendrograms (Section 5.6) and various factor rotations including varimax (discussed in a different context in Section 13.6) can be employed for simplifying the data.

13.4 OTHER INFORMATION

NIR, whilst a classical method for analysing food, is by no means the only approach that benefits from chemometrics. NIR is though so ingrained that I have come across some statistically minded colleagues who often confuse chemometrics with NIR, such is the established success of this technique.

13.4.1 Spectroscopies

Other closely related spectroscopies such as mid infrared and Raman can be employed, the weakness being that the spectroscopies are often slightly less stable and instrumentation less well established than NIR, but pose similar challenges. In Section 8.2 the differences between spectroscopies in the context of reaction monitoring is discussed, some of the same principles relate to the analysis of food.

Fluorescence spectroscopy however whilst not common holds special interest as an extra dimension of data can be obtained if excitation/emission spectroscopy is performed meaning that an excitation/emission matrix (EEM) can be obtained for each sample. Whereas gaining an extra dimension is common in coupled chromatography such as gas chromatography – mass spectrometry or diode array high performance liquid chromatography, there are two advantages of fluorescence spectroscopy. The first is that the wavelength ranges are generally quite stable, unlike elution times in chromatography. The second is that the analysis is fairly fast. The drawback is that the data now require more elaborate methods for handling, but three-way methods, especially PARAFAC (Section 5.11) and normal PLS (Section 6.8) can be employed. One example involves using fluorescence EEM to determine dioxins in fish oils [4], where three-way methods are compared with two-way PLS of the fluorescence spectral profiles, and another involves determining the ash content in sugars [5]. One obvious disadvantage is there is as yet limited commercial software for multiway calibration and pattern recognition, so one has to be an enthusiast and usually a Matlab user to successfully apply these techniques. Nevertheless, multiway methods are slowly but surely gaining acceptance in the chemometrics community, especially in the area of food analysis, where there is a more established tradition of separate statistical data analysis teams that have strong computing expertise, than in mainstream chemistry where the experimenters often demand highly user friendly software as they want to do the data analysis themselves.

13.4.2 Chemical Composition

Sometimes it is possible to obtain chemical information about the composition of food. This could be done using chromatography or other tests. Pattern recognition can be performed on these data. In such cases the aim may not be so much to develop a fast method for screening but to provide insights into whether there are differences between products or grades, and why. This type of pattern recognition is primarily exploratory in nature.

An example involves looking at 120 Tuscan olive oils using 29 chemical descriptors such as the amount of free fatty acids, the ratio of C26/C24 chains and the ratio of triterpene alcohols to total terpene alcohols [6]. The principle is borrowed from Quantitative Structure–Activity Relationships (QSAR), where typically molecular descriptors are employed to see whether there are patterns in data. Using pattern recognition, four groups of samples are found, and the samples can be related roughly to geographical origin. It is also possible to see which variables are correlated and obtain some insight into why there are differences from a chemical perspective.

Often this chemical information may come not from descriptors but from chromatographic peak areas, such as in GC, particularly if the identities of specific peaks are known. Section 13.2 discusses an application from GC, but the potential is very wide.

13.4.3 Mass Spectrometry and Pyrolysis

Both head space and pyrolysis mass spectrometry has a role to play in food analysis. Pyrolysis data are so complex that it is hard to determine patterns within some chemometric data analysis. The *Journal of Analytical and Applied Pyrolysis* contains numerous examples. Section 10.8.4 describes applications of pyrolysis in the context of biological pattern recognition, and similar principles apply to food chemistry.

13.5 SENSORY ANALYSIS: LINKING COMPOSITION TO PROPERTIES

Multivariate methods have a long history in sensory statistics. To the food scientist it is important to be able to relate the results from a taste panel to the nature of consumer products, often their chemistry or composition. The brewing industry has employed sensory panels for many years, and it is possible to obtain huge quantities of information, so we will illustrate this section by reference to beers, however the same principles apply to most food products. Graphical presentation employing multivariate methods is especially important and allows large quantities of data to be displayed in a readily comprehensible format.

13.5.1 Sensory Panels

A typical example is where a sensory panel is asked to assess the quality of a series of beers. It is not easy work being a member of such a panel, as there are often tens or even more than a hundred possible criteria. These can be grouped into different categories, for example, odour, taste, mouthfeel and aftertaste each of which have several descriptors attached to them. Many descriptors relate to more than one category, a beer can have a sweet odour, sweet taste and sweet aftertaste for example, so saying a beer is 'sweet' is not necessarily very specific, whereas some descriptors such as oily or sticky refer only to one category, e.g. mouthfeel. Each beer is scored for every criterion, on a scale, an example between 8 (strong) and 0 (none). Each member of the panel has to fill in a form, as shown

Sensory Panel Member : John Smith
Beer Number : 8
Descriptors

	0	1	2	3	4	5	6	7	8
Odour									
Worty	☐	☐	☐	☐	☐	☐	☐	■	☐
Cheesy	☐	■	☐	☐	☐	☐	☐	☐	☐
Fresh	☐	☐	☐	☐	☐	☐	■	☐	☐
	☐	☐	☐	☐	☐	☐	☐	☐	☐

Figure 13.1 Typical portion of a form completed by a taste panel member

in Figure 13.1. Typically an individual beer is tasted several times on different days, by several panel members. If a beer is tasted on four occasions by 15 panel members, this makes 60 different sets of scores, all consisting of values between 0 and 8, so the quality of a beer for each of the categories will be represented by a summed value between 0 and 480. A well designed tasting session would present beers in random order, perhaps in dark containers to disguise their colour (unless this is specific factor that is of interest).

The result of the sensory studies can then be presented as a multivariate matrix, whose rows relate to the different beers and whose columns relate to the overall sensory scores. If 50 beers are studied for 100 characteristics a 50 × 100 matrix, as illustrated in Figure 13.2, is obtained. Each cell represents an overall sensory panel value for a specific descriptor, for example the worty odour of beer 8, which in the example above will be represented by a number between 0 and 480. As we know, there is lots that can be done with this information.

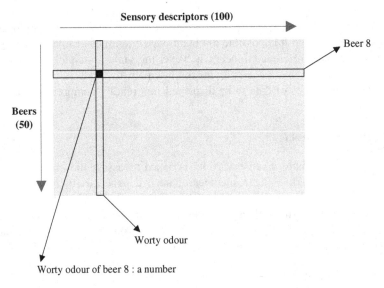

Figure 13.2 Typical multivariate matrix where 100 sensory descriptors are obtained on 50 beers

13.5.2 Principal Components Analysis

The simplest procedure is to perform PCA and then examine the first few PCs visually using PC plots. Sometimes the data are first standardized (Section 5.5) prior to PCA but because the variables are recorded on similar scales in the case described in Section 13.5.1, this is not always necessary; centring the columns however is usual. There are several ways of displaying the information some of which have been discussed previously (Section 5.3).

A scores plot examines the relationship between beers, for example, are beers from a certain brewer clustered together, or is the type of beer, e.g. stouts, bitters and lagers more significant? Normally the latter clustering will be more significant. However what sometimes happens is that certain later PCs relate to different factors so the less important trends are clearer in plots of later PCs. A loadings plot looks at the variables, in this case sensory characteristics. Often there are quite a large number of such variables, but an experienced sensory analyst can usually interpret which are significant for any specific combination of components. Characteristics in the middle of the loadings plot are probably not very useful as they have low values, so are probably not useful sensory descriptors for the main features in the corresponding scores plot. Remember that all the variables are usually measured on roughly similar scales, even if not standardized. The central variables will be different for each combination of PCs, for example there will be different variables in a graph of PC2 versus PC1 compared with a graph of PC3 versus PC5. One of the most important tools in sensory analysis is a biplot (Section 5.4) in which scores and loadings are superimposed. This allows the beers to be related to the main sensory characteristics, so the closer a beer or cluster of beers is to a characteristic the more strongly they are related. Therefore we could for example, state that a specific group of beers have a worty aroma.

13.5.3 Advantages

It may at first appear that this information could be obtain using simpler means not just by inspecting the survey data, but the advantage of multivariate methods is that a huge amount of information can be compared graphically. If we study 100 characteristics from 50 beers that are tasted on four occasions by 15 panel members, there are about 300000 sensory measurements available. To wade through this information manually would take a long time, and yet the numbers obtained from the surveys can quite readily be entered into a computer and informative diagrams produced that provide guidance especially about the relationship between beers and their sensory characteristics.

If specific beers have certain characteristics, it is possible to then see whether there are specific chemicals that result in these characteristics and so develop new processes or even blends that optimize certain qualities of the product or appeal to certain niche markets.

13.6 VARIMAX ROTATION

PCA is sometimes called 'abstract factor analysis' (Section 5.2). PCs are mathematical entities, that are the solution to best fit equations. There is no physical significance in the components. However in areas such as food research we may want to produce pictures similar to PC plots that we can interpret in physical terms. For example it might be desirable to produce a loadings plot for a series of food whose axes relate primarily to sweetness and texture. These two properties are often referred to as physical (or chemical) factors. The texture of a food may be measured by a variety of different sensory statistics obtained from a large number of measurements. We could then obtain an equivalent scores plot, which allows us to match the food to the sensory properties, so we can see which foods are most and least sweet pictorially – remember that this assessment may involve the summary of large amounts of raw data and not be obvious from the tables of numbers. The beauty of multivariate methods is that a huge amount of data can be presented in a graphical form, which would be very hard to do using tables of numbers or univariate statistics.

In order to illustrate the method we will use a simple dataset obtained from *Applied Multivariate Statistical Analysis* by Härdle and Simar [7]. Although the example is not chemical in nature, it does focus on food and illustrate the methods quite well. The data were collected in France in the early 1980s and reflect the average monthly expenditure (in French francs) of several groups of families on various different types of food. The families are categorized into three employment classes, namely MA (manual workers), EM (employees) and CA (managers), and also the number of children (between two and five). Therefore MA3 relates to average expenditure of families of manual workers with three children. The information is presented in Table 13.1.

Varimax rotation is a method that converts PCs into what are often called factors (note that there are many definitions in the literature as to what is really meant by a factor, we will not enter the debate in this book). Mathematically, however, a rotation or transformation converts PC scores to factor scores and PC loadings to factor loadings. It converts something quite abstract to something that is often physically interpretable. It can be expressed mathematically by:

$$S = T \cdot R$$

and

$$Q = R^{-1} \cdot P$$

where T and P are the PC scores and loadings, S and Q the corresponding factor scores and loadings, and R is called a rotation or transformation matrix. We have come across these concepts in several other situations in this book (see Sections 7.9 and 8.3) although

Table 13.1 Average monthly expenditure (in French Francs) on food for 12 families. MA, manual workers; EM, employees; CA, managers; the number of children being indicated

	Bread	Vegetables	Fruits	Meat	Poultry	Milk	Wine
MA2	332	428	354	1437	526	247	427
EM2	293	559	388	1527	567	239	258
CA2	372	767	562	1948	927	235	433
MA3	406	563	341	1507	544	324	407
EM3	386	608	396	1501	558	319	363
CA3	438	843	689	2345	1148	243	341
MA4	534	660	367	1620	638	414	407
EM4	460	699	484	1856	762	400	416
CA4	385	789	621	2366	1149	304	282
MA5	655	776	423	1848	759	495	486
EM5	584	995	548	2056	893	518	319
CA5	515	1097	887	2630	1167	561	284

Data taken from Härdle and Simar [7].

the criteria are quite different. For users of Matlab, there is public domain code for this; a good Website is available [8] which gives some code and further information.

The aim of varimax rotation is to find factors that are interesting, which are factors often defined as those that maximize the variance of the loadings. These new factors will usually consist of variables with a few large loadings, as these contribute to high sums of squares for the variance of the loadings, rather than several comparably sized loadings.

Now let us pass on from the technical details and look at our example. Before analysing the data we first standardize it (Section 5.5) so all the variables are effectively on a similar scale. The next step is to perform PCA, and then look at plots of scores and loadings (Section 5.3). The plots for PC2 versus PC1 are given in Figure 13.3 and for PC3 versus

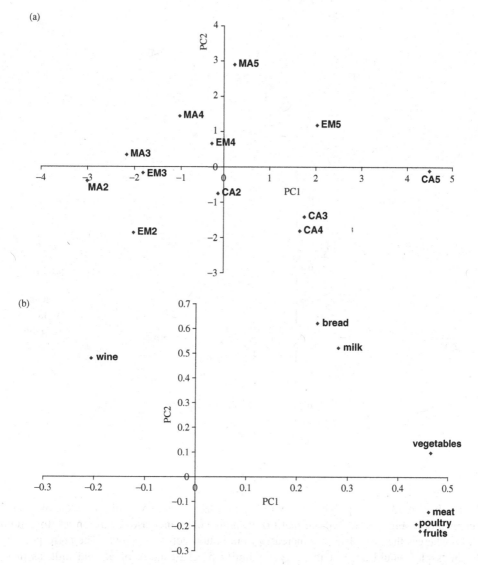

Figure 13.3 Graph of PC2 versus PC1 for data in Table 13.1. (a) Scores; (b) loadings

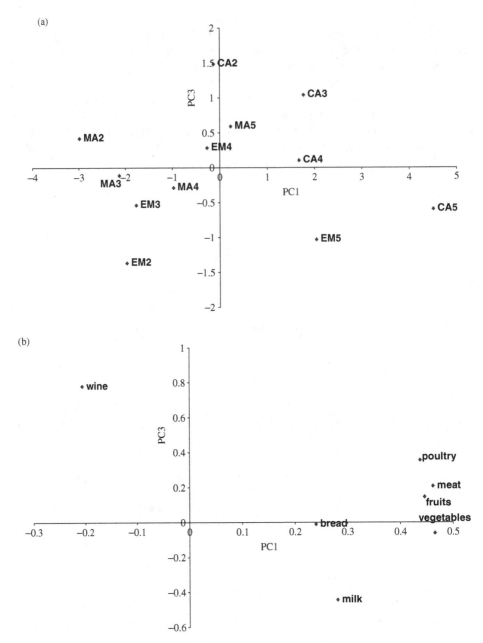

Figure 13.4 Graph of PC3 versus PC1 for data in Table 13.1. (a) Scores; (b) loadings

PC1 in Figure 13.4. It is possible to try to interpret these graphs visually, for example, in the scores plot of Figure 13.3 we see that the economic class increases from top left to bottom right, whereas the size of family increases from bottom left to top right. The corresponding loadings plot would suggest that larger families purchase more bread and milk as these variables are in the top right, but that the lower the economic class the greater the purchase of wine as this variable is in the top left. The highest purchase of meat, poultry, vegetable and fruits corresponds to CA5, the largest families of the highest social class. So from the PC plots we can obtain quite a lot of information.

What happens if we perform varimax rotation? In this example we will retain six PCs and then determine of the rotation matrix, meaning that the first three varimax factors are the result of transforming or rotating the first six PCs. Figure 13.5 illustrates the scores and loadings plots for factor 2 versus 1, and Figure 13.6 for factor 3 versus 1. In the scores plot

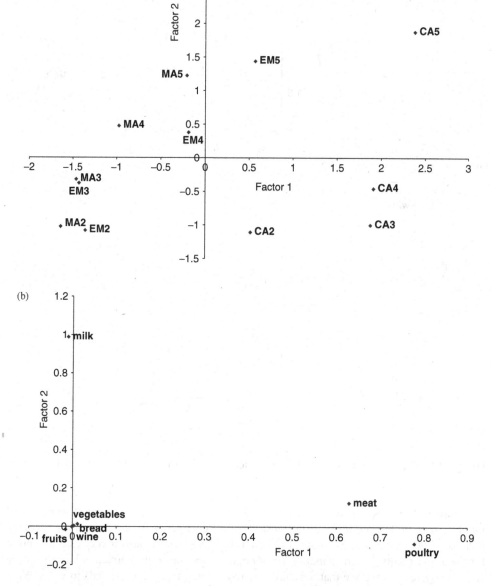

Figure 13.5 Graph of varimax factor 2 versus factor 1 for data in Table 13.1. (a) Scores; (b) loadings

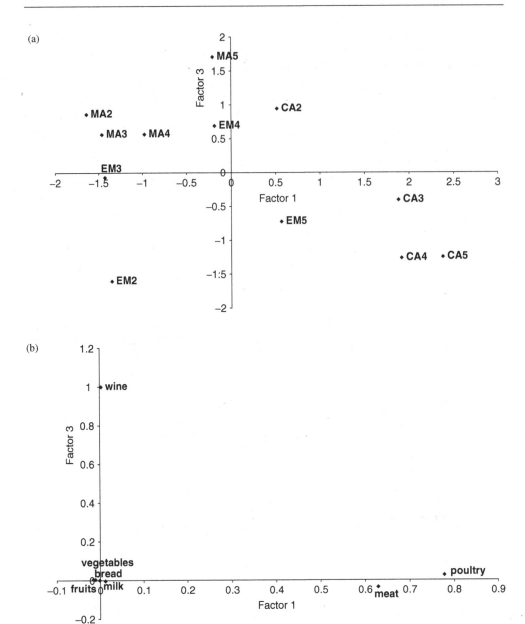

Figure 13.6 Graph of varimax factor 3 versus factor 1 for data in Table 13.1. (a) Scores; (b) loadings

of Figure 13.5, we approximately see that economic class increases from left to right and family size from bottom to top. The most dramatic changes, however, are in the loadings plot, as it is these that the rotation tries to alter directly. There is now a very clear direction for meat and poultry, corresponding to economic class, and milk corresponding to family size. In Figure 13.5 we see that the third varimax rotated factor corresponds to wine. What has happened? Each factor is now primarily characterized by a small number of variables.

This example has been a simple one, but generally several variables group together as in factor one (meat and poultry in our case). The variables in the centre of the graph can then be largely ignored. This means that a particular graph is mainly based on a subset of the original variables, often those that are highly correlated, and the factors can usually be interpreted in physical terms. In a more complex situation, such as in food chemistry, we might measure 50 to 100 sensory variables, and each factor might represent a grouping, say of 10 or so variables, that are linked together. This then tells us a lot about the variables but also about the objects, as it is common to compare scores and loadings plots.

Varimax rotation is one of several methods of factor analysis that is often employed for the simplification and interpretation of PC plots and is probably the most widely used.

13.7 CALIBRATING SENSORY DESCRIPTORS TO COMPOSITION

In food chemistry, it is often important to relate the composition of a food to its appearance and taste. In Sections 13.5 and 13.6 we described PC based approaches in which two graphs, for example a scores plot and a loadings plot, are compared, often visually.

An alternative approach is to use calibration to form a model between the two types of information. PLS (Section 6.6) is a popular method that is often chosen for this purpose. In chemometrics jargon, the aim is to relate two 'blocks' of information, sometimes called an x block (the ingredients) and a c block (the descriptors). Note that there is sometimes different terminology according to authors. Some people refer to the y rather than the c block. This problem is analogous to a spectroscopic one of relating an observed spectrum (x) to the concentrations of compounds in the underlying mixture (c), and similar types of approach can be employed.

We will take as an example, information obtained from Folkenberg *et al.* [9] in which six sensory descriptors are obtained from eight blends of cocoa as presented in Table 13.2. Can we relate the blend to the sensory descriptors? There are numerous ways of doing this, but if we use PLS, we try to form a relationship between these two types of information. At this point, it is important to recognize that there are several variants on PLS in the literature. Some newcomers to chemometrics get very confused by this. In the old adage

Table 13.2 Cocoa example

Sample	Ingredients			Assessments					
	%Cocoa	%Sugar	%Milk	Lightness	Colour	Cocoa odour	Smooth texture	Milk taste	Sweetness
1	20.00	30.00	50.00	44.89	1.67	6.06	8.59	6.89	8.48
2	20.00	43.30	36.70	42.77	3.22	6.30	9.09	5.17	9.76
3	20.00	50.00	30.00	41.64	4.82	7.09	8.61	4.62	10.50
4	26.70	30.00	43.30	42.37	4.90	7.57	5.96	3.26	6.69
5	26.70	36.70	36.70	41.04	7.20	8.25	6.09	2.94	7.05
6	26.70	36.70	36.70	41.04	6.86	7.66	6.74	2.58	7.04
7	33.30	36.70	30.00	39.14	10.60	10.24	4.55	1.51	5.48
8	40.00	30.00	30.00	38.31	11.11	11.31	3.42	0.86	3.91

'all roads lead to Rome', similar answers are usually obtained from most algorithms, but each has different features. Some people get confused as to the difference between *methods* and *algorithms*. There are basically two methods of PLS appropriate to this problem. PLS1 models each variable separately. In this case there are three c variables, namely, cocoa, sugar and milk, so three separate calculations are required. PLS2 models all the variables at the same time, so only one calculation is required. Whereas PLS2 is quicker and, in fact, results in less and, therefore, simpler, output, in most cases it performs less well, so in practice most people use PLS1. An algorithm is a means for calculating PLS components. There are several different algorithms in the literature. The way in which the scores and loadings of each component is determined differs, and even the definition of scores and loadings can change according to the paper or book or package you are using. However the estimates of the c block are a product of two or more matrices and it normally makes very little difference which algorithm is employed. Here we use PLS1. Since intermediate steps of the calculations are not presented, results will be fairly similar no matter which algorithm is used. Section 6.6 expands on this theme.

In PLS, we can predict the value of each descriptor using successively more and more PLS components. The more we use, the closer the prediction. Table 13.3 is of the prediction of the percentage of cocoa, using a PLS1 model of the six sensory descriptors and using autoprediction on the training set. Similar calculations could be performed using the other two descriptors. The more the components, the better the prediction if the calculation is performed on the training set (obviously different results will be obtained for cross-validation or independent test sets: see Section 6.7). However, the table demonstrates that it is possible to use sensory descriptors to predict the composition, which has two benefits. First of all one can examine an unknown sample and from its taste, odour and texture, predict its composition. Second one can determine a relationship between composition and sensory attributes. Ultimately this allows the possibility of relating chemistry to consumer acceptability.

The first two PLS components account for 98 % of what chemometricians often call the variability of the data in this case, so we can describe the original information quite well using two components. Neglecting later components loses us only 2 % of the original information, which may be accounted for by noise. We can also produce PLS1 loadings and scores plots (Figure 13.7) for the first two components; for the scores plot, the points are labelled according to percentage cocoa in the samples. Note that there will be different sets of graphs for performing PLS1 calibration for each of the three variables, cocoa, milk and

Table 13.3 Use of PLS1 to predict the percentage of cocoa

True %cocoa	Predicted %cocoa		
	1 component	2 components	3 components
20	17.87	19.79	20.38
20	19.73	18.99	19.19
20	21.67	19.79	20.19
26.7	26.23	27.32	26.44
26.7	28.16	28.03	27.66
26.7	27.45	26.98	26.15
33.3	34.47	34.18	34.37
40	37.82	38.32	39.03

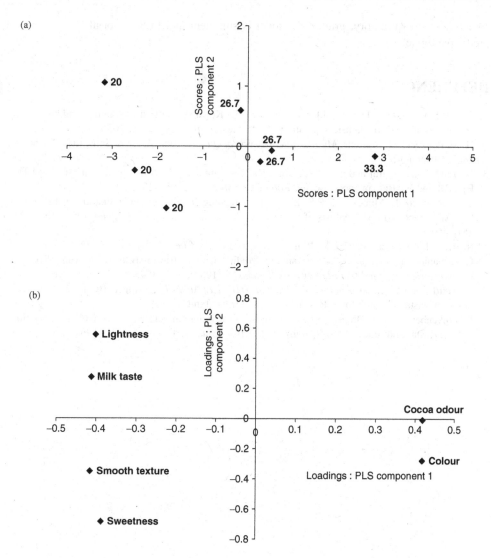

Figure 13.7 (a) score plot for %cocoa and (b) loadings plot for descriptors corresponding to data in Table 13.2

sugar, which differs from the case of PCA where there is one single set of loadings and scores plots. The two plots can, however, be compared, and it is not surprising that colour and cocoa odour are associated with high percentage of cocoa.

Note that another important issue is to determine how many significant components can be used to characterize the data. Obviously, the more PLS components employed, the closer the sensory descriptors are predicted. However, this does not mean that the model is correct. The reason is that the later PLS components may model noise, and so although it appears the prediction is good on the original dataset, in fact it is misleading. There are many ways to check this such as cross-validation and the bootstrap as discussed elsewhere. However sometimes an exploratory, graphical, model is sufficient and PLS is not always used in a predictive mode.

REFERENCES

1. B.E.H. Saxberg, D.L. Duewer, J.L. Booker and B.R. Kowalski, Pattern-recognition and blind assay techniques applied to forensic separation of whiskies, *Anal. Chim. Acta.*, 103 (1978), 201–212
2. H. Martens and M. Martens, *Multivariate Analysis of Quality*, John Wiley & Sons, Ltd, Chichester, 2001, Chapter 12
3. H. Yang, J. Irudayaraj and M.M. Parakdar, Discriminant analysis of edible oils and fats by FTIR, FT-NIR and FT-Raman spectroscopy, *Food Chemistry*, 93 (2005), 25–32
4. D.K. Pedersen, L. Munck and S.B. Engelsen, Screening for dioxin contamination in fish oil by PARAFAC and N-PLSR analysis of fluorescence landscapes, *Journal of Chemometrics*, 16 (2002), 451–460
5. R. Bro, Multiway calibration. Multilinear PLS, *Journal of Chemometrics*, 10 (1996), 47–61
6. C. Armanino, R. Leardi, S. Lanteri and G. Modi, Chemometric analysis of Tuscan olive oils, *Chemometrics and Intelligent Laboratory Systems*, 5 (1989), 343–354
7. W. Härdle and L. Simar, *Applied Multivariate Statistical Analysis*, Springer, Berlin, 2003
8. http://www.stat.ufl.edu/~tpark/Research/about_varimax.html
9. D. Folkenberg, WL.P. Bredie and M. Martens, What is mouthfeel? Sensory-rheological relationships in instant hot cocoa Drinks, *Journal of Sensory Studies*, 14 (1999), 181–195

Index

accuracy   65, 85
adulteration of foods   351, 355
agglomerative clustering   169–70
alloys   331–2
Alternating Least Squares (ALS)   247–8, 251, 261
   and image analysis   345–7
amino acids
   letter notation   321
   scoring matrix   321
   sequences   320–2
analysis of variance (ANOVA)   19
arthritis   313
auto-correlograms   113–5
   and cyclical features   113–4
   and noise   114, 116
autoprediction   186–7, 212–3, 215–6
   and model calibration   214
average experimental value   22–3

bacteria *see* micro-organisms
bar chart   70–1
Bayes' theorem   124–5, 302
   definitions   124
Bayesian methods   124–5, 300–3
   contingency tables   302–3
   in medicine   301–2
   posterior probability calculation   303
best fit coefficient *see* coefficient
best fit line   93, 97–9, 101
bioinformatics   3
biological classification *see* taxonomy
biological macromolecules *see*
      macromolecules
biology   4, 287–318
biplot   160, 359

bivariate discriminant models   175–7
   class distance   175–8
   projection   175–6
blob chart   69
bootstrap   173, 217
brewing industry   353–4, 357–9

calibration   193–220
   model appropriateness   214–7
   model effectiveness   214–7
      autoprediction   214
      bootstrap   217
      cross-validation   215
      test sets   215–6
   and Multiple Linear Regression   206–8
   multivariate   58–62, 194–5, 202–6
   multiway calibration   217–20
   Partial Least Squares   211–4
   Principal Components Regression   208–11
   of spectra   207–8
   univariate   194–202
calibration set *see* training set
calibration transfer   254–5
central composite designs   39–44
   experimental conditions calculation   43
   four factor designs   43
   rotatable   42–3
   terminology   43
   three factor designs   40–1
chemometrics
   applications   3–4
   development   1–3
chi-square distribution   95–6
chromatograms
   composition areas   228–30
   embedded peaks   229

Applied Chemometrics for Scientists   R. G. Brereton
© 2007 John Wiley & Sons, Ltd

chromatograms (*continued*)
 impurity monitoring 230
 overlapping peaks 229–30
 tailing peaks 230
chromatography 112
 brewing industry 353–4
 derivatives 138–42, 239–41
 and Multiple Linear Regression 207–8
 peak shapes 134–8
 and Principal Components Analysis 148
 see also coupled chromatography
cladistics 325–6
cladogram 170, 325–6
class distance 175–8
 determination 297–300
  Euclidean distance 297–8
  Mahalanobis distance 298–9
 in SIMCA 180
classification *see* supervised pattern recognition;
  taxonomy
cluster analysis 146, 167–71
 agglomerative clustering 169–70
 dendrogram 170
 and DNA/protein sequencing 322
 molecular biology 327–9
 in numerical taxonomy 290–1
 similarity determination 168
 terminology 329
 'ultrametric' 328
 UPGMA 327–9
 usage 167–8
coefficient 24–5
 calculation 27
 interpretation 27
 significance 25
 size interpretation 31
 usefulness 25
column scaling 223–4
Component Detection Algorithm (CODA) 226
composition areas 228–30
confidence
 in linear regression 89–93
 in the mean 93–5
 in predictions 64
  and experimental region 57
 of parameters 92–3
 in standard deviation 95–6
confidence bands 54, 91
 graphical representation 91–2
confidence interval calculations 90–2
confidence limits 93–4
constrained mixture designs 47–9

bounds 47–9
contingency table 302–3
control charts
 Cusum chart 106–8
 range chart 108
 Shewhart charts 104–6
convolution 131, 134
correlation coefficient 73, 168–9, 234–5
correlation structure 100
correlograms 112–6
 auto-correlograms 113–6
 cross-correlograms 115
 multivariate 115
 principle 113
coupled chromatography 221–48
 applications 222
 data preparation 222–8
 deconvolution 241–2
 derivatives 239–41
 Evolving Factor Analysis 236–9
 iterative methods 246–8
 metabolomics 314–7
 noniterative methods for resolution 242–6
 Orthogonal Projection Approach (OPA) 236
 peak tables 314–6
 and Principal Components Analysis 150–4
 purity curves 230–4
  peak purity 234, 239–40
 similarity methods 234–6
 SIMPLISMA 236
 Window Factor Analysis 236–9
covariance 72
cross-correlograms 115
cross-validation 172–3, 186–7, 215–6
cumulative frequency 76–8
 computation 77–8
cumulative frequency measurement 69–70
Cusum chart 106–8
 calculation 107

data compression 143
data description 67–73
 correlation coefficient 73
 covariance 72
 graphs 69–72
 statistical 68–9
data fit 31
data interpretation 31–2
data matrix 151–2, 185, 228–9
data preparation 222–8
 preprocessing 222–4
 variable selection 224–8

datasets
  chromatographic 2–3
  component numbers 185–7
deconvolution 221–2
  and filtering 140
  Fourier self deconvolution 130–1, 133
  spectral mixture analysis 241–2
degrees of freedom 12–6
  tree 14–16, 40, 42
dendrogram 170, 290, 325
derivatives 138–42
  calculation 138–41
    disadvantages 140
  in coupled chromatography 239–41
  of Gaussian 139
  micro-organism investigations 309–11
  peak purity determination 141
  peak resolution 138–40
  spectral following 239–40
    method implementation 240
design matrix 22–5, 31, 38
  calculation 23–4
diode-array HPLC 226–8, 231–4
Discrete Fourier Transform (DFT) 120–1, 123
discriminant function 291–2, 293–7,
    299–300, 354
discriminant models 175–8
  extensions 178
  see also bivariate discriminant models
Discriminant Partial Least Squares (D-PLS)
    306–7
  calibration model 307
  graphical information 308
  multi-class 308
  two classes 307
  validation 307
discrimination methods 291–7
  class numbers 293–6
  discriminant analysis 291, 293
  discriminant function 291–7, 299–300
  Fisher score 292, 295
  Fisher's iris data 291, 293–4
  limitations 296–7
  modifications 296–7
  parameter combination 293
  pooled variance covariance matrix 292
  in taxonomy 291–7
disease diagnosis 288
distribution curve 70–2
Dixon's Q-statistic 100–1, 108–9
DNA 319–20
dummy factors 36–8

eigenvalues 152, 185–6, 212
  in Window Factor Analysis 237–8
entropy 125–8
  see also maximum entropy
environmental studies 4, 111
equilibria studies
  multivariate matrices 249
  profiles 249–51
  spectroscopic 249–52
errors 14, 64–7
  behaviour 106
  distribution 73
  estimation 212–4
  and instrumental noise 67
  mistakes v. errors 65
  in real world 65
  sample preparation 66–7
  sampling 65–6
  significance 19
  size determination 17
  sources 67
Euclidean distance 168, 178, 235–6, 292
  in taxonomy 297–8
evolutionary theory 325–6, 329
Evolving Factor Analysis (EFA) 236–9, 244
  see also Window Factor Analysis
experimental designs 9–62, 272–3
  compound selection 11
  descriptor selection 11
  orthogonality 33, 35–6, 272
  rationale 9–12
experiments
  number reduction 32–5
  subsets 32–3
Exploratory Data Analysis (EDA) 145–6

face centred cube 40
factor analysis 208–9, 346
factorial designs 25–32
  experiment numbers 32
  extension of factors 28
  levels 28–9
    multilevel 28–9
    two levels 26–8
    limitations 27–8, 34–5
  practical example 29–32
  weakness 32
factors 28
  interaction 34, 269–70
  number reduction 32
  relevant 32
  and varimax rotation 360–5

false colour technique 337
filters 116–20
  linear 117
  moving average 117–9
  Multivariate Image Analysis 335–7
  Savizky–Golay 117–9
Fisher discriminant function 291–2, 293–7,
    299–300, 354
Fisher score 292, 295
Fisher's iris data 289–1, 293–4
fluorescence spectroscopy 356
food chemistry 3, 49, 109, 351–68
  chromatography 353–4
  component significance 368
  fluorescence spectroscopy 356
  food adulteration 351, 355
  food composition 356–7
  image analysis 352
  ingredient measurement 352
  mid infrared spectroscopy 356
  near infrared spectroscopy 193, 354–6
    calibration 354–5
    exploratory methods 355–6
    food classification 355
    protein in wheat 354–5
  origin determination 353–4
  Partial Least Squares 365–8
    data variability 366
    PLS1/PLS2 365–8
  Principal Components Analysis 359
  product origin 353–4
  product quality 352
  Raman spectroscopy 356
  sensory studies 352, 357–9
    and composition 365–8
  varimax rotation 359–361, 363–5
    factors 360–5
    loadings/scores plot 359–62, 364
Fourier filter 128–34
  convolution 131, 134
  double exponential 129–30, 132
  Fourier self deconvolution 130–1, 133
  and noise 128–30
  peak resolution 128–30
  of time series 130–1, 134
Fourier transforms 112–3, 120–4
  Discrete Fourier Transform 120–1
  domain spectra 122
  Fast Fourier Transform 124
  FTNMR 120–1
  peak shapes 122
  phase errors 122

time to frequency 120–1
fractional factorial designs 32–5, 273, 276
  limitations 34–5
  rules 32–3
frequency histogram 70–1
F-test 19, 81, 85–9
  and accuracy 85
  critical value tables 86–9
  and one-tailed test 86–9
  and two-tailed test 86–9
  and variance 85–6
FTNMR 120–1
fuzzy membership functions 97–8

gas chromatography (GC) 353
gene sequences 319
genomics 287–8
geological processes 111, 123
graphs 69–72
  bar chart 70–1
  blob chart 69
  cumulative frequency measurement 69–70
  distribution curve 70–2
  frequency histogram 70–1
Grubb's test 100–1

hat matrix 54–5
high performance liquid chromatography
    (HPLC) 1, 3–4, 226–8
hyphenated chromatography 221
hypothesis testing 64, 80–1

image analysis
  Alternating Least Squares 345–7
  factor analysis 346–7
  in food industry 352
  multiway methods 347–9
    complex images 347–8
    PARAFAC 347–8
    three way images 347
    Tucker3 decomposition 347–8
image enhancement 337–40
  false colour technique 337
  Principal Components Analysis 337–40
  unfolding 339
  use of contour plots 337–9
image filtering 335–7
  digital representation 335
  grid representation 335–6
  missing pixels 337
  nonlinear transformation 337

image regression 340–5
  calibration testing 343
  and partial information 345
  reference images 342–3
  residuals 343
  two images comparisons 340–1
image representation 332
image scaling 333–5
  multiway preprocessing 334–5
  spatial variables 334
  spectral variables 334
image smoothing 335–7
image unfolding 333, 339
indicator functions 186
industrial process control 111
interactions 9–11, 269–70
  selection 34
  terms 22–3
intercept term 17
inverse matrix 20–1, 25, 39
iterative methods 246–8
  Alternating Least Squares 247–8
  Iterative Target Transform Factor Analysis 247
  'needle' search 247
Iterative Target Transform Factor Analysis (ITTFA) 247, 261
IUPAC notation 20

K Nearest Neighbour (KNN) method 182–5
  implementation 183
  limitations 184
  modelling methods compared 184–5
kernel methods 305–6
kinetics 257–60
  applications 257–8
Kolmogorov–Smirnov test 78–9

lack-of-fit error 19, 86
leukaemia 313–4
leverage 54–8
  calculated 54–6
  and experimental designs 54–7
  measures of confidence 54, 57
  and outliers 103–4
linear regression 89–93
  linear calibration 90–2
  and outliers 97, 101–2
  see also Multiple Linear Regression
linear terms 22–3
literature sources 5–7

loadings
  case studies 154–6
  graphical representation 154–9
  and mean centring 162–4
  normalization 167
  and parameter information 157–8
  plots 157–9, 233
  and standardization 164–5

macromolecules 319–29
  sequence alignment 320–2
  sequence similarity 322–4
  tree diagrams 324–9
Mahalanobis distance 108, 178
  in taxonomy 298–301
Manhattan distance 168
mass spectrometry (MS) 224–6
Matlab 2
matrix 20–1
  applications 20–1
  hat matrix 54–5
  identity 20
  inverse 20–1, 25, 39
  in molecular biology 322–3, 327
  multivariate 249
  pseudoinverse 21, 25
  rank 152, 228–9
  and spectroscopy of mixtures 206
  transpose 20
maxent *see* maximum entropy
maximum entropy 125–8
  computational approach 127–8
  and modelling 126–8
  optimum solution 127–8
  and probability calculations 125–6
mean 68
  comparisons 81–5
  and outliers 84–5
  t-test 82–4
mean centring 162–4
median 68
medicine 4
  diagnoses 313–4
metabolomics 287–8, 314–7
  coupled chromatography 314–7
    peak tables 314–6
    total chromatographic profile 316–7
  Nuclear Magnetic Resonance 120–1, 317
metabonomics *see* metabolomics
micro-organisms 308–9
  growth curves 311
  mid infrared spectroscopy 309–11

micro-organisms (*continued*)
    baseline problems 309–10
    derivatives (use of) 309–11
    spectral regions 309–11
  near infrared spectroscopy 311–2
  pyrolysis mass spectrometry 312–3
  Raman spectroscopy 311–2
mid infrared (MIR) spectroscopy 252–4, 256, 356
  of micro-organisms 309–11
mixtures
  composition determination 203–6
    chromatography 207–8
    Multiple Linear Regression 206–8
    Principal Components Regression 208–11
    spectroscopy 203–5
  defined 44
  experimental designs 44–9
    applications 51
    constrained mixture designs 47–9
    four component 49–51
    practical example 49–51
    simplex centroid designs 45–6
    simplex lattice designs 47
    three component 44
  resolution
    iterative methods 246–8
    noniterative methods for resolution 242–6
  variables 278–86
    minor constituents 285
    models 285–6
    and process variables 285
    ratios 283–4
    simplex designs 278–83
modelling 12–6, 260
  of experimental data 22
  and maximum entropy 126–8
  of reactions *see* reaction modelling
models 53–8
  calibration 214–7
  confidence bands 54
  leverage 54–8
  of mixture experiments 285–6
    Sheffé model 286
  prediction 53–4
moving average (MA) filter 117
Multiple Linear Regression (MLR) 206–8, 242–4
  and spectroscopy of mixtures 206–8
  see also linear regression
multivariate analysis
  outliers 99–100

pattern analysis 159–60
multivariate calibration 58–62, 194–5, 202–6
  designs 61–2
  mixture composition 203–6
  models 58–9, 178, 263–4
  software 58, 60
  and spectroscopy 203–5
multivariate correlograms 115
multivariate curves 234–6
Multivariate Image Analysis (MIA) 331–49
Multivariate Statistical Process Control (MSPC) 108–9
multiway methods 187–90, 217–20, 347–9
  calibration 217–20
    N-PLSM 219–20
    trilinear PLS1 218–9
    unfolding 217–8
  pattern recognition 187–90, 347–9
    PARAFAC 189–90, 347–8
    three-way chemical data 187–8
    Tucker3 models 188–9, 347–8
    unfolding 190

N-PLSM 219–20
  dimensionality 219
near infrared (NIR) spectroscopy 193, 252, 254–5, 257
  in food industry 354–6
  historical growth areas 1, 4
  of microbes 311–2
noise 116, 120
  and Fourier filter 128–30
noniterative methods for resolution 242–6
  diagnostic spectroscopic variables 243
  Multiple Linear Regression 243–4
  partial selectivity 244–6
  Principal Components Regression 244
  selective/key variables 242–3
normal distribution 64, 73–6
  applications 75–6
  error distribution 73
  and hypothesis test 80–1
  and outliers 100–1
  probability density function 74
  t-statistic compared 84
  tables 74–5
  verification 76–80
    consequences 79–80
    cumulative frequency 76–8
    Kolmogorov–Smirnov test 78–9
Normal Operating Conditions (NOC) 108–9
normalization 166–7

Nuclear Magnetic Resonance (NMR) 120–1, 228, 317
null hypothesis 80
numerical taxonomy 290–1
Nyquist frequency 123

olive oil blending 49–51
one-tailed test 81, 83–4
  and F-test 86, 89
optimization 9–11, 39, 51–3
Orthogonal Projection Approach (OPA) 236, 247
orthogonality 33, 35–6, 153–4, 272
  and multivariate calibration designs 60–1
outliers 64, 79, 96–100
  and best fit line 97–8, 101
  and calibration 197–8
  causes 96
  detection 100–4, 174, 216
    linear regression 101–2
    multivariate calibration 103–4
    normal distribution 100–1
    use of residuals 101–2
  and filters 119
  in multivariate analysis 99–100
  overcoming 97–9
    removal 97
    using median of regression parameters 98–9
    using membership functions 97–8
  and t-tests 84–5
over-fitting 206, 215

Parallel Factor Analysis (PARAFAC) 189–90, 347–8
Partial Least Squares (PLS) 3, 108, 193, 211–4, 306–8
  compared with PCA 211–2
  data modelling 212–3
  in food chemistry 365–8
  in image analysis 342, 346
  nonlinearity 213
  PLS1 103–4, 212, 218–9, 365–8
  PLS2 212–3, 365–6
  presentation 211–2
  reaction modelling 260
  in supervised pattern recognition 213–4
  weights matrix 212
partial selectivity 244–6
  use of PCA/PCR 245
  zero concentration/composition 0 window 245

pattern recognition 145–91, 353
peak purity 141, 234, 239–40
peak shapes 134–8
  applications 138
  asymmetric 137–8
  characteristics 135–6
  Gaussian 136–7
  Lorentzian 136–7
  resolved using derivatives 138–42
peak tables 314–16
  peak intensity integration 316
  peak matching 315–6
  peak picking 315
pH profiles 249–50
pharmaceutical industry 2–4, 109, 230
  chromatographic pattern recognition 2
  drug costs 268–9
  economics 267–8
  Multivariate Image Analysis 331
  patenting 268
  and Process Analytical Technology 263
  yield optimization 269
phylogenetic trees 327–9
phylogram 170, 326–7
pixels 332–3, 336–7
Plackett–Burman designs 35–6, 273–4
  application 37–8
  coefficient interpretation 39
  design matrix 38
  factor significance 39
  generators 273–4
PLS1 103–4, 212, 218–9, 365–8
PLS2 212–3, 365–6
pooled variance 83
prediction 53–4, 194–5
preprocessing 160–7, 222–4
  baseline correction 222–3
  column scaling 223–4
  example 160–2
  importance 160, 166
  and mean centring 162–4
  row scaling 166, 223
Principal Components (PCs) 148–50, 212
  conversion to factors 360–1
  in dataset characterization 185–7
  extensions 159
  loadings plots 157–8, 233
  and mixture analysis 209
  model characteristics 185
  scores plots 156–7, 233
  in SIMCA 180

Principal Components Analysis (PCA) 108,
  147–54, 221
  autoprediction 186–7
  and chromatography 148
  column determination 185
  component identification 148–9
  component number determination 152–3,
    185–6
  and coupled chromatography 150–4, 244
  cross-validation 186–7
  data matrix 151–2, 185
  eigenvalues 152, 185–6
  error matrix 186
  food chemistry 359
  historical aspects 147
  image analysis 337–40, 345–6
  loadings 153–9, 162–5, 167
  measurement error 151
  and mixture analysis 209
  multivariance 148
  orthogonality 153–4
  and partial selectivity 245
  and reaction modelling 261
  scores 150, 153–8, 162–5, 167
Principal Components Regression (PCR)
    208–11, 242
  and mixture analysis 208–11, 244
probability density function (pdf) 74
  normal distribution 74
Process Analysis 2
Process Analytical Technology (PAT) 3,
    263–5
  continuous improvement 264–5
  knowledge management 264–5
  multivariate models 263–4
  process analysers 264
  process control tools 264
process improvement 267–86
process variables 275–8
procrustes analysis 160
proteins
  sequence 319
    alignment 320–2
  in wheat 354–5
proteomics 287
pseudoinverse 21, 25
psychometrics 147
purity curves 226–8, 230–4
  drug detection 230–1
  elution profiles 231–2
  limitations 234
pyrolysis mass spectrometry 312–3, 357

Q-statistic 100–1, 108–9
quadratic terms 22–3
quality control 64, 104–6, 352
quantitative model 39–40
quartiles 69

Raman spectroscopy 256, 311–2, 356
rate constant 259, 262–3
reaction modelling 257–60
  analysis 259–60
    component resolution 259–60
    concentration profiles 259–60
    PLS 260
  constraints 261
  data merging 262
  information combining 261
  kinetics 258–9
    first order reaction 258, 262
    rate constant 259
    second order reaction 258–9
  three-way analysis 262–3
reaction monitoring 252–7
  data analysis 252, 256–7
  methods summarized 256–7
  mid infrared spectroscopy 252–4
  near infrared (NIR) spectroscopy 254–5
  Raman spectroscopy 256
  UV/visible spectroscopy 255–6
reaction performance
  experiment numbers 275–7
  experimental designs 272–3
  factors
    coefficients 274
    interaction 269–70
    limits 272–3
    relate to response 273–4
    screening 271–5
    types 271–2
  fractional factorial designs 273, 276
  improvement 269–71
    grid search 270–1
    response surface 270
  interactions 269–70
  mathematical modelling 276–8
  optimization
    prediction 278
    process variables 275–8
    yield 271–4
regression *see* linear regression; Multiple Linear
    Regression; Principal Components
    Regression
replicates 14, 16

reroughing 120
residuals
   and Multivariate Image Analysis 343
   and outlier detection 101–2
root mean square error 214
rotatable central composite designs 42–3
row scaling 166, 223

sample correlation 216
sample standard deviation 69
sample variance 69
sampling 63–4
   errors 65–6
      and measurement deviation 66
      and sample preparation 66–7
Savitzky–Golay method 117–19, 140–2
scaling
   of images 333–5
   in preprocessing 166, 223–4
scores 154–8
   case studies 154–6
   graphical representation 154–9
   and mean centring 162–4
   normalization 166–7
   plots 150, 156–7, 233
   and standardization 164–5
semiconductors 331–2
sensory analysis 357–9
sequence alignment 320–2
   DNA 320
   proteins 320–2
sequence similarity
   cluster analysis 322
   indicator function 323–4
   similarity matrix 322–3
   statistical measures 323
sequential data 111–2, 228–30
   chromatograms
      overlapping peaks 229–30
      regions 228–9
sequential methods 111–44
Sheffé models 46, 286
Shewhart charts 104–6
   interpretations 105–6
   and limits 105–6
   principle 105
   purpose 104
   and quality control 104–6
significance tests 78–80, 82–3
SIMCA 178–82
   and class distance 180
   modelling power 180

soft v. hard modelling 178–9
   variables 181
similarity methods 168, 234–6
   correlation coefficients 234–5
   distance measures 235–6
   macromolecule investigations 322–4
   matrices 322–3, 327
   phylogenetic trees 327
simplex 45, 51–3, 278–83
   constraints 280–2
   defined 45, 52
   designs 282–3
      centroid 45–6, 278–9, 281
      Sheffé models 46
      lattice 47, 279–80
   limits 280–2
   optimization 51–3
      limitations 53
      rules 52–3
SIMPLISMA 236, 247
smoothing 116–20
   and correlograms 114–5
   method choice 120
   moving average 117–8
   Multivariate Image Analysis 335–7
   reroughing 120
   Running Median Smoothing functions 119
   Savitzky–Golay filter 117–9
   and wavelets 143
soft v. hard modelling 178–9
species 289
spectroscopy 112
   derivatives 138–42
   equilibria studies 249–52
   medical diagnosis 313–4
   of mixtures 203–5
   and multivariate calibration 58–9
   peak shapes 134–8
   reaction monitoring 252–7
standard deviation 68–9, 82, 163–4
   accuracy 95–6
   confidence 93–6
standardization 163
   uses 165–6
'star' points 40, 41
statistical classification techniques 174–82
   bivariate discriminant models 175–7
   multivariate discriminant models 177
   SIMCA 178–82
   statistical output 182
   univariate classification 175

statistics
  descriptive 68–9
  outliers 64
  terminology 43
  uses
    descriptive 63
    distributions 63–4
    hypothesis testing 64
    prediction confidence 64
    quality control 64
sum of squares 17–9
supervised pattern recognition 146–7,
    171–4
  bootstrap 173
  cross-validation 172–3
  industrial process control 174
  method stability 174
  model application 174
  model improvement 173
  model optimization 173–4
  and Partial Least Squares 213–4
  and prediction 171–3
  test set 172–4
  training set 171–4
Support Vector Machines (SVMs) 303–6
  boundary establishment 304–6
  kernel methods 305–6
  linearity vs. nonlinearity 304–6
  multiclass problems 306
  optimization 306

Taguchi designs 36–7
taxonomy 288–313
  algorithmic methods 288–9
  Bayesian methods 300–3
  class distance determination 297–300
  defined 289
  Discriminant Partial Least Squares 306–8
  discrimination methods 291–7
  Fisher's iris data 289–90, 291, 293–4
  micro-organisms 308–12
  numerical 290–1
  Support Vector Machines 303–6
test set 172, 215–6
  and compound correlations 216
  dataset representation 215
  and outlier detection 216
three-way chemical data 187–8
time series 130–1, 134

time series analysis 111–2
Total Ion Chromatogram (TIC) 316–7
training set 58–60, 171
  and calibration experiments 205–6
  and classification 171–3
  designs 216
tree diagrams 14–6, 40, 42, 324–9
  branches 324–5
  cladistics 325–6
  dendograms 325
  evolutionary theory 325–6
  nodes 324–5
  phylograms 326–7
  representation 324–5
trilinear PLS1 218–9
  data types 219
t-statistic 83–4, 94
t-test 81–5, 94–5
  one-tailed 83–4
  principle 82
  statistics tabulated 83–4
  uses 83–4
  two-tailed 81
Tucker3 models 188–9, 347–8

unfolding 190, 217–8
  disadvantages 218
  of images 333, 339, 341–2
univariate calibration 195–202
  classical 195–7
  equations 198–9
  extra terms 199
  graphs 199–202
  inverse 196–7
  matrix 198–9
  outliers 197–8
  terminology 195
univariate classification 175
unsupervised pattern recognition 146,
    167–71
  see also cluster analysis
UPGMA 327–8
UV/visible spectroscopy 138, 224, 249, 252,
    255–7

variable combination 228
variable selection 224–8
  noise 225–6
  optimum number 225–6

variance   19, 68–9
  comparisons   85–6
  and *F*-test   85–6
varimax rotation   247
  food chemistry   359–65

wavelets   142–3
  applications   143
  in datasets   142–3
  defined   142

weights   188
Window Factor Analysis   236–9
  eigenvalues   237–8
  expanding windows   237
  fixed size windows   238–9
  variations   239

yield improvement   267–86
  optimization   269, 271–4

Printed in the United States
By Bookmasters